theclinics.com

NEUROLOGIC CLINICS

Current Issues in Clinical Neurovirology: Pathogenesis, Diagnosis and Treatment

GUEST EDITORS
Christopher Power, MD
Richard T. Johnson, MD

CONSULTING EDITOR
Randolph W. Evans, MD

August 2008 • Volume 26 • Number 3

SAUNDERS

An Imprint of Elsevier, Inc.
PHILADELPHIA LONDON TORONTO MONTREAL SYDNEY TOKYO

W.B. SAUNDERS COMPANY
A Division of Elsevier Inc.

1600 John F. Kennedy Blvd., Suite 1800, Philadelphia, PA 19103-2899

http://www.theclinics.com

NEUROLOGIC CLINICS
August 2008
Editor: Donald Mumford

Volume 26, Number 3
ISSN 0733-8619
ISBN-13: 978-1-4160-5820-5
ISBN-10: 1-4160-5820-6

Neurologic Clinics (ISSN 0733-8619) is published quarterly by Elsevier Inc., 360 Park Avenue South, New York, NY 10010–171. Months of issue are February, May, August, and November. Business and editorial offices: 1600 John F. Kennedy Blvd., Suite 1800, Philadelphia, PA 19103-2899. Customer Service Office: 6277 Sea Harbor Drive, Orlando, FL 32887–4800. Accounting and circulation offices: 6277 Sea Harbor Drive, Orlando, FL 32887-4800. Periodicals postage paid at New York, NY, and additional mailing offices. Subscription prices are $218.00 per year for US individuals, $344.00 per year for US institutions, $109.00 per year for US students, $273.00 per year for Canadian individuals, $404.00 per year for Canadian institutions, $285.00 per year for international individuals, $404.00 per year for international institutions, and $145.00 for Canadian and foreign students/residents. To receive student/resident rate, orders must be accompanied by name of affiliated institution, date of term, and the *signature* of program/residency coordinator on institution letterhead. Orders will be billed at individual rate until proof of status is received. Foreign air speed delivery is included in all *Clinics* subscription prices. All prices are subject to change without notice. POSTMASTER: Send address changes to *Neurologic Clinics*, Elsevier Periodicals Customer Service, 6277 Sea Harbor Drive, Orlando, FL 32887-4800. **Customer Service: 1-800-654-2452 (US). From outside of the US, call 1-407-563-6020. Fax: 1-407-363-9661. E-mail: JournalsCustomerService-usa@elsevier.com.**

Reprints. For copies of 100 or more of articles in this publication, please contact the Commercial Reprints Department, Elsevier Inc., 360 Park Avenue South, New York, New York, 10010-1710; Tel.: (+1) 212-633-3812; Fax: (+1) 212-462-1935, and E-mail: reprints@elsevier.com.

Neurologic Clinics is also published in Spanish by Nueva Editorial Interamericana S.A., Mexico City, Mexico.

Neurologic Clinics is covered in *Current Contents/Clinical Medicine, MEDLINE/PubMed (Index Medicus), EMBASE/Excerpta Medica*, and *PsycINFO*, and *ISI/BIOMED*.

Printed in the United States of America.

CURRENT ISSUES IN CLINICAL NEUROVIROLOGY: PATHOGENESIS, DIAGNOSIS AND TREATMENT

CONSULTING EDITOR

RANDOLPH W. EVANS, MD, Clinical Professor of Neurology, Baylor College of Medicine, Houston, Texas

GUEST EDITORS

CHRISTOPHER POWER, MD, Professor of Neurology, Psychiatry, Medical Microbiology and Immunology, Canada Research Chair in Neurological Infection and Immunity, Faculty of Medicine and Dentistry, Departments of Medicine (Neurology), Medical Microbiology and Immunology, and Psychiatry, University of Alberta, Edmonton, Alberta, Canada

RICHARD T. JOHNSON, MD, Distinguished Service Professor of Neurology, Microbiology and Neuroscience, Department of Neurology, Johns Hopkins University School of Medicine and Bloomberg School of Public Health, Baltimore, Maryland

CONTRIBUTORS

J. RICHARD BARINGER, MD, Professor of Neurology and Pathology, Emeritus, Department of Neurology, University of Utah Health Sciences Center, Salt Lake City, Utah

J. DAVID BECKHAM, MD, Instructor in Medicine, Division of Infectious Diseases, University of Colorado Health Sciences Center; and Research Service, Denver Veterans Affairs Medical Center, Denver, Colorado

LYSA BOISSÉ, MD, MSc, BSc, Division of Neurology, Queen's University, Kingston, Ontario, Canada

BRUCE J. BREW, MBBS, MD, FRACP, Professor of Medicine (Neurology), University of New South Wales, New South Wales, Australia; and Head of Neurology Program Director (Medicine), Department of Neurology, St Vincent's Hospital, Darlinghurst, Sydney, New South Wales, Australia

RANDALL J. COHRS, PhD, Professor, Department of Neurology, The University of Colorado School of Medicine, Denver, Colorado

STUART D. COOK, MD, Ruth Dunietz Kushner and Michael Jay Serwitz Professor of Neurology and Neurosciences, Department of Neurology and Neuroscience; Director, Multiple Sclerosis Center at UMDNJ-New Jersey Medical School, Newark, New Jersey

LARRY E. DAVIS, MD, Chief of Neurology Service, New Mexico Veterans Affairs Health Care System; and Distinguished Professor of Neurology and Molecular Genetics and Microbiology, University of New Mexico School of Medicine, Albuquerque, New Mexico

DEREK EMERY, MD, Department of Radiology and Diagnostic Imaging, University of Alberta, Edmonton, Alberta, Canada

DONALD H. GILDEN, MD, Professor and Chairman, Department of Neurology, The University of Colorado School of Medicine; Professor, Department of Microbiology, The University of Colorado School of Medicine, Denver, Colorado

M. JOHN GILL, MBBS, Departments of Medicine and Microbiology and Infectious Diseases, University of Calgary, Calgary, Alberta, Canada

ALEJANDRA GONZALEZ-DUARTE, MD, NeuroAIDS Fellow, Mount Sinai School of Medicine, New York

KAROLINA HANKIEWICZ, MD, Hospital Clinico y Provincial, Barcelona, Spain

DAVID N. IRANI, MD, Associate Professor of Neurology, Department of Neurology, Holtom-Garrett Program in Neuroimmunology, University of Michigan Medical School, Ann Arbor, Michigan

ALAN C. JACKSON, MD, Departments of Internal Medicine (Neurology) and of Medical Microbiology, University of Manitoba, Winnipeg, Manitoba, Canada

STEVEN JACOBSON, PhD, Viral Immunology Section, Neuroimmunology Branch, National Institute of Neurological Diseases and Stroke, Bethesda, Maryland

RICHARD T. JOHNSON, MD, Distinguished Service Professor of Neurology, Microbiology and Neuroscience, Department of Neurology, Johns Hopkins University School of Medicine and Bloomberg School of Public Health, Baltimore, Maryland

JOHN A. LINCOLN, MD, PhD, Neuroimmunology Fellow, Department of Neurology and Neuroscience, Multiple Sclerosis Center at UMDNJ-New Jersey Medical School, Doctors Office Center, Newark, New Jersey

RAVI MAHALINGAM, PhD, Professor, Department of Neurology, The University of Colorado School of Medicine, Denver, Colorado

NIKLAUS H. MUELLER, PhD, Fellow, Department of Neurology, The University of Colorado School of Medicine, Denver, Colorado

MARIA A. NAGEL, MD, Instructor, Department of Neurology, The University of Colorado School of Medicine, Denver, Colorado

FARSHID NOORBAKHSH, MD, PhD, Postdoctoral Fellow, Department of Medicine, University of Alberta, Edmonton, Alberta, Canada

UNSONG OH, MD, Viral Immunology Section, Neuroimmunology Branch, National Institute of Neurological Diseases and Stroke, Bethesda, Maryland

CHRISTOPHER POWER, MD, Professor of Neurology, Psychiatry, Medical Microbiology and Immunology, Canada Research Chair in Neurological Infection and Immunity, Faculty of Medicine and Dentistry, Departments of Medicine (Neurology), Medical Microbiology and Immunology, and Psychiatry, University of Alberta, Edmonton, Alberta, Canada

JESSICA ROBINSON-PAPP, MD, Assistant Professor of Neurology and Pathology, Mount Sinai School of Medicine, New York

DAVID M. SIMPSON, MD, Professor of Neurology, Director, Clinical Neurophysiology Laboratories and NeuroAIDS Program, Mount Sinai School of Medicine, New York

KENNETH L. TYLER, MD, Reuler-Lewin Family Professor of Neurology, Professor of Medicine, and Microbiology, University of Colorado Health Sciences Center; and Chief, Neurology Service, Denver Veterans Affairs Medical Center, Denver, Colorado

THOMAS WEBER, MD, Professor of Neurology, Neurologische Klinik, Marienkrankenhaus Hamburg, Academic Teaching Hospital, University of Hamburg, Hamburg, Germany

STEVE L. WESSELINGH, BMBS, PhD, FRACP, Professor, Dean, Faculty of Medicine, Nursing, and Health Sciences, Monash University, Melbourne, Victoria, Australia; and The Alfred Hospital, Melbourne, Victoria, Australia

EDWINA J. WRIGHT, MBBS, FRACP, The Alfred Hospital Infectious Diseases Unit, Burnet Institute, Melbourne, Victoria, Australia; The Burnet Institute Melbourne, Victoria, Australia; and Faculty of Medicine Nursing and Health Sciences, Monash University, Melbourne, Victoria, Australia

CONTENTS

FORTHCOMING ISSUES

November 2008

Stroke
Philip Gorelick, MD and
Sean Ruland, MD, *Guest Editors*

February 2009

Neuroimaging
Laszlo Mechtler, MD, *Guest Editor*

May 2009

Primary Headache
Randolph W. Evans, MD, *Guest Editor*

RECENT ISSUES

May 2008

Neurologic Critical Care
Romergryko G. Geocadin, MD
and Robert D. Stevens, MD,
Guest Editors

February 2008

Sports Neurology
Cory Toth, MD, FRCPC,
Guest Editor

November 2007

Brain Tumors in Adults
Patrick Y, Wen, MD and
David Schiff, MD, *Guest Editors*

ELSEVIER
SAUNDERS

Neurol Clin 26 (2008) xiii–xv

NEUROLOGIC
CLINICS

Preface

The Guest Editors (L-R) Christopher Power and Richard T. Johnson with longstanding neurovirological colleague and friend, Avi Nath, on a dog sled trip in northern Manitoba, Canada.

During the past 40 years, there has been a steadily burgeoning appreciation for the importance of viral infections of the nervous system in terms of clinical features, epidemics, and their global impact. Indeed, the pace has escalated in recent years with the introduction of West Nile virus into North America, the recognition of multiple disease neurologic phenotypes associated with Varicella zoster virus infection, increased survival among patients who have HIV infection, and the growing number of global infections. Neurovirologic infections represent particularly challenging problems for clinicians and their patients because of the inherent complexity in diagnosing neurologic disorders, the nervous system's particular vulnerability to infection because of limited host immune defenses within the brain, the relative inaccessibility of the nervous system tissue for diagnosis, and a fixed intracranial volume. In addition, delivery of drugs to the brain is a major

doi:10.1016/j.ncl.2008.06.003

obstacle in many cases, even when suitable drugs are available. Viral resistance to various drugs also has become an important issue in treating neurologic infections, especially herpes virus and retrovirus infections, just as it is for systemic infections. Future treatment strategies (eg, the use of stem cells from human embryos or animal organs for xeno-transplantation) also might introduce new infections.

A rapidly expanding repertoire of diagnostic tools has become the cornerstone of controlling disease outbreaks, monitoring ongoing epidemics, and permitting expedient clinical diagnoses. A greater understanding of host susceptibility to infection also has arisen through the identification of specific genotypes associated with disease, including single-nucleotide polymorphisms. Moreover, neurologic diagnostic and therapeutic options have evolved with the advent of improved neuroimaging, neurocognitive testing, and a wider availability of specific drugs for neurologic infections.

This issue of the *Neurologic Clinics* includes articles by authors who are recognized internationally for their expertise in areas related to both the clinical and experimental aspects of neurovirologic infections and their management. The overall strategy for the monograph is to provide essential background understanding of different disease phenotypes and their management associated with viral infections of the nervous system, in articles by Wright and colleagues and by Irani, followed by descriptions of specific infections. The three subsequent articles concentrate on herpes virus infection, with articles on Herpes simplex virus by Baringer, on Varicella zoster virus by Mueller and colleagues, and on Epstein-Barr virus by Lincoln and colleagues. Articles on rabies by Jackson and on arbovirus infections by Davis and colleagues supply comprehensive overviews of these changing fields. Because acute disseminated encephalomyelitis often is included in the differential diagnosis of a neurologic infection, this issue includes an article on this topic by Noorbakhsh and colleagues. Human retroviral infections of the central and peripheral nervous systems constitute serious disease burdens globally and are described herein: Oh and Jacobson discuss Human lymphotropic virus I, and Boisse and colleagues and Gonzalez-Duarte and colleagues discuss HIV-1. An article by Weber discusses progressive multifocal leukoencephalopathy, which has evolved from being a rare infection to a comparatively common entity, particularly among patients who have HIV/AIDS. Finally, Johnson and Power review entities contributing to new or emerging neurovirologic infections, highlighting several recent examples such as enterovirus 71, West Nile virus, Nipah virus, and a future concern, hepatitis C virus, together with new diagnostic and therapeutic approaches.

We thank for the authors for their substantial efforts invested in assembling their respective articles, Leah DeBlock for her assistance in handling and editing each article, and Donald Mumford at Elsevier for his commitment to this project. We hope the information provided in this issue will permit the reader in neurology, neurosurgery, infectious diseases, public health, pediatrics, and related experimental disciplines to glean more insights into viral infections of the nervous system and will lead to the improved care of patients afflicted by neurologic infections.

Christopher Power, MD
Departments of Medicine (Neurology)
Medical Microbiology
Immunology, and Psychiatry
University of Alberta, 6-11 Heritage Medical Research Centre
Edmonton, Alberta, T6G 2S2, Canada

E-mail address: chris.power@ualberta.ca

Richard T. Johnson, MD
Department of Neurology, Pathology 627
The Johns Hopkins Hospital
Baltimore, MD 21287, USA

E-mail address: rtj@jhmi.edu

ELSEVIER
SAUNDERS

Neurol Clin 26 (2008) xvii

NEUROLOGIC
CLINICS

Erratum

In the May 2008 issue of *Neurologic Clinics*, Volume 26, number 2, Dr. Leonardo Rangel-Castillo's name was misspelled. The correct spelling is Leonardo Rangel-Castilla.

doi:10.1016/j.ncl.2008.06.004 ***neurologic.theclinics.com***

ELSEVIER
SAUNDERS

NEUROLOGIC CLINICS

Neurol Clin 26 (2008) 617–633

Pathogenesis and Diagnosis of Viral Infections of the Nervous System

Edwina J. Wright, MBBS, FRACP[a,b,c],
Bruce J. Brew, MBBS, MD, FRACP[d,e],
Steve L. Wesselingh, BMBS, PhD, FRACP[a,c,*]

[a]*The Alfred Hospital Infectious Diseases Unit, 2nd Floor, Burnet Institute, 85 Commercial Road, Melbourne, Victoria, Australia, 3004*
[b]*The Burnet Institute, 85 Commercial Road, Melbourne, Victoria, Australia, 3004*
[c]*Faculty of Medicine Nursing and Health Sciences, Monash University, Wellington Road, Melbourne, Victoria, Australia, 3800*
[d]*Faculty of Medicine, University of New South Wales, Australia, NSW, 2052*
[e]*Department of Neurology, St Vincent's Hospital, 390 Victoria Street, Darlinghurst, Sydney, NSW, Australia, 2010*

The diagnosis of some viral infections of the nervous system may be made by the clinician after only a simple glance such as the early vesicles of zoster while others are much more challenging or rare and are made only after extensive investigations and considerable diagnostic vexation, for example measles subacute sclerosing panencephalitis. The humble zoster and uncomplicated viral meningitis are the commonest viral infections of the nervous system that the clinical neurologist is likely to encounter during his or her career. Nonetheless, this article seeks to provide a pathway that can be negotiated to help the clinician understand the pathogenesis and thereby the clinical features and relevant diagnostics of both common and obscure viral infections of the nervous system.

The host

Humans have encountered viruses with a predilection for the nervous system for thousands of years: human T-lymphotropic virus (HTLV), a ubiquitous virus that causes spastic paraparesis, is presumed to have entered human populations in Africa between 5000 and 23,000 years ago

* Corresponding author. Monash University, Bldg 64-Wellington Road, Melbourne, Victoria 3800, Australia.

E-mail address: steve.wesselingh@med.monash.edu.au (S.L. Wesselingh).

0733-8619/08/$ - see front matter
doi:10.1016/j.ncl.2008.03.006 *neurologic.theclinics.com*

[1]. Smallpox, which can cause encephalitis, was found in the mummy of Rhameses V in approximately 1140 BC [2]. More recently it has been surmised that the death of Alexander the Great in 323 BC, whose terminal illness included flaccid paralysis and was preceded by the death of ravens in the city of Babylon where he succumbed, was due to West Nile virus [3]. As a corollary, there is strong evidence to show that the immune system of mammals including humans has been broadly shaped by its interaction with viruses over thousands of years [4]. In this section, we discuss the cells that comprise the healthy nervous system and the specialized nature of the brain's immune system.

Cells of the central nervous system

The brain

The healthy brain is rich in two differentiated cell types—neurons and neuroglial cells—with the ratio of the former to the latter being approximately 1:10. The neuroglia (glia) comprise astrocytes, microglia, oligodendrocytes and ependymal cells and their collective purpose is to provide structural, functional, and nutritional support to the neuronal network.

Neurons. Neurons possess the unique property of being able to transmit, conduct, and receive signals and, in doing so, afford the host the capacity to experience the world through sensation, movement, thought, and emotion and then to integrate these experiences and retain them in memory. Hence, for the host, the integrity of neurons and neuronal pathways is vital, and although there are over 100 billion neurons in the basal ganglia and cerebral and cerebellar cortices, infection or insult, even to limited areas of the brain, may threaten host survival.

Neuroglial cells. Of the neuroglial cells, astrocytes are the most abundant and are critical to overall brain function wherein they (1) act as a de facto connective tissue system that supports neurons, (2) mechanically contribute to the blood brain barrier, (3) metabolize neurotransmitters notably the excitatory transmitter glutamate, (4) provide nutritional and metabolic support for neurons, and (5) play an important role in the host's innate immune response to viral infection. In the event of neuronal death or damage, local scar production occurs via the proliferation of astrocytes.

Microglial cells are monocyte-derived cells of bone marrow origin and are found throughout the brain parenchyma. Microglia are the resident antigen-presenting cells (APCs) of the CNS and play a key role in both the innate and adaptive CNS immune response. Oligodendrocytes are the cells responsible for the production of myelin and are found in the white matter of the brain. Myelin insulates neuronal axons and permits neuronal electrical signal conduction. Ependymal cells are those cells that line the ventricles within the brain and the central canal of the spinal cord.

Other CNS cells. Further, there are other important nonglial cells within the brain that can be targeted by viruses and these include perivascular and choroid plexus endothelial cells, macrophages and dendritic cells, and cells of the leptomeninges. Cerebrovascular endothelial cells (CVEs) are vital components of the blood brain barrier (BBB) (see later in this article) and line the luminal surface of bloods vessels within the brain. CVEs may play a role in the innate immune response to viral infection [5]. Finally, there is ongoing traffic of blood-derived leukocytes throughout the nervous system that maintains homeostasis and acts as a surveillance mechanism for foreign pathogens.

The spinal cord

The outer segments of the spinal cord are composed of myelinated descending (motor) and ascending (sensory) axonal tracts. The center of the spinal cord contains gray matter with anterior and posterior horns.

Cells of the peripheral nervous system

The dorsal root ganglia (DRG) lie just outside the spinal cord and contain the cell bodies of the sensory nerves. There are 31 pairs of spinal nerves that arise from the spinal cord (8 cervical, 12 thoracic, 5 lumbar, and 1 coccygeal) and each of these nerves is composed of anterior and dorsal (posterior) roots with the dorsal root carrying sensory fibers from the periphery. On the dorsal root one finds a swelling known as the dorsal (or posterior) root ganglion. The cellular constituents include neurons and their processes, resident macrophages, Schwann cells that can produce myelin, and closely related satellite cells that are found in close proximity to neuronal sources within the DRG.

Neuroimmunology

The healthy central nervous system

The human brain is thus the complex functional integration of these different cell types. However the brain is vulnerable both because of its innate complexity and its core value to the host, hence the brain is protected through mechanical means. It is the mechanical boney encasement of the skull and the thick dura mater that protect the brain from direct traumatic injury. However, by being protectively enclosed thus, if trauma or infection do occur, the brain has a limited capacity to swell with impunity in response to these insults: an immune inflammatory response to infection or trauma that might be "normal" in other organs may be excessive in the brain and lead to injury, death, or inappropriate repair of crucial neurons and neuronal pathways. It is surmised, therefore, that because of the protective mechanical confinement of the skull that the brain has had to develop an attenuated inflammatory response to infection and injury and, as such, it is "an immunologically specialized organ" [5].

Evidence for this is as follows: first, the brain possesses an anatomic blood brain barrier that limits cellular trafficking into the brain by cells of the peripheral immune system. Second, the brain has only a rudimentary lymphatic system and therefore lacks the efficiently streamlined immune arc that other organs enjoy in the setting of infection: typically APCs capture the antigen within an organ, travel to the draining lymph nodes, and present the antigen to T cells within the lymph node; this activates the T cells and stimulates a proinflammatory T-cell response that is directed against the offending antigen within the organ. Third, the brain has an actively suppressed response to inflammatory stimuli relative to other tissues: the resident APCs of the brain-microglia and perivascular macrophages in the healthy brain have much lower surface expression of the major histocompatibility complex (MHC) molecules compared with APCs of other organs. $CD8^+$ and $CD4^+$ T cells can only recognize an antigen that has been digested by APCs and then re-presented in fragments at the cell's surface within the binding clefts of class I and class II MHC molecules, respectively. Hence, this lower expression of MHC molecules means that there is less opportunity for a vigorous inflammatory response within the brain [5,6]. Despite these circumstances, the central nervous system (CNS), when confronted with infection, mounts dynamic innate and adaptive immune inflammatory responses (discussed later in this article) albeit carefully titrated to minimize damage to and protect the resident CNS cells [5,7].

The virus

There is a panoply of viruses that can infect and cause disease within the nervous system and in Table 1 a number of these leading viruses are grouped according to whether they are RNA, DNA, or retroviruses. Viruses contain their genetic material in the form of either RNA or DNA: DNA viruses require DNA polymerase, found in the host cell's nucleus, to reproduce themselves (replicate) and therefore must enter the host cell's nucleus to access this enzyme; RNA viruses do not replicate using a DNA intermediate form and therefore replicate in the cell's cytoplasm. Retroviruses are an exception wherein their genetic material is in the form of RNA, but they replicate using a DNA intermediate: upon entering a cell they use their own reverse transcriptase enzyme to assume a cDNA form, which then enters the cell's nucleus and integrates into the host's chromosomal DNA, as a provirus and thence replication begins.

By reviewing Table 1 it becomes clear that, broadly speaking, RNA viruses are the commonest cause of acute meningitis and encephalitis while the DNA and retroviruses are more typically associated with chronic disease forms including spinal cord disease and dementia, respectively. In the case of the DNA viruses herpes simplex virus (HSV) 1 and 2 and varicella zoster virus (VZV) and the retrovirus human immunodeficiency virus type 1(HIV-1), this is in part due to the capacity for these viruses to establish

Table 1
Neurologic disorders and causative RNA, DNA, and retroviruses

Neurologic disorders	RNA viruses	DNA viruses	Retroviruses
CNS			
Meningitis	Coxsackie Echovirus *Rare* Mumps, LCMV	HSV-2 (*associated with primary genital herpes*)	HIV-1 (*acute during seroconversion illness and chronic during course HIV disease*)
Encephalitis	JE virus* Coxsackie Echovirus Influenza (*children > adults*) *Other arthropod-borne viruses *West Nile virus, Eastern, Western, Venezuelan equine viruses, St Louis, Murray Valley, Rift Valley fever, California, La Crosse virus (children > adults)* *Rare* Rabies, Nipah, Hendra (henipahviruses), Dengue	HSV-1 (*adults*) HSV-2 (*neonates*) Adenoviruses *Rare* VZV HHV6 CMV	HIV-1 (*seroconversion illness*)
Brainstem encephalitis (Rhombencephalitis)	West Nile Virus	HSV-1 *Rare* HSV-2 EBV	
Dementia	Measles -Subacute sclerosing panencephalitis (*children*) Rubella -Progressive rubella panencephalitis (*young adults*)		HIV-1 (*adults and children*)
Spinal cord			
Myelopathy		HIV-1 HTLV-1 HTLV-2 (*rare*)	
Myelitis	West Nile virus	HSV-1 VZV Polio virus	
PNS			
Neuropathy			HIV-1

Abbreviations: EBV, Epstein Barr virus; HHV, human herpes virus; HIV, human immunodeficiency virus; HLTV, human T-lymphotropic virus; HSV, herpes simplex virus; JE virus, Japanese encephalitis virus; LMCV, lymphochoriomeningitis virus; MCV, cytomegalovirus; VZV, varicella zoster virus.

latency within host cells and therein become relatively impervious to immune system detection and eradication.

Modes of exposure and viral transmission

In Table 2 we have provided the possible routes of exposure employed by viruses that infect the nervous system. Although Table 2 is discussed in more detail in the section "The diagnostic approach to viral infections of the nervous system" it is raised here to illustrate the numerous modus operandi by which viruses with a predilection for the nervous system can infect the host. Apropos of the discussion earlier, it is interesting to note that DNA and retroviruses are more likely to be spread following sexual, breastfeeding, peripartum, or blood transfusion exposures as opposed to the RNA viruses that are typically spread via oral, vector, or airborne routes.

Viral entry into the nervous system

Viral entry into the nervous system is manifold: hematogenous spread occurs when viruses directly enter the brain through blood vessel walls; viruses also gain direct access to the cerebrospinal fluid (CSF) via infection of the cells of the choroid plexus; some viruses enter the brain "hidden" within cells of the peripheral immune system, for example, HIV enters the brain within infected monocytes and $CD4^+$ T cells and also enters the brain as a result of turnover of perivascular macrophages from the bone marrow; JC virus (JCV), the agent that causes PML (progressive multifocal leucoencephalopathy), probably enters the brain within infected B cells. HSV-1 also exhibits transneuronal spread wherein following a period of latency within the geniculate ganglion it appears to enter the temporal and frontal lobes via the axon in a retrograde manner; primary HSV-1 infection of the brain may also occur following infection of the nasal mucosa and direct spread into the anterior fossa through the cribiform plate [8], or perhaps occurs following latent infection of the brain parenchyma itself. Recently West Nile virus has been shown to enter the CNS via transneuronal spread [9].

Viral entry into target cells

We have listed and provided examples of the cell types infected or affected by the viruses outlined in Box 1. Of note, some viruses have a proclivity for infecting only one cell type, for example, Japanese encephalitis (JE) virus only infects neurons; polio virus infects the motor cells found in the anterior horn of the spinal cord. However, other viruses, like HIV-1 and cytomegalovirus (CMV), can infect several cell types within the brain: HIV infects macrophages, microglia, astrocytes, and possibly neurons although *productive* HIV-1 infection occurs only within brain macrophages and microglial cells, although the latter cell type's involvement is debated [10].

Table 2
Types of exposure to viruses that cause neurologic disease

Exposure type	
Sexual	HIV-1 and -2, HSV-1 and 2, HTLV-1and -II
Breast feeding	HIV-1 and -2, HTLV-1 and -II
Blood transfusion	HIV-1, West Nile virus, HTLV-1 and -II
Vertical transmission	HIV-1 and -2, HTLV-1 and -II, CMV
Oral	Echovirus, coxsackie virus, polio virus, EBV
Respiratory	HSV-1, measles, mumps, rubella, PML, VZV, adenovirus
Percutaneous	Rabies
Vector-borne	Mosquitoes: JE virus, West Nile virus, Eastern equine encephalitis, Western equine encephalitis, Venezuelan equine encephalitis, St Louis encephalitis virus, Rift Valley fever virus, California encephalitis viruses, Murray Valley encephalitis. Ticks: Powassan virus, Colorado tick fever, tick-borne encephalitis (central European and Russian subtypes)
Animal exposures	Rabies (dogs and other mammals), Bat Lyssa virus (bats), Hendra virus (horses), Menangle virus and Nipah virus (pigs), Herpes B virus (monkeys)
Vaccine-related	Yellow fever, polio, mumps, small pox
Tissue grafts, organ donations	West Nile virus, rabies, LCMV, HHV-6
Immunocompromised patients	CMV, PML, HSV-1 & -2, VZV, HHV-6,
Children	HSV-2 (neonates), JE virus, HIV-1, coxsackie and enteroviruses, La Crosse virus, measles, mumps, rubella, influenza virus
Regional	
North America	Across the continent: West Nile virus, St Louis encephalitis, Jamestown Canyon, HIV-1; Eastern USA: La Crosse, eastern equine encephalitis; Midwest and western USA: Western equine encephalitis
Canada	Southern Canada: St Louis encephalitis; Across the continent: Eastern equine encephalitis, western equine encephalitis, HIV-1
South America	Central and South America: St Louis encephalitis; Brazil: Dengue, West Nile virus, Venezuelan equine encephalitis, Chikungunya, eastern equine encephalitis, JE virus, St Louis encephalitis, HIV-1
Africa	HIV-1 and -2, ebola, polio, rabies, Rift Valley fever
Europe	Rabies, HIV-1, West Nile virus, tick-borne encephalitis subtypes (central Europe and previous USSR)
Asia	JE virus, West Nile virus, Nipah virus, HIV-1, Chikungunya virus,
Australia	Murray Valley encephalitis, Hendra virus, Australia Bat Lyssa virus, Kunjin, JE virus (Northern Australia), HIV-1

Abbreviations: CMV, cytomegalovirus; EBV, Epstein Barr virus; HIV, human immunodeficiency virus; HHV-6, human herpes virus-6; HLTV, human T-lymphotropic virus; HSV, herpes simplex virus; JE virus, Japanese encephalitis virus; LCMV, lymphocytic choriomeningitis virus; PML, progressive multifocal leucoencephalopathy; VZV, varicella zoster virus.

Viruses may infect their target cells through specific receptors: HIV-1 enters microglia and macrophages in the brain through use of the $CD4^+$ receptor and use of one or both of its co-receptors, CCR5 and CXCR4 [11]; JCV enters oligodendrocytes via the serotonin receptor 5HT2A [12]

Box 1. Cell types of the nervous system and examples of viruses that may infect them

Neurons
HSV-1, JE virus, coxsackie virus, echo virus, West Nile virus
Motor neurons
Polio virus, coxsackie virus, echo virus, West Nile virus
Microglia, perivascular macrophages and astrocytes
HIV-1
Oligodendrocytes
JCV
Meningeal cells including meningeal macrophages
Coxsackie virus, echovirus, HSV-2, HIV-1, HIV
Dorsal root ganglia
VZV, HSV

Abbreviations: HSV, herpes simplex virus; JE, Japanese encephalitis; HIV, human immunodeficiency virus; JCV, JC virus; VZV, varicella zoster virus.

and probably other receptors. HSV-1 and rabies viruses and the polio virus [13] can replicate and infect other neurons via trans-synaptic spread [14]. Cell-to-cell transmission also occurs.

Mechanisms of damage

Viral infections of neurons may cause neuronal injury through direct damage to the cell itself, often through lysis or impaired metabolic function. Virus-encoded proteins may also cause injury to the neuronal support cells, microglia and astrocytes, resulting in their dysfunction and their concomitant failure to support neuronal function.

The host and the virus

In this section we provide an overview of the host's immune response to viral infections of the nervous system.

The CNS innate immune response to viral infection

The innate immune response is the host's immediate, first-line defense against pathogens. Pathogens are recognizable to the host via their molecular patterns, known as pathogen-associated molecular patterns (PAMPs), examples of which include single- and double- stranded viral RNA and bacterial lipopolysaccharides [5]. The innate immune response is triggered when PAMPs are recognized by the host's Toll-like receptors (TLRs), 10 of which have been identified to date in the mammalian host [15]. Within the brain, TLRs are richly distributed on microglia and astrocytes, as these cells are the chief resident CNS cells involved in the innate immune response [7].

The triggering of TLRs on microglia and astrocytes by viruses and other pathogens transposes these cells from a quiescent to an activated state and thereby affords them a number of properties. First, they produce the antimicrobial agents nitrous oxide (NO) and type 1 interferons (IFN-α and IFN-β). IFN-β is the principal IFN produced in the CNS and is less neurotoxic than IFN-α; it has a direct antiviral effect and mitigates viral spreading and replication. Neurons infected with viruses also produce IFN-β and can signal their distress to and activate nearby microglia and astrocytes. Second, viral infection within the CNS triggers the expression of MHC on microglia, perivascular macrophages, and astrocytes. Third, activated microglia and astrocytes produce both cytokines and chemokines. Cytokines are proteins and peptides that send a broad array of proinflammatory signals to different cell types; these include interleukin-1 and -6 and tumor necrosis factor-α (TNF-α). Chemokines are small cytokines that send chemotactic signals to migratory cells such as macrophages, T cells, and monocytes. Chemokines are produced by activated microglia, and astrocytes and include CCL2 (or MCP-1, monocyte chemotactic protein-1), CCL3 (or MIP-α, macrophage inflammatory proteinα), CCL4 (or MIP-β, macrophage inflammatory proteinβ), CCL5 (or RANTES), and CXCL10 (or IP10, interferon-γ-induced protein of 10 kDa).

These and other products of the inflamed CNS act as signals to the peripheral immune system that a CNS viral infection is present. In response to the cytokines listed above, CVEs produce intercelluar adhesion molecules, vascular cell adhesion molecules, and matrix metalloproteinases, which increase the permeability of the BBB to facilitate the entry of cells of the peripheral immune system into the brain at around day 3 or 4 following CNS infection [5,16]. In sum, these events herald the onset of the adaptive immune response.

The CNS adaptive immune response to viral infection

The cells from the peripheral immune system that arrive and infiltrate the CNS in response to cytokine and chemokine signaling from the CNS antigen-presenting cells are typically made up of a mixture of natural killer cells, antigen-specific $CD8^+$ and $CD4^+$ T cells, B cells, and monocytes/macrophages, and typically arrive in that order [16]. Antigen is presented to the $CD8^+$ and $CD4^+$ T cells, antigen-specific antibodies are produced by B cells, and monocytes/macrophages are able to engulf debris.

Clearance of RNA viruses

Griffin [16] reviewed the mechanisms by which RNA viruses are cleared from the nervous system. Broadly, the clearance of RNA viruses from neurons is somewhat dependent on the type of viral infection present: antibodies are certainly key to controlling neuronal viral infections in a manner that precludes the risk of cytolysis of important neurons. Although the manner by which antibodies achieve viral clearance or at least replicative control within neurons is not fully understood, the model of Sindbis virus infection of mouse neurons shows that bivalent antibodies crosslink viral proteins present on the surface of infected neurons and thereby allow restoration of key neuronal functions [17]. This process may be augmented by type 1 IFNs and activated $CD8^+$ cells [16]. Production of IFN-γ by $CD4^+$ and $CD8^+$ cells is also key to control of viral infections of neurons that are richly imbued with IFN-γ receptors. Neuroglial cells are more typically the recipients of assistance from T cells for viral clearance, which may involve lysis of infected glial cells. Notably however, oligodendrocytes do require IFN-γ for viral clearance [18]. Virus-specific antibodies produced by B cells may assist in the long-term prevention of viral replication in neuroglial cells.

Clearance of DNA and retroviruses

In some neuronal DNA infections, notably HSV-1–activated $CD8^+$ cells are the key contributors to neuronal viral clearance and control but the mechanisms by which they achieve this, in a noncytolytic manner, are not well understood and ultimately HSV-1 can remain latent in the geniculate ganglion out of reach of the immune system.

Similarly, viral clearance of the retrovirus HIV-1 from its cellular reservoirs within the CNS (perivascular macrophages and possibly astrocytes) is not (yet) possible but HIV-1 infection of the CNS can be regulated and controlled through both antibody and cell-mediated immunity. CSF plasmablast levels correlate strongly with CSF HIV viral load and disappear from the CSF as the CSF HIV viral load falls following commencement of antiretroviral infection, suggesting that there is a localized humoral immune response directed against HIV within the CNS [19]. Evidence of cellular immune control of HIV-1 being integral to control of CNS infection is gleaned from a small number of histopathological studies undertaken in patients with early and asymptomatic HIV infection wherein cellular cytotoxic infiltrates, increased chemokine expression, and up-regulation of microglial activation markers have been observed. McCrossan and colleagues [20] found that there was a significant positive correlation between the proportion of $CD8^+$ T cells and the HIV proviral load in brain tissues, suggesting that the level of immune surveillance increases as the level of HIV proviral DNA increases. The authors propose that with progressive peripheral immunosuppression, the integrity of the CNS surveillance afforded by cellular immunity wanes and hence the brain is vulnerable to up-regulated local CNS infection and increasing numbers of infected monocyte/macrophages trafficking into the CNS [20].

The host immune response may be neuroprotective

Recently, Crack and Bray [7] highlighted the salutary role that activated brain macrophages and microglia may play as agents of neuroprotection in CNS trauma and stroke. Further they noted that TLRs may also play a role in regulating the degree of immune activation of microglia and astrocytes. Recently CD4+CD25 T regulatory cells have been shown to afford neuroprotection in Parkinson's disease via a modulatory effect upon microglia [21]. Human interferon-β has been reported to be neuroprotective following stroke and cord contusion in animal models [22]. Currently there are few data to demonstrate what neuroprotective properties the cytokines and chemokines involved in neuroinflammation secondary to viral infection may have, but it is anticipated that studies will be forthcoming.

Host genetics and viral infections of the nervous system. Host genetics appear to play an as yet confined role in increased risk for some viral infections of the nervous system, notably HSV-1. UNC-93B is an endoplasmic reticulum protein that is involved in activation of TLRs 3, 7, and 9 [23]. Autosomal recessive deficiency of UNC-93B leads to impaired cellular production of interferon-α/β and -γ and has been linked to HSV-1 encephalitis in two children with this autosomal condition [23]. Further, autosomal dominant Toll-like receptor 3 (TLR3) deficiency has also been linked to an increased risk for HSV-1 encephalitis [24]. Gonzalez and colleagues [25] have demonstrated that HIV seropositive patients with the monocyte chemotactic protein-1 (MCP-1) 2578G allele have a 4.5-fold increased risk of HIV-associated dementia.

Clinical manifestations of viral diseases of the nervous system. As shown in Table 1, viruses are associated with a number of different neurologic presentations that broadly reflect the underlying cell type involved.

Meningitis. Acute viral meningitis is a viral infection of the cells of the meninges. Cells targeted by viruses include meningeal macrophages and ependymal cells. The host immune response leads to increased permeability of the BBB, a rise in CSF IL-6, IFN-γ, and IL1-β [26–28], the influx of activated T cells, presumably sensitized to the antigen through contact with it in peripheral tissues and organs, and a B cell infiltrate. The host inflammatory response and the resultant meningeal inflammation manifests as headache, fever, and neck stiffness in the patient. Cranial nerve meninges may also become inflamed, hence patients may develop cranial nerve palsies; this may be seen with a chronic form of HIV-1 meningitis following HIV-1 seroconversion, but is generally uncommon with viral meningitis. Raised intracranial pressure (ICP) that leads to a change in mentation may occur in viral meningitis, but is uncommon.

In addition to the neurologic features of viral meningitis, the patient may show other signs and symptoms suggestive of other organ system involvement: patients with enteroviral meningitis may have had a gastrointestinal illness preceding their meningitis, rash may be present, or occasionally pericarditis or pleurisy are noted; children may have features of hand-foot-and-mouth disease. Patients with acute HIV seroconversion illness and meningitis may give a history of recent HIV exposure [29].

Encephalitis. Encephalitis denotes inflammation of the brain parenchyma; neurons are the target cells in this disease. As discussed in an earlier section, the host's innate and adaptive immune responses lead to production of antiviral cytokines and a cellular immune response within the brain parenchyma. Patients with encephalitis present with a change in their consciousness that occurs early in the illness and may manifest as confusion, which may progress to stupor, then coma. Meningoencephalitis may also occur wherein both the meninges and the brain parenchyma are involved in the infective process. These patients may present with signs and symptoms of both encephalitis and meningitis, which may be diagnostically challenging.

Depending on the area of the brain involved, patients may present with focal signs and seizures. HSV-1, one of the hallmark viruses that causes encephalitis at a global level, infects the medial temporal lobe and the orbital surface of the frontal lobe hence patients present with bizarre behavior, speech disturbance and hallucinations that signify the nature of the lobes involved [8]. Japanese B encephalitis virus has a penchant for the thalamus and parts of the basal ganglia: this gives the clinician a clue wherein approximately 30% of patients infected with this virus will evince signs of extrapyramidal involvement with tremor, rigidity, and abnormal movements [30]. Brainstem encephalitis, or "rhombencephalitis," is a rare infection of the brainstem and the virus that typically causes it is HSV-1. Patients present with cranial nerve palsies and signs of cerebellar and long tract involvement. Finally, raised ICP can occur in viral encephalitis, notably in HSV-1 encephalitis and Japanese B encephalitis, and may lead to signs of incipient brain herniation followed by death.

Myelitis. Myelitis denotes inflammation of the spinal cord and may involve sensory and motor tracts and neuronal cells within the anterior and posterior horns. Patients with myelitis secondary to polio or the other enteroviruses—coxsackie and echoviruses—develop a flaccid paralysis without sensory signs. Patients with transverse myelitis develop a sensory level, limb weakness, back pain, and bladder and bowel disturbance. Encephalomyelitis may also occur, hence patients present with features of encephalitis and myelitis.

The diagnostic approach to viral infections of the nervous system.

Clinical history. While the clinician should not struggle to make a clinical diagnosis of meningitis or encephalitis, the challenge comes in determining the infectious etiology of the presenting illness: virus or bacteria (or occasionally fungus)? To this end, Table 2 may be useful to the clinician, ie, what sort of exposure has the patient had to a possible causative virus?

Travel history is vital and if the patient has not traveled, is there an endemic virus in the area in which you practice that could be responsible for the illness? A history of exposure to animals, tick bites, or mosquito bites can be useful clues in the differential diagnosis. The season during which the patient presents should not escape the clinician, for example enteroviruses are the commonest cause of childhood viral meningitis and typically occur during summer and fall. Finally, clinicians must consider the patient before them: if a child, then certain viruses are commoner or evince more severe neurologic disease in children than in adults. The immunocompromised patient with a neurologic infection is always diagnostically challenging, but the list of leading suspect viruses in this patient group is relatively well defined (see Table 2). The clinical neurologist may be asked to see a patient with an unusual or rapid-onset neurologic disease following organ donation; although very rare, organ and tissue donations are increasingly being reported as vectors of serious viral infections of the nervous system, including rabies and West Nile virus [31,32]. Hence, before embarking on the tangible diagnostics described as follows, much can be gained from a detailed history.

Neuroimaging. The benefits of performing a CT or MRI scan of the brain in patients with features of meningitis, encephalitis, or both are twofold: the scan may provide some (typically nonspecific) evidence suggestive of an underlying viral process, eg, changes in the temporal lobe suggestive of HSV-1 encephalitis or changes in the basal ganglia and thalamus as seen in both West Nile Virus and Japanese B encephalitis [33]. More importantly, the scan can rule out the presence of a space-occupying lesion with edema and mid-line shift, or features of obstructive hydrocephalus, both of which make performance of a lumbar puncture dangerous. Hence, CT or MRI brain scans are routinely performed before undertaking a lumbar puncture in patients with suspected meningitis or encephalitis. In patients with myelitis, neuroimaging may be useful and may show cord lesions and inflammation of contiguous spinal nerve roots.

Lumbar puncture and CSF analysis. Obtaining CSF for analysis is one of the most valuable steps that the clinician can undertake on this diagnostic pathway. CSF analysis provides the opportunity to distinguish between viral, bacterial, and fungal causes of meningitis and encephalitis. CSF should be sent for a total white cell count, glucose and protein, Gram stain, bacterial culture, and, where indicated, fungal and mycobacterial cultures. In Table 3, the typical CSF findings in patients with acute bacterial and viral meningitis and viral encephalitis are provided.

CSF nucleic acid amplification using PCR testing is useful and sensitive for some of the more common viruses including the enteroviruses, herpes viruses (HSV-1 and 2, VZV, CMV, EBV), and the less common viruses JC virus, mumps, and influenza virus [34]. Polymerase chain reaction (PCR) testing for the West Nile virus is currently approved for screening of blood before blood or organ donation only [35], although recently universal primers for the flavivirus group to which West Nile belongs were

Table 3
Typical patterns of cerebrospinal fluid parameters in acute meningitis and encephalitis

CSF parameter	Bacterial meningitis	Viral meningitis	Viral encephalitis
White cells			
Total count (N: <5 cells/mm^3)	1000–5000 cells/mm^3	Typically <500 cells/mm^3	Variable; may be less than 10 cells or as high as 2000 cells/mm^3
Differential (normally lymphocytes; no neutrophils or red cells)	Neutrophils predominate but lymphocytes may be present early in illness	Lymphocytes predominate but neutrophils may be present	Lymphocytes predominate and red cells may be present
Glucose concentration (NR: 2.8–4.4 mmol/L or 50.4–79.2 mg/dL)	Decreased	Normal	Normal
CSF: serum glucose ratio (N: >0.60)	Decreased	Normal	Normal
Protein concentration (N: <0.45 g/L or <45 mg/dL)	Elevated	Normal or slightly increased	Normal or slightly increased
Gram's stain	In 70%–90% cases it will be positive	Negative	Negative

identified that allow for accurate one-step PCR testing [36]. Of note, the current recommendation for diagnosis of West Nile CNS viral infections is by detection of West Nile virus–specific IgM antibody in CSF using an ELISA technique [37]. Japanese B virus detection using PCR is sensitive and well established [38]. Unfortunately, PCR techniques for other vector-borne encephalitic viruses outlined in Table 2 are not as well optimized [34].

Other techniques for viral isolation. In some patients, throat and nasopharyngeal swabs and stool samples may be sent for viral culture and/or PCR, notably patients with suspected enteroviral or influenza infections of the CNS.

Serology. Serology may by useful in the setting of suspected HIV-1 or West Nile virus infections. Acute and convalescent serology may be useful for those infections where CSF PCR is negative including enteroviruses, measles, mumps, rubella, and influenza: in these instances a fourfold rise is requisite for diagnosis.

EEG testing. EEG testing is an adjunctive test and is most notably performed in the setting of suspected HSV-1 encephalitis where changes of periodic sharp and slow wave complexes are seen localized over the fronto-temporal regions of the brain.

Brain biopsy. The interposition of a brain biopsy in the diagnostic process implies three things. First, the patient has a serious neurologic illness for which a diagnosis is required for either therapeutic or prognostic reasons, or both. Second, the patient's illness has not been ascertained through the neuroradiological or CSF analysis metrics discussed above. Third, the risk of an adverse outcome secondary to the biopsy is outweighed by the perceived benefits of a histopathological diagnosis. The overall risk for a major morbidity rate following a stereotactic brain biopsy is 3.5% and the operative mortality rate is 0.7% [39], but the risk may be higher in different subgroups [40]. The diagnostic yield has been cited to be as high as 91% [39].

Once conceded as necessary and consented to by the patient, a successful brain biopsy requires a facilitated process between the clinician, the radiologist, the neurosurgeon, and the neuropathologist. In the setting of diffuse encephalitis, as opposed to a discrete lesion that can be specifically targeted, the brain biopsy is typically taken from the nondominant frontal lobe. Tissue should be obtained for culture, frozen section, and formalin fixation. Appropriate stains and routine immunoperoxidase tests may yield a diagnosis, or PCR analysis for individual viruses may need to be performed on paraffin-embedded tissue.

Summary

Through an understanding of the make-up of both the healthy nervous system and the refined nuances of its immune system, a fuller appreciation of the pathogenesis and thereby the clinical history and presentation and the appropriate investigations of the patient with a viral infection of his or her nervous system is afforded to the clinician.

References

[1] Van Dooren S, Salemi M, Vandamme AM. Dating the origin of the African human T-cell lymphotropic virus type-i (HTLV-I) subtypes. Mol Biol Evol 2001;18(4):661–71.
[2] Leslie KS, Levell NJ. Skin disease in mummies. Int J Dermatol 2006;45(2):161–3.
[3] Marr JS, Calisher CH. Alexander the Great and West Nile virus encephalitis. Emerg Infect Dis 2003;9(12):1599–603.
[4] Beutler B, Eidenschenk C, Crozat K, et al. Genetic analysis of resistance to viral infection. Nat Rev Immunol 2007;7(10):753–66.
[5] Bailey SL, Carpentier PA, McMahon EJ, et al. Innate and adaptive immune responses of the central nervous system. Crit Rev Immunol 2006;26(2):149–88.
[6] Carson MJ, Doose JM, Melchior B, et al. CNS immune privilege: hiding in plain sight. Immunol Rev 2006;213:48–65.
[7] Crack PJ, Bray PJ. Toll-like receptors in the brain and their potential roles in neuropathology. Immunol Cell Biol 2007;85(6):476–80.
[8] Gilden DH, Mahalingam R, Cohrs RJ, et al. Herpesvirus infections of the nervous system. Nat Clin Pract Neurol 2007;3(2):82–94.
[9] Samuel MA, Wang H, Siddharthan V, et al. Axonal transport mediates West Nile virus entry into the central nervous system and induces acute flaccid paralysis. Proc Natl Acad Sci U S A 2007;104(43):17140–5.
[10] Gonzalez-Scarano F, Martin-Garcia J. The neuropathogenesis of AIDS. Nat Rev Immunol 2005;5(1):69–81.
[11] Gorry PR, Sterjovski J, Churchill M, et al. The role of viral coreceptors and enhanced macrophage tropism in human immunodeficiency virus type 1 disease progression. Sex Health 2004;1(1):23–34.
[12] Elphick GF, Querbes W, Jordan JA, et al. The human polyomavirus, JCV, uses serotonin receptors to infect cells. Science 2004;306(5700):1380–3.
[13] Daley JK, Gechman LA, Skipworth J, et al. Poliovirus replication and spread in primary neuron cultures. Virology 2005;340(1):10–20.
[14] Sik A, Cote A, Boldogkoi Z. Selective spread of neurotropic herpesviruses in the rat hippocampus. J Comp Neurol 2006;496(2):229–43.
[15] Bowie A, O'Neill LA. The interleukin-1 receptor/Toll-like receptor superfamily: signal generators for pro-inflammatory interleukins and microbial products. J Leukoc Biol 2000; 67(4):508–14.
[16] Griffin DE. Immune responses to RNA-virus infections of the CNS. Nat Rev Immunol 2003; 3(6):493–502.
[17] Despres P, Griffin JW, Griffin DE. Effects of anti-E2 monoclonal antibody on sindbis virus replication in AT3 cells expressing bcl-2. J Virol 1995;69(11):7006–14.
[18] Parra B, Hinton DR, Marten NW, et al. IFN-gamma is required for viral clearance from central nervous system oligodendroglia. J Immunol 1999;162(3):1641–7.
[19] Cepok S, von Geldern G, Nolting T, et al. Viral load determines the B-cell response in the cerebrospinal fluid during human immunodeficiency virus infection. Ann Neurol 2007; 62(5):458–67.
[20] McCrossan M, Marsden M, Carnie FW, et al. An immune control model for viral replication in the CNS during presymptomatic HIV infection. Brain 2006;129(Pt 2):503–16.
[21] Reynolds AD, Banerjee R, Liu J, et al. Neuroprotective activities of CD4+CD25+ regulatory T cells in an animal model of Parkinson's disease. J Leukoc Biol 2007; 82(5):1083–94.

[22] Gok B, Okutan O, Beskonakli E, et al. Effect of immunomodulation with human interferon-beta on early functional recovery from experimental spinal cord injury. Spine 2007;32(8): 873–80.
[23] Casrouge A, Zhang SY, Eidenschenk C, et al. Herpes simplex virus encephalitis in human UNC-93B deficiency. Science 2006;314(5797):308–12.
[24] Zhang SY, Jouanguy E, Ugolini S, et al. TLR3 deficiency in patients with herpes simplex encephalitis. Science 2007;317(5844):1522–7.
[25] Gonzalez E, Rovin BH, Sen L, et al. HIV-1 infection and AIDS dementia are influenced by a mutant MCP-1 allele linked to increased monocyte infiltration of tissues and MCP-1 levels. Proc Natl Acad Sci U S A 2002;99(21):13795–800.
[26] Frei K, Leist TP, Meager A, et al. Production of B cell stimulatory factor-2 and interferon gamma in the central nervous system during viral meningitis and encephalitis. Evaluation in a murine model infection and in patients. J Exp Med 1988;168(1):449–53.
[27] Chonmaitree T, Baron S. Bacteria and viruses induce production of interferon in the cerebrospinal fluid of children with acute meningitis: a study of 57 cases and review. Rev Infect Dis 1991;13(6):1061–5.
[28] Ramilo O, Mustafa MM, Porter J, et al. Detection of interleukin 1 beta but not tumor necrosis factor-alpha in cerebrospinal fluid of children with aseptic meningitis. Am J Dis Child 1990;144(3):349–52.
[29] Brew BJ. HIV neurology. 1st edition. New York: Oxford University Press; 2001.
[30] Diagana M, Preux PM, Dumas M. Japanese encephalitis revisited. J Neurol Sci 2007; 262(1–2):165–70.
[31] Srinivasan A, Burton EC, Kuehnert MJ, et al. Transmission of rabies virus from an organ donor to four transplant recipients. N Engl J Med 2005;352(11):1103–11.
[32] CDC. West Nile virus infections in organ transplant recipients—New York and Pennsylvania, August–September, 2005. MMWR Morb Mortal Wkly Rep 2005;54:1–3.
[33] Petropoulou KA, Gordon SM, Prayson RA, et al. West Nile virus meningoencephalitis: MR imaging findings. AJNR Am J Neuroradiol 2005;26(8):1986–95.
[34] Cinque P, Bossolasco S, Lundkvist A. Molecular analysis of cerebrospinal fluid in viral diseases of the central nervous system. J Clin Virol 2003;26(1):1–28.
[35] FDA. FDA approves second West Nile virus screening test for donated blood and organs. FDA News 2007. Available at: http://www.fda.gov/bbs/topics/NEWS/2007/NEW01691.html. Accessed April 24, 2008.
[36] Maher-Sturgess SL, Forrester NL, Wayper PJ, et al. Universal primers that amplify RNA from all three flavivirus subgroups. Virol J 2008;5(1):16.
[37] CDC. West Nile Virus: information and guidance for clinicians. Available at: http://www.cdc.gov/ncidod/dvbid/westnile/clinicians/. Accessed February 3, 2008.
[38] Santhosh SR, Parida MM, Dash PK, et al. Development and evaluation of SYBR green I-based one-step real-time RT-PCR assay for detection and quantitation of Japanese encephalitis virus. J Virol Methods 2007;143(1):73–80.
[39] Hall WA. The safety and efficacy of stereotactic biopsy for intracranial lesions. Cancer 1998; 82(9):1749–55.
[40] McGirt MJ, Woodworth GF, Coon AL, et al. Independent predictors of morbidity after image-guided stereotactic brain biopsy: a risk assessment of 270 cases. J Neurosurg 2005; 102(5):897–901.

ELSEVIER
SAUNDERS

Neurol Clin 26 (2008) 635–655

NEUROLOGIC
CLINICS

Aseptic Meningitis and Viral Myelitis

David N. Irani, MD

Department of Neurology, Holtom-Garrett Program in Neuroimmunology, University of Michigan Medical School, BSRB Room 4007, 109 Zina Pitcher Place, Ann Arbor, MI 48109–2200, USA

Viral infections of the central nervous system (CNS) encompass both acute and chronic conditions caused by a broad range different pathogens. As a group, these diseases can have a complex and variable pathogenesis that is influenced by host, viral, and environmental factors. In terms of frequency, the number of cases of viral meningitis that occurs annually exceeds the total number of meningitis cases caused by all other etiologies combined. This disorder results following widespread viral dissemination to the meningeal coverings of the CNS. Focal CNS infections such as occur in the spinal cord with viral myelitis, however, are much less common and may be confused with noninfectious disorders that cause acute flaccid paralysis (AFP). This article reviews some of the important clinical features, epidemiology, diagnostic approaches, and management strategies for patients with aseptic meningitis and viral myelitis. Particular focus is placed on the diseases caused by enteroviruses (EVs), which as a group account for most aseptic meningitis cases and many focal infections of the spinal cord.

Aseptic meningitis

The term "aseptic meningitis" describes a clinical syndrome characterized by meningeal inflammation not caused by an identifiable bacterial pathogen in the cerebrospinal fluid (CSF) [1]. It also distinguishes a group of disorders that do not typically cause notable parenchymal involvement of the brain (encephalitis) or spinal cord (myelitis). Implicit in the definition of aseptic meningitis is a somewhat more benign clinical course compared with either bacterial meningitis or the hybrid syndromes of meningoencephalitis or encephalomyelitis. As might be anticipated, however, there is sufficient enough clinical overlap among the infectious causes of aseptic meningitis,

Supported by Grant No. AI057505 from the National Institutes of Health.
E-mail address: davidira@med.umich.edu

doi:10.1016/j.ncl.2008.03.003 ***neurologic.theclinics.com***

meningoencephalitis, and encephalomyelitis to require that all etiologic agents be considered in an individual patient. It also bears remembering that aseptic meningitis can also occur in the setting of an underlying connective tissue disorder or malignancy, or following certain drug ingestions or administrations, which necessitates a search for the noninfectious etiologies of this disorder. The aseptic meningitis syndrome is reviewed here, with particular emphasis on the viral causes of this disorder.

Clinical features

Clinical disease observed in patients with viral meningitis can vary with the host's age and underlying immune status, and can span the spectrum of an asymptomatic CSF pleocytosis to an illness causing an alarming degree of neurologic impairment. Despite this heterogeneity, however, most patients with aseptic meningitis present with fever accompanied by complaints of headache, stiff neck, malaise, anorexia, and vomiting. In neonates, CNS involvement may or may not be evidenced by overt signs of meningeal inflammation (nuchal rigidity, bulging of the anterior fontanelle), but in the setting of infections caused by EVs, the neonatal population often shows evidence of major systemic involvement in the form of hepatic necrosis, myocarditis, and necrotizing enterocolitis [2]. Indeed, such multiorgan failure may closely resemble overwhelming bacterial sepsis. The CNS disease caused by EVs in newborns can be more overt and progress to a more encephalitic picture, however, with the appearance of seizures and focal neurologic deficits. Death in this setting, however, is still more likely to occur as the result of hepatic or cardiac failure [2,3]. The morbidity and mortality that accompany perinatal EV infections has been estimated to be as high as 70% and 10%, respectively [2–5].

In older infants and children, EV meningitis is rarely fatal but may still be complicated by significant short-term morbidity and prolonged clinical recovery. These patients present with the abrupt onset of fever to 38°C to 40°C; temperature curves are often biphasic with the first peak accompanying systemic constitutional symptoms and the second one reappearing with the onset of meningeal signs [2,6]. If old enough to report them, affected children complain of headache, photophobia, and myalgias, and many also experience vomiting, diarrhea, cough, sore throat, and rash [2,6]. Some EV serotypes also produce distinctive clinical stigmata. During a large EV-71 outbreak in Taiwan, young patients had a characteristic hand-foot-and-mouth disease with vesicular lesions erupting over these body regions [7]. Febrile seizures may accompany EV meningitis in children without other evidence of parenchymal CNS involvement, and the syndrome of inappropriate antidiuretic hormone secretion can occasionally be seen [8]. Overall, older infants and children often are symptomatic for more than a week with these infections, causing significant economic impact on care providers [9]. Eventually, however, full recovery ensues, and despite lingering suspicions otherwise, there is no convincing evidence that these infections lead to any subsequent neurodevelopmental abnormalities [10]. One exception to this rule occurs in young patients with congenital hypo- or agammaglobulinemia; the antibody-dependent mechanisms critical for EV clearance from the CNS mean that a chronic meningitis or meningoencephalitis can develop in this immunologic setting, often with a fatal outcome [11]. Survivors of chronic EV infections can develop superimposed rheumatologic disorders, commonly in the form of dermatomyositis, which are believed to result from lingering extracerebral tissue infection [11].

Older infants and children are also susceptible to other forms of viral meningitis beyond those caused by EVs, albeit at much lower rates. The mosquito-borne flaviviruses and bunyaviruses (especially St. Louis encephalitis virus and La Crosse virus), mumps virus (either with or without an accompanying parotitis), and various members of the herpesvirus family (herpes simplex virus [HSV]-1 and -2, varicella-zoster virus [VZV], and human herpesvirus type 6) all can cause a meningitic disease in this age group. In most cases a distinguishing clinical feature that identifies a particular pathogen is not observed, and a specific diagnosis rests heavily on epidemiologic and laboratory data (reviewed later). Occasionally, however, a highly distinctive rash (VZV) or the occurrence of parotitis (mumps) can render a specific diagnosis clear. Although cases of infections caused by all of these pathogens where a more encephalitic picture is followed by an adverse outcome are reported, the aseptic meningitis caused by these agents is typically self-limited and followed by full recovery in most patients [2].

Adults with EV meningitis may have symptoms that persist for several weeks, although the overall severity of these illnesses tends to be somewhat less than in children [12]. Other forms of viral meningitis in adults are, for the most part, similarly benign. Primary genital infections caused by HSV-2 are accompanied by meningitis in more than one third of women and 11% of men [13], although many cases of both HSV-1 and HSV-2 meningitis are reported without any recent occurrence of genital lesions [14,15]. These two pathogens must also be kept in mind in adults with recurrent episodes of aseptic meningitis; so-called "Mollaret's meningitis" has been linked by polymerase chain reaction (PCR) with HSVs, particularly HSV-2, in the absence of signs or symptoms of concurrent genital infection [16]. In the clinical setting of aseptic meningitis accompanied by acute weakness of the extremities occurring in the summer or fall months, West Nile virus (WNV) has become the modern day version of paralytic poliomyelitis as the most common infectious cause of AFP in the Western hemisphere [17]. This diagnosis must be considered in any patient with a CSF pleocytosis and clinical or electrophysiologic findings consistent with lower motor neuron involvement.

Epidemiology

Because more than 90% of all viral meningitis cases are caused by EVs, patterns associated with the occurrence of this syndrome at a population level are driven largely by the epidemiology of these pathogens. Overall EVs occur in a worldwide distribution [18], although only a handful of specific serotypes predominate in a particular part of the world in any given year [19]. Humans are their only natural reservoir, and they are transmitted primarily by fecal-oral contamination and less commonly in respiratory secretions [18]. As such, EVs exhibit a summer-to-fall seasonality in temperate climates and a high year-round incidence in tropical and subtropical areas (ostensibly because of sparse clothing and lower hygiene standards among children in these environments) (Fig. 1). Indeed, although EVs are still the most common cause of viral meningitis in adults [12], most cases occur in children under the age of 5 years [2–6,18]. In the United States, the 15 or so most common serotypes that cause disease cycle with varying periodicity, likely reflecting the birth of new susceptible hosts (ie, nonimmune children) within a given community [19]. Occasional outbreaks in adults are caused by those serotypes that have not been present in a community for some time, again because a pool of susceptible hosts without pre-existing immunity needs a longer time to develop [20]. Among the many EV serotypes that cycle from year to year, certain ones are also more associated with the development of aseptic meningitis than others. A more detailed review of these pathogens is covered elsewhere in this issue.

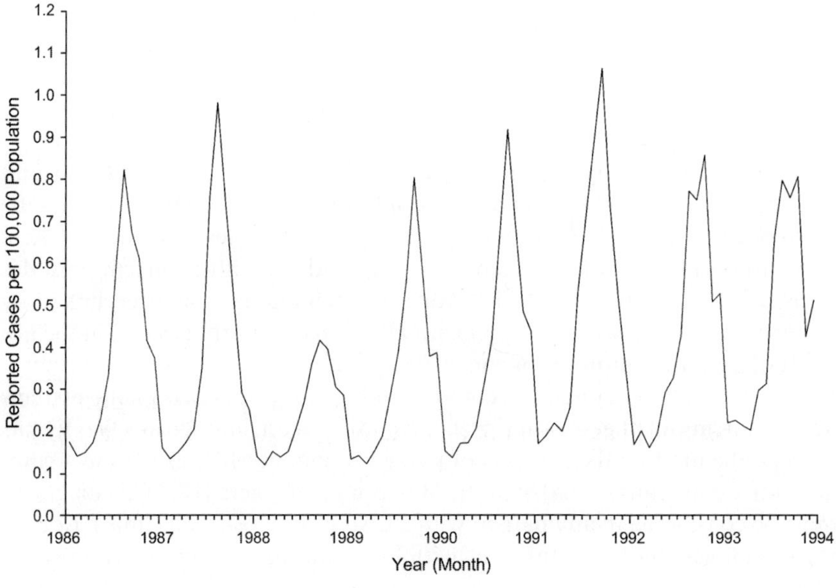

Fig. 1. The seasonal occurrence of aseptic meningitis in the United States from 1986 to 1994, as reported to the Centers for Disease Control and Prevention. The striking peak of cases during the summer and early fall months of each year reflects the predominance of enteroviruses as the main etiologic agents of this disorder. (*From* Centers for Disease Control. Summary of notifiable diseases, United States, 1993. MMWR Morb Mortal Wkly Rep 1994;42:1–73.)

Many non-EV cases of viral meningitis also demonstrate seasonality, even as much less common causes of this syndrome. For vector-borne pathogens, such as St. Louis encephalitis virus, WNV, and La Crosse virus, the resultant disease occurs predominantly during the summer and fall months in geographic regions where infected mosquitoes are abundant. Often only a small proportion ($<5\%$) of these arthropod-borne infections actually produce symptoms, and even when they do, febrile illnesses may or may not have any accompanying signs of meningitis. St. Louis encephalitis virus was first identified in the Midwestern United States, but it is now recognized to cause disease throughout Central and North America. The clinical picture with symptomatic infections caused by this pathogen is more meningitic in children but tends to be more encephalitic in older patients [21]. Paradoxically, exposure to infected vectors among all age groups can occur indoors as often as outdoors, and open windows are a known predisposing factor [21]. For WNV, neuroinvasive disease occurs in only 1% to 2% of all infections, but nearly half of these cases (more than 20,000 since 1999) are manifest primarily as aseptic meningitis clustering from March through October [17]. Each year since 1999 has brought clusters of WNV cases to different regions of the United States, Canada, and Mexico. La Crosse virus is endemic in forested regions of the United States, especially around the Great Lakes, and mostly causes aseptic meningitis in children [2].

Mumps virus is transmitted by respiratory secretions and clinical infections predominate in the winter and early spring months when person-to-person transmissions are most common. With the widespread use of the attenuated live-virus vaccine, however, mumps has gone from being one of the most common identifiable causes of meningitis to being only rarely diagnosed in the United States today [22]. When it occurs, symptomatic mumps infection is twice as frequent in males as it is in females, and neurologic involvement is three times more common in the male population [2]. These unusual gender differences in the pathogenesis of mumps infection have never been adequately explained.

Differential diagnosis

Patients who develop the syndrome of acute aseptic meningitis may or may not become ill as a result of an underlying infection. Of primary importance is the exclusion of partially treated bacterial meningitis that can occur in the setting of preceding antibiotic exposure and that requires further antibiotic therapy. Other bacteria, such as mycoplasma, spirochetes (borrelia, syphilis, and leptospira), mycobacteria, and brucella, along with various fungi, can occasionally present as acute aseptic meningitis. Noninfectious etiologies are rare, but include certain drugs (especially nonsteroidal anti-inflammatory drugs and intravenous immunoglobulin); vaccines; systemic illnesses including connective-tissue disorders (notably systemic lupus erythematosus); vasculitides (eg, Kawasaki disease); granulomatous conditions (mostly sarcoidosis); neoplasms that have spread to the meninges; and rarely, migraine. The broad differential diagnosis of the acute aseptic meningitis syndrome segregated based on the relative frequency of each disorder in the general population is outlined in Table 1.

Pathophysiology and pathogenesis

Viruses that cause meningitis spread to the CNS by extracerebral routes. For EVs, it is likely that the initial inoculum is swallowed and passes into the lower intestinal tract where it infects enterocytes [23]. The virus then traverses the intestinal wall and moves into gut-associated lymphoid tissue, such as Peyer's patches, where primary replication occurs. A viremia ensues, and multiple tissues (liver, lungs, heart, CNS) can be seeded [23]. Even if the

Table 1
Causes of the aseptic meningitis syndrome segregated based on their relative frequency in the general population

Common	Uncommon	Rare
Viruses	*Viruses*	*Viruses*
Enteroviruses	Mumps	HSV-1
Arboviruses[a,b]	LCMV	VZV
HSV-2	HIV	CMV
	HHV6	EBV
		Measles
		Influenza A and B
		Parvovirus B19
Bacteria	*Bacteria*	
Borrelia burgdorferi[b]	*Mycobacterium tuberculosis*	
Partially treated bacterial meningitis (common agents)	*Leptospira* spp[b]	
	Mycoplasma pneumoniae	
Parameningeal bacteria infection (sinusitis, otitis, mastoiditis)		
	Other	*Other*
	Fungi[b]	*Brucella* spp
		Fungi
		Autoimmune diseases (lupus, Sjögren's)
		Drugs (NSAIDs, IVIg)
		Malignancy
		Vasculitis (Kawasaki)

Abbreviations: CMV, cytomegalovirus; EBV, Epstein-Barr virus; HHV, human herpesvirus; HSV, herpes simplex virus; LCMV, lymphocytic choriomeningitis virus; VZV, varicella-zoster virus.

[a] Arthropod-borne viruses (including alphaviruses, flaviviruses, bunyaviruses, and reoviruses).

[b] Incidence varies significantly with geographic region.

Data from Rotbart HA. Viral meningitis and the aseptic meningitis syndrome. In: Scheld WM, Whitley RJ, Durack DT, editors. Infections of the central nervous system. Second edition. Philadelphia: Lippincott-Raven; 1997. p. 23–46.

meninges do not become infected at this stage, further replication at these extracerebral sites can produce a second viremia that then causes neurologic involvement [23]. The pathogenesis of routine EV infections within the CNS is not very well understood because pathologic data are scarce and fatal cases invariably reflect a more severe encephalitic syndrome rather than more typical meningitis. Still, a single case of a child who died of coxsackievirus B5 myocarditis and who also had meningitis has been reported; here, prominent inflammation of the choroid plexus and ependymal lining of the ventricles was described [24]. Fibrotic changes in the basal leptomeninges were also seen, but there was limited direct parenchymal involvement by the pathogen [24].

For the vector-borne viruses, local subcutaneous replication after the virus is inoculated leads to regional lymph node spread, viremia, and dissemination to the CNS by a hematogenous route [25]. In some cases, prominent infection of olfactory neurons suggests that CNS entry can be achieved by this pathway [26]. Mumps is contracted by respiratory secretions and the virus replicates initially in the upper respiratory epithelium. Local invasion of the parotid gland causes the prototypical parotitis, but viremia and CNS spread can occur in the absence of overt parotid involvement [27]. The passage of infected mononuclear cells across the choroid plexus into the CSF is believed to result in meningoencephalitis [28]. With lymphocytic choriomeningitis virus, an arenavirus previously known to be a major known cause of aseptic meningitis, human transmission occurs by inhalation or ingestion of contaminated rodent urine and feces [29]. Spread to the CNS occurs after replication in the lungs and hilar lymph nodes, likely by the bloodstream [29]. Aseptic meningitis caused by infections from the herpesviruses can occur with either primary or reactivated infection; both hematogenous and intraneuronal routes of CNS dissemination are described.

Common etiologic agents

Enteroviruses

The EVs constitute more than 60 viral serotypes within the picornavirus family; they are subdivided into the polioviruses, the coxsackieviruses, the echoviruses, and the newer numbered EVs. Although genetic variability in the coding region of the capsid protein produces the many serotypes, all EV virions consist of a small icosahedral capsid that surrounds a single strand of positive-sense RNA [18]. Viral particles bind to specific receptors on target cells, allowing for the direct release of the viral genome into the cytoplasm. A single polyprotein is translated directly from this RNA strand, and its rapid cleavage produces individual products that regulate viral RNA transcription and others that create new capsids [18]. Viral replication causes a rapid shut off of host cell protein synthesis, and the cell becomes a factory for new viral production. Finally, direct lysis causes the release of infectious virions that target adjacent cells.

Flaviviruses

St. Louis encephalitis virus and WNV are 2 of 69 known flaviviruses. These pathogens are also small, enveloped viruses containing a spherical ribonucleoprotein core with a single-stranded, positive-sense RNA genome. A membrane and an envelope protein are present in the outer lipid envelope of the virion; variability in the latter, in particular, distinguishes the many viral subtypes, and epitopes therein are the main targets of host immunity. Although virions are inactivated by high temperature, ultraviolet light, gamma-irradiation, and various disinfectants, virus-containing aerosols are stable at room temperature for up to 6 hours making this route of transmission conceivable in certain settings [30]. Cells infected in vitro demonstrate reduced macromolecular synthesis late in the replication process, around the time when cytopathic effects appear. Only some of these viruses are neurotropic in vivo, whereas others infect visceral organs and cause fatal hemorrhagic fevers.

Mumps

Mumps virus is a paramyxovirus that exists as a single serotype. Its virions are larger enveloped structures that contain a ribonucleoprotein core with a single strand of negative-sense RNA. Proteins associated with the core assist with genomic replication and maintenance of the core structure. Two surface glycoproteins mediate neuraminidase, hemagglutination, and fusion functions that allow virus adsorption to host cells and penetration of the genome into the cytoplasm. Mumps RNA replication occurs by the production of a positive-strand intermediate, which serves as both the template for negative-sense RNA replication and messenger RNA for protein translation [31].

Herpesviruses

Herpesviruses are large, enveloped viruses with an icosadeltahedral capsid that surrounds a complex, double-stranded DNA genome. Eight of the nearly 100 known herpesviruses cause observable disease in humans, either as a primary infection or by means of reactivation. Virus assembly takes place in the nucleus, and the production of progeny virus invariably destroys the infected target cell. Although some human herpesviruses (HSV-1, HSV-2) infect a wide range of host cells in vivo, others (Epstein-Barr virus [EBV], human herpesvirus type 6) are much more restricted in their cellular targets [32]. Latency is another property of herpesviruses that influences the type of disease these pathogens can produce, and the nervous system is a common site of this event. As such, neurologic involvement may or may not be accompanied by signs of systemic infection.

Nonviral pathogens

More fastidious bacteria and a variety of fungi usually cause chronic meningitis, but CNS involvement by these pathogens can occasionally present as an acute illness. A specific microbiologic diagnosis can be difficult in this setting because of a lack of reliable culture methods or the very slow growth of these organisms. PCR assays applicable to CSF are being investigated for some of these pathogens (tuberculosis, Lyme disease), but none have achieved routine clinical use at present. Many patients with these infections have extracerebral foci of involvement at the same time of their meningitis, however, and a search for systemic involvement is often warranted. Even if such a diagnostic approach does not confirm one of these infections, it can sometimes uncover a previously undocumented malignancy or granulomatous disease that could explain the CNS symptoms.

Noninfectious etiologies

Aseptic meningitis can occur in 2% to 4% of patients with systemic lupus erythematosus, usually as part of ongoing disease evolution but rarely as an initial manifestation of the disorder [33]. Leptomeningeal metastasis has been estimated to occur in up to 5% of all patients with cancer, and unlike those with solid tumors, patients with leukemias or lymphomas can develop neoplastic meningitis without evidence of systemic disease, during periods of remission, and even rarely at clinical presentation [34]. An occasional patient taking nonsteroidal anti-inflammatory drugs may develop aseptic meningitis without any other underlying explanatory cause, and a handful of individuals have developed this syndrome in the setting of intravenous immunoglobulin infusions. Here, most symptoms occur within 24 hours of completing an infusion, may recur with reinfusion, and always recover spontaneously in less than a week after their onset [35].

Diagnostic approach

A diagnosis of aseptic meningitis begins with the identification of a CSF pleocytosis (either mononuclear or polymorphonuclear cell predominant) in a patient with the appropriate clinical features. Bacterial smears and cultures of the CSF are uniformly negative. Variability from patient to patient in the total CSF cell count (usually 100–1000 cells/mm^3), protein concentration (usually normal to mildly elevated), and glucose level (usually normal to slightly depressed) is the norm, even during confirmed outbreaks caused by a single viral serotype [2]. Much effort has gone into the development of predictive algorithms to analyze routine CSF parameters in the setting of acute meningitis, but they retain little practical significance for use at the bedside. Indeed, this difficulty distinguishing between aseptic and partially treated bacterial meningitis on clinical and routine laboratory grounds alone means that most patients should be treated for the latter disorder until an alternative cause is uncovered or the patient demonstrates significant clinical improvement.

For EV meningitis, a specific diagnosis traditionally has depended on the recovery of virus from the CSF using cell culture techniques. Older studies conducted in the pre-PCR era for all patients with aseptic meningitis identified a specific viral pathogen in 55% to 70% of cases, and they consistently confirmed that EVs were the predominant virus isolated using these laboratory methodologies (Table 2). Still, the mean time to a positive viral culture can range from 3.7 to 8.2 days [48], often well after a final CSF bacterial culture result has been available and clinical outcome has been determined, making the practical clinical significance of these techniques unclear. Application of PCR methods to the analysis of CSF samples has resulted in higher rates of EV isolation in much shorter periods of time; more than two thirds of what turn out to be culture-negative specimens are positive for an EV using these techniques in less than 1 day [49]. This means that EVs now account for 85% to 90% of all cases of aseptic meningitis where an etiologic agent is identified [2]. EVs have been found by PCR in only about one half of CSF samples tested in large retrospective cohorts [50,51], however, raising some question about whether hospital laboratories should routinely use EV PCR for the evaluation of aseptic meningitis and at what frequency the assay should be performed. In one series, patients with positive CSF EV PCR assays had shorter durations of antibiotic coverage and hospitalization compared with patients with negative results or in whom the assay was not performed [51]. Prospective studies are needed to determine the optimal cost-benefit strategy for diagnostic testing in patients with aseptic meningitis.

Management

Management of patients with acute aseptic meningitis is largely supportive in nature. Appropriate interventions include aggressive fluid, electrolyte,

Table 2
Viral causes of aseptic meningitis as reported in selected case series

		Viruses identified (%)								
Years	Cases	P	NPE	Arbo	Mumps	HSV	LCMV	other[a]	None	Ref
1941–1946	374				15.3		11.2	No	73.5	[36]
1947–1952	480				13.3	5.3	9.7	No	74.8	[37]
1953–1958	430	8.8	29.8	0.7	15.8	1.4	8.8	No	29.0	[38]
1955–1958	407	7	41		8			No	46.0	[39]
1958	368	2	57		9	1		Yes	31.0	[40]
1958–1963	374	4.8	38.5	0.8	7.5	0.5	1.9	Yes	43.5	[41]
1972–1979	2382	0.5	24	1.4	1.2	2.7	0.5	Yes	68.3	[42–46]
1986–1990	274	0.007	61.3					Yes	38.4	[47]

Abbreviations: Arbo, arthropod-borne viruses; HSV, herpes simplex virus; LCMV, lymphocytic choriomeningitis virus; NPE, nonpolio enteroviruses; P, polioviruses.

[a] Rare cases of adenovirus, measles virus, Epstein-Barr virus, influenza A virus, cytomegalovirus, varicella-zoster virus, rubella virus, influenza B virus, parainfluenza type-3 virus, respiratory syncytial virus.

Data from Rotbart HA. Viral meningitis. Semin Neurol 2000;20:277–92.

and pain management, and close observation for potential neurologic and neuroendocrine sequelae (seizures, brain edema, syndrome of inappropriate antidiuretic hormone secretion). Antiviral therapies active against many of the causative agents underlying viral meningitis either do not exist (arthropod-borne viruses, mumps virus, lymphocytic choriomeningitis virus), or have not been shown to accelerate clinical recovery in placebo-controlled trials (herpesviruses). This latter point must not, however, be confused with the established benefit of acyclovir in patients with herpes simplex encephalitis. Two therapies recently used against EVs are immune serum globulin and pleconaril, the latter being a newly developed antipicornaviral agent. Both have relatively limited data behind their use; immune serum globulin has been given therapeutically in neonates and prophylactically in immunocompromised hosts, whereas pleconaril has been studied in only a few placebo-controlled trials. In neonates with systemic EV infection, one study showed that immune serum globulin treatment produced lower serum viral titers, but it was too small to demonstrate any clinical benefit to this approach [52]. In agammaglobulinemic adults given regular doses of immune serum globulin, the incidence of chronic EV meningoencephalitis seems to have fallen, and the course of such infections in these patients may be attenuated [53]. The therapeutic benefit of such treatment in established cases of chronic EV meningoencephalitis has also been anecdotally reported [2].

Pleconaril blocks EV attachment to cellular receptors and inhibits the viral uncoating process. The drug has broad-spectrum anti-EV activity in vitro and high oral bioavailability in vivo [2]. Unfortunately, despite promising initial results, the drug was recently found to have no detectable clinical or virologic benefit in infants with EV meningitis compared with a placebo [54]. In two randomized studies with more than 600 combined patients, pleconaril only modestly shortened the course of the illness in a subgroup of adults with EV meningitis who experienced more severe symptoms and who were treated early in disease [55]. Applications for federal approval of the drug have not yet been accepted based on these data, and it remains unclear what future the drug has in these diseases.

Viral myelitis

Myelitis (inflammation of the spinal cord) can have numerous causes, including direct viral infection of the neural elements at this site. In some forms of acute myelitis caused by viral pathogens, infection of anterior horn cells in the gray matter of the spinal cord produces a clinical picture of AFP without significant accompanying sensory or autonomic (ie, bowel and bladder) disturbances. Alternatively, more prominent infection of spinal white matter tracts can result in the syndrome of acute transverse myelitis (ATM), where affected individuals develop motor, sensory, and autonomic disturbances that extend up to a defined rostral level of the spinal cord. Chronic viral infection of the spinal cord, however, is usually the result of underlying retroviral infection caused by either HIV or human T-cell lymphotropic virus type I. Finally, myelitis can rarely occur as a postinfectious process on the heels of a systemic viral illness, sometimes without any concomitant brain involvement. Chronic retroviral infections and postinfectious disorders affecting the spinal cord are covered elsewhere in this issue. This article focuses on the diagnosis, pathogenesis, and management of the acute viral myelitides.

Clinical features

Acute viral myelitis can be subdivided into gray matter syndromes causing AFP (reminiscent of poliomyelitis) and partial or complete white matter syndromes with or without simultaneous gray matter involvement (ATM). Patients with virus-induced AFP usually develop headache, fever, and meningeal signs typical of aseptic meningitis. Flaccid weakness of one or more extremities then becomes evident within a few days, but affected individuals generally do not report prominent bladder symptoms or manifest overt sensory impairment (even in the setting of subjective sensory complaints). In contrast, patients who develop ascending sensory deficits and urinary retention over several days, along with progressive weakness caudal to the sensory level, may have ATM caused by viral infection of the spinal white matter. Here, both halves of the spinal cord are commonly involved to produce relatively symmetric deficits with a defined rostral border. For those patients where the signs and symptoms of myelopathy evolve over a period of minutes to a few hours, a vascular or compressive etiology is much more likely than a focal viral infection. Conversely, chronic viral myelitis is associated with a slow and progressive evolution of clinical deficits over a period of months to years.

Systemic features can occasionally highlight particular causes of viral myelitis. Although fever and nonspecific respiratory symptoms often herald the onset of this disorder regardless of its cause, a distinct dermatomal rash in the preceding 1 to 2 weeks is the hallmark of VZV-induced myelitis in both immunocompetent and immunocompromised patients [56]. During EV-71 infections, the characteristic hand-foot-and-mouth disease causes vesicular lesions to erupt over these particular body regions right before or directly at the time that neurologic deficits appear [7]. Coxsackievirus A16 is another cause of hand-foot-and-mouth disease, although this virus is less neurovirulent than EV-71 and much less likely to cause myelitis [57]. For EBV-associated myelitis, symptoms typical of infectious mononucleosis (fatigue, pharyngitis, cervical lymphadenopathy) often precede the neurologic involvement by several weeks. In immunocompromised patients, cytomegalovirus (CMV) is a primary diagnostic consideration in a patient where the clinical picture of ATM develops [58]. These patients commonly have pain and objective sensory loss in perianal regions, along with depressed deep tendon reflexes indicative of concomitant lumbosacral nerve root involvement [58]. Finally, any history of antecedent illness or vaccination in the preceding few weeks can suggest a postinfectious or a postvaccinal cause.

Epidemiology

AFP of viral causes has been estimated to occur with an annual worldwide incidence of 4 cases per 100,000 individuals [59], whereas ATM of all causes occurs at a rate of 1 to 4 cases per million population per year [60]. Polioviruses were previously the world's most common cause of AFP (>350,000 confirmed cases across 125 countries as recently as 1988), but because of an aggressive vaccination campaign championed by the World Health Organization (WHO), there has been a more than 99.8% reduction in wild-type (ie, nonvaccine-associated) polio cases [59]. Indeed, the WHO certified the Americas "polio free" in 1994, followed thereafter by the western Pacific region (2000) and the European region (2002). The 1996 polio cases reported in 2006 occurred in 16 countries in Africa, Southeast Asia, and the Middle East, and as a group, poliovirus infections now account for only approximately 5% of all AFP cases reported to the WHO each year [59]. In the Western hemisphere, emerging infections, such as WNV, have become the foremost cause of acute spinal cord disease caused by viruses, and North America has witnessed a rising number of WNV infections since the first cases were identified in 1999 [17]. Despite the fact that AFP has been associated with only 2% to 3% of all confirmed WNV infections over the past few years, the syndrome is documented in nearly one third of hospitalized patients [61]. Although many of these more severely affected patients had some degree of simultaneous encephalitis, others had neuromuscular weakness to the point of requiring intubation and mechanical ventilation without overt brain involvement [62]. For unexplained reasons, AFP cases tend to be younger than patients with WNV encephalitis, a situation analogous to the hosts that are most susceptible to polioviruses. Because of the required vector transmission of WNV, most viral-induced AFP cases in the United States occur in the summer and fall months [17].

Differential diagnosis

Any patient with an acute myelopathy must first have a compressive spinal cord lesion urgently excluded by MRI. Assuming that such a structural lesion is ruled out and evidence to suggest an inflammatory, intramedullary spinal cord lesion is found, other potential diagnostic considerations in this setting include nonviral infections of the spinal cord and various autoimmune and vascular disorders. Metabolic and hereditary conditions should also be kept in mind, even though they are less likely to cause acute illnesses and overt lesions on spinal MRI scans. The nonviral pathogens that cause noncompressive myelopathies include bacteria (*Mycoplasma pneumoniae*, *Borrelia burgdorferi*, *Treponema pallidum*, *Mycobacterium tuberculosis*); fungi (*Actinomyces*, *Blastomyces*, *Aspergillus*); and parasites (*Shistosoma mansoni*) [57]. Serologic testing and CSF cultures, for the most part, help to identify these agents. Connective tissue disorders (systemic lupus erythematosus, Sjögren's syndrome, mixed connective tissue disease, antiphospholipid syndrome) form another significant group of causes that underlie the inflammatory myelopathies, and granulomatous conditions (sarcoidosis) and the vasculidites (Wegener's granulomatosis, Behçet's disease, giant cell arteritis, primary arteritis of the CNS) are other known predisposing conditions [57,60]. A more complete nosology of these disorders is reviewed elsewhere [60].

For isolated subacute or chronic myelopathies without overt brain involvement, the differential diagnosis is broader and it can be somewhat harder to distinguish inflammatory from noninflammatory disorders. Many of the previously referenced infectious and inflammatory etiologies still come into play, as do other inflammatory and demyelinating disorders, such as neuromyelitis optica and progressive multiple sclerosis. Here, one is also obliged to consider more strongly metabolic derangements (vitamin B_{12} or E deficiency) and certain hereditary disorders (Friedrich's ataxia, other spinocerebellar ataxias). Another dilemma can arise in the setting of radiographic evidence of significant degenerative disk or bone disease in the cervical region; it can be very difficult to determine what proportion of a patient's signs and symptoms are caused by compressive myelopathy versus intrinsic cord dysfunction before any surgical intervention. Finally, a sizable proportion of patients with chronic myelopathies never have an underlying etiology uncovered and are labeled as having chronic idiopathic inflammatory processes.

Pathogenesis

The pathogenesis of virus-induced AFP is likely multifactorial in nature. In WNV-induced disease, electrophysiologic studies have confirmed denervating changes attributable to the presence of significant anterior horn cell injury in affected patients [63]. Neuropathologic study of fatal WNV cases associated with severe AFP have shown widespread motor neuron destruction in the spinal cord [63], and it is generally assumed that direct viral infection and cytolysis of these cells has occurred. Interestingly, however, other cases of WNV-associated weakness have revealed motor deficits that are transient and completely reversible in nature [64]. The exact pathophysiology in this setting is unclear, because many patients do not show electrophysiologic evidence of motor neuron involvement [64]. Likewise, the reversibility of these motor deficits strongly implies that actual motor neuron destruction does not occur. Demyelinating polyradiculoneuropathies have been observed with WNV infection, although electrophysiologically confirmed cases of Guillain-Barré syndrome in this setting are very rare [64]. Direct muscle involvement or transiently disrupted electrical signaling in the central motor pathways are both unproved alternative possibilities.

Less virulent or even noncytolytic viruses can also cause ATM, and in this setting the associated host response within the spinal cord may underlie disease pathogenesis. Certainly, disease mechanisms in postinfectious myelitis are primarily immunologic in nature, and this feature has obvious implications for both treatment and the potential reversibility of deficits. A viral etiology is uncovered in only a small subset of ATM cases, and proper controlled clinical trials proving that immunotherapy is beneficial to outcome have not been conducted. Overall, it has been estimated that about one third of these patients make good recoveries with few residual sequelae, one third are left with moderate deficits, and one third make little to no improvement and are left with profound neurologic impairment [60]. The presence of detectable 14-3-3 protein (a cytoplasmic-axonal protein of neurons) in CSF at the time of clinical nadir is strongly associated with a poor clinical outcome [65].

Common etiologic agents

Herpesviruses

HSV-2 more commonly causes myelitis in adults, whereas HSV-1 is the main etiologic agent of this syndrome in children. Both forms of disease can vary from mild involvement with full recovery to a severe necrotizing process with devastating sequelae [66,67]. Patients with HSV-2 myelitis in one series generally did not have known genital herpes, but lesions, when present, preceded spinal cord involvement by several days [68]. Patients with VZV myelitis often have some underlying immunologic deficit, and spinal cord dysfunction typically follows the zoster rash by 10 to 14 days (range, 0–94 days) and evolves over 1 to 3 weeks [56,66]. CMV involvement of the spinal cord is a disease primarily of HIV-infected patients, presenting either as pure ATM or a cord syndrome accompanied by radicular or peripheral nerve involvement. A distinguishing feature of this disorder is the common occurrence of a neutrophil-predominant pleocytosis in the CSF of up to 1000 cells/mm^3 [69]. The radicular component is often discernible by electrophysiologic studies, and may show both reduced conduction velocities indicative of demyelination and diminished sensory and motor amplitudes that reflect axonal injury [69]. ATM as a manifestation of EBV infection is rare, but may occur 1 to 2 weeks after infectious mononucleosis and present with combined spinal cord and radicular features [66,70].

Polioviruses

AFP caused by polioviruses can occur as either wild-type infections in non–polio free parts of the world, or as a rare complication of the live attenuated ("Sabin") polio vaccine. Most cases occur with a preceding meningitis, and in the 1% to 2% of patients who go on to develop further neurologic signs and symptoms, asymmetric flaccid weakness with areflexia develops within a few days and is followed within a week or two by discernible muscle atrophy [66]. Unlike other EVs, polioviruses are often not cultured from the CSF and a diagnosis may hinge on the demonstration of a fourfold increase in serum, virus-specific antibodies. PCR on CSF can be used to confirm the presence of an EV, but most diagnostic laboratories do not carry primers that specifically discriminate the three poliovirus serotypes from the many nonpolio EV strains. Patients with paralytic poliomyelitis typically manifest some recovery within a few weeks of infection that can reach a plateau by 6 months of onset [71].

West Nile virus

WNV has spread throughout North and Central America since the original New York outbreak in 1999. Serologic surveys suggest that asymptomatic infections outnumber clinically overt ones by more than 100:1, and even then, some patients only develop a febrile illness without any overt neurologic involvement. Still, the diagnosis should be considered in any patient with AFP and a CSF pleocytosis. It is confirmed primarily by means of serologic assays; detection of virus-specific IgM by capture ELISA in CSF is the most sensitive and specific method, although demonstration of both antiviral IgM and IgG in a single serum sample is also considered confirmatory. Reverse transcriptase PCR is highly specific for WNV sequences, but at present seems to be less sensitive than the serologic assays. The virus is only rarely cultured from CSF samples.

Diagnostic approach

Spinal MRI imaging begins the evaluation of any patient with an acute myelopathy to exclude an extramedullary or extradural compressive lesion. In cases of acute viral myelitis, spinal MRI scans can reveal a wide range of findings from a normal-appearing spinal cord to a swollen, enhancing cord lesion that extends over multiple spinal levels. Myelitis is usually distinguished from spinal multiple sclerosis plaques on radiographic grounds by virtue of involving a greater cross-sectional area of the cord and by extending over more than two contiguous spinal levels. Viral myelitis, however, can be very difficult to distinguish from the spinal onset of neuromyelitis optica based on MRI findings. Contrast enhancement of the lesion is usually seen in the earliest stages of disease, but may disappear within days to a few weeks. Hyperintensity of the cord on noncontrast T1-weighted sequences suggests the presence of hemorrhagic necrosis. Imaging of the brain is also commonly advised to exclude a multifocal process, such as acute disseminated encephalomyelitis or multiple sclerosis.

The presence of inflammation is suggested by gadolinium enhancement on MRI and confirmed by lumbar puncture and CSF analysis. A mononuclear cell pleocytosis with elevated total protein content are the expected findings in viral myelitis, but this profile does not exclude other nonviral or noninfectious etiologies. Furthermore, some confirmed cases of viral myelitis are accompanied by a high proportion of neutrophils in the CSF (CMV and WNV, in particular), whereas others (approximately 3%–5%) do not show any evidence of a pleocytosis. Hypoglycorrhachia ($<$40 mg/dL) is uncommon with all CNS viral infections, but it can occasionally be seen in those disorders that also elicit a neutrophilic pleocytosis. Direct isolation of a viral pathogen from the CSF is rarely accomplished in either ATM or AFP. Instead, PCR for DNA viruses and reverse transcriptase PCR for RNA viruses are now the standard methods used in the rapid identification of a specific viral pathogen from clinical samples. These assays are routinely available in most institutions for HSV, CMV, VZV, EBV, EVs, and HIV, and are becoming more commonplace for WNV and human T-cell lymphotropic virus type I. One important general limitation of these PCR tests is that viral replication often peaks early in CNS viral infection and then declines rapidly to undetectable levels; one study found that the incidence of a positive PCR assay was highest when CSF samples were obtained 5 days after symptom onset [72]. A diagnosis can also be confirmed by detecting virus-specific IgM in CSF, because these large molecules do not readily cross the blood-brain barrier and indicate intrathecal synthesis. Measurement of serum IgM titers can also sometimes be useful, as is the demonstration of a fourfold higher IgG titer during the acute compared with the convalescent phase of disease. This latter approach is only useful in establishing a retrospective diagnosis, because convalescent samples should be obtained at least 6 weeks after the acute illness.

Management

Treatment of viral myelitis focuses on the use of specific antiviral agents, when available. Although formal clinical trial evidence of efficacy is lacking given the rarity of these diseases, it is advisable to administer intravenous acyclovir, 10 to 15 mg/kg every 8 hours for 10 to 14 days, to patients with HSV and VZV myelitis as soon as a pathogen is confirmed in CSF by PCR or even empirically if the clinical suspicion is high enough (ie, recent zoster rash or recurrent genital herpes outbreaks) [57]. Ganciclovir (5 mg/kg every 12 hours) and foscarnet (90 mg/kg every 12 hours) both have been used in cases of CMV myelitis with some effectiveness, although overall outcome from this disorder remains poor even when drug treatment is continued over 2 to 3 weeks [57,58,60]. Spinal EBV infections have not responded well to either of these two therapies. There is no proved antiviral therapy for spinal poliomyelitis, although pleconaril can be made available from the manufacturer (ViroPharma) on a compassionate use basis. The drug was believed to be of some use in two of three cases of vaccine-associated poliomyelitis in an open-label study [73].

Another approach being taken in patients with neuroinvasive WNV infections is the administration of intravenous immunoglobulin pooled from donors where the virus is endemic and high titers of neutralizing antibodies are found. Such passively transferred antibodies would be presumed to cross the blood-brain barrier and facilitate viral clearance from the CNS, as it does in animal models of this infection [17]. Unfortunately, however, results of a phase I and II clinical trial have not yet reported any efficacy compared with a placebo. Other immunotherapeutic interventions are also conceivable in the setting of efficacy in animal models of alphavirus and flavivirus myelitis [74,75], but these similarly require further testing before any use in humans is advised. The application of high-dose corticosteroids, either as an anti-inflammatory intervention or through actions by some other mechanism, has not been systematically studied in viral myelitis, although anecdotal evidence has reported benefit in four patients with EV-71 infections extending into the brainstem to ameliorate long-term deficits [76].

Summary

Aseptic meningitis is frequently caused by viral pathogens and remains the most common form of CNS viral infection. Acute myelitis is a distinctly unusual manifestation that occurs with viral invasion of the CNS and is frequently confused with other nonviral and noninfectious causes of acute myelopathy. Both syndromes can result following EV infection of the CNS, although myelitis in particular can have other viral etiologies. Patients with acute viral infections of the spinal cord recover to widely variable degrees from their illnesses, whereas those individuals with aseptic meningitis typically have a much more favorable long-term outcome. Therapies for both these disorders are still in their infancy, but as pathogenesis studies in animal models unravel disease mechanisms, novel treatment interventions are anticipated to follow. In this regard, it seems conceivable that both the pathogens themselves and the host responses elicited by these infections may be drug targets.

References

[1] Wallgren A. Die atiologie der enzephalomeningitis bei kindern, besonders des syndromes der akuten abakteriellen (aseptichen) meningitis. Acta Paediatr Scand 1951;40:541–65.

[2] Rotbart HA. Viral meningitis. Semin Neurol 2000;20:277–92.

[3] Abzug MJ, Levin MJ, Rotbart HA. Profile of enterovirus disease in the first two weeks of life. Pediatr Infect Dis J 1993;12:820–4.

[4] Modlin JF. Perinatal echovirus infection: insights from a literature review of 61 cases of serious infection and 16 outbreaks in nurseries. Rev Infect Dis 1986;8:918–26.

[5] Abzug MJ. Perinatal enterovirus infections. In: Rotbart HA, editor. Human enterovirus infections. Washington, DC: ASM Press; 1995. p. 221–38.

[6] Wilfert CM, Lehrman SN, Katz SL. Enteroviruses and meningitis. Pediatr Infect Dis 1983;2: 333–41.

[7] Ho M, Chen E-R, Hsu K-H, et al. An epidemic of enterovirus 71 infection in Taiwan. N Engl J Med 1999;341:929–35.

[8] Chemtob S, Reece ER, Mills EL. Syndrome of inappropriate secretion of antidiuretic hormone in enteroviral meningitis. Am J Dis Child 1985;139:292–4.
[9] Pichichero ME, McLinn S, Rotbart HA, et al. Clinical and economic impact of enterovirus illness in private pediatric practice. Pediatr 1998;102:1126–34.
[10] Rorabaugh ML, Berlin LE, Rosenberg L, et al. Absence of neurodevelopmental sequelae from aseptic meningitis. Pediatr Res 1992;30:177A.
[11] McKinney RE, Katz SL, Wilfert CM. Chronic enteroviral meningoencephalitis in agammaglobulinemic patients. Rev Infect Dis 1987;9:334–56.
[12] Rotbart HA, Brennan PJ, Fife KH, et al. Enterovirus meningitis in adults. Clin Infect Dis 1998;27:896–8.
[13] Corey L, Adams HG, Brown ZA, et al. Genital herpes simplex virus infections: clinical manifestations, course, and complications. Ann Intern Med 1983;98:958–72.
[14] Schlesinger Y, Tebes P, Gaudreault-Keener M, et al. Herpes simplex virus type 2 meningitis in the absence of genital lesions: improved recognition with use of the polymerase chain reaction. Clin Infect Dis 1995;20:842–8.
[15] Hartford CG, Wellinghoff W, Weinstein RA. Isolation of herpes simplex virus from the cerebrospinal fluid in viral meningitis. Neurology (Minneap) 1975;25:198–200.
[16] Tedder DG, Ashley R, Tyler KL, et al. Herpes simplex virus infection as a cause of benign recurrent lymphocytic meningitis. Ann Intern Med 1994;121:334–8.
[17] Davis LE, DeBiasi R, Goade DE, et al. West Nile virus neuroinvasive disease. Ann Neurol 2006;60:286–300.
[18] Melnick JL. Enteroviruses: polioviruses, coxsackieviruses, echoviruses, and newer enteroviruses. In: Fields BN, Knipe DM, Howley PM, editors. Fields virology. 3rd edition. New York: Lippincott-Raven Press; 1996. p. 655–712.
[19] Strikas RA, Anderson LJ, Parker RA. Temporal and geographic patterns of isolates of nonpolio enterovirus in the United States, 1970–1983. J Infect Dis 1986;153:346–51.
[20] Kinnunen E, Hovi T, Stenvik M, et al. Localized outbreak of enteroviral meningitis in adults. Acta Neurol Scand 1987;74:346–51.
[21] Tsai TF, Canfield MA, Reed CM, et al. Epidemiological aspects of a St. Louis encephalitis outbreak in Harris County, Texas, 1986. J Infect Dis 1988;157:351–6.
[22] Centers for Disease Control and Prevention. Mumps: United States, 1985–1988. MMWR Morb Mortal Wkly Rep 1989;38:101–5.
[23] Rotbart HA. Viral meningitis and the aseptic meningitis syndrome. In: Scheld WM, Whitley RJ, Durack DT, editors. Infections of the central nervous system. 2nd edition. Philadelphia: Lippincott-Raven Publishers; 1997. p. 23–46.
[24] Price RA, Garcia JH, Rightsel WA. Choriomeningitis and myocarditis in an adolescent with isolation of Coxsackie B5 virus. Am J Clin Pathol 1970;53:825–31.
[25] Johnson RT. Viral infections of the nervous system. 2nd edition. Philadelphia: Lippincott-Raven Publishers; 1998.
[26] Monath TP, Cropp CB, Harrison AK. Mode of entry of a neurotropic arbovirus into the central nervous system: reinvestigation of an old controversy. Lab Invest 1983;48: 399–410.
[27] Kilham L. Mumps meningoencephalitis with and without parotitis. Am J Dis Child 1949;78: 324–33.
[28] Wolinsky JS, Klassen T, Baringer JR. Persistence of neuroadapted mumps virus in brains of newborn hamsters after intraperitoneal inoculation. J Infect Dis 1976;133:260–7.
[29] Peters CJ, Buchmeier M, Rollin PE, et al. Arenaviruses. In: Fields BN, Knipe DM, Howley PM, editors. Fields virology. 3rd edition. New York: Lippincott-Raven Press; 1996. p. 1521–51.
[30] Monath TP, Heinz FX. Flaviviruses. In: Fields BN, Knipe DM, Howley PM, editors. Fields virology. 3rd edition. New York: Lippincott-Raven Press; 1996. p. 961–1034.
[31] Wolinsky JS. Mumps virus. In: Fields BN, Knipe DM, Howley PM, editors. Fields virology. 3rd edition. New York: Lippincott-Raven Press; 1996. p. 1243–66.

[32] Roizman B. Herpesviridae. In: Fields BN, Knipe DM, Howley PM, editors. Fields virology. 3rd edition. New York: Lippincott-Raven Press; 1996. p. 2221–30.
[33] Johnson RT, Richardson EP. The neurological manifestations of systemic lupus erythematosus. Medicine 1968;47:337–69.
[34] Chamberlain MC, Nolan C, Abrey LE. Leukemic and lymphomatous meningitis: incidence, prognosis, and treatment. J NeuroOncol 2005;75:71–83.
[35] Hamrock DJ. Adverse events associated with intravenous immunoglobulin therapy. Int Immunopharmacol 2006;6:535–42.
[36] Rasmussen Jr AF. The laboratory diagnosis of lymphocytic choriomeningitis and mumps. Presented at the Rocky Mountain Conference on Poliomyelitis, Denver, CO, December 16, 1946.
[37] Adair CV, Gauld RL, Smadel JE. Aseptic meningitis, a disease of diverse etiology: clinical and etiologic studies on 854 cases. Ann Intern Med 1953;39:675–704.
[38] Meyer HM Jr, Johnson RT, Crawford IP, et al. Central nervous syndromes of viral etiology. Am J Med 1960;29:334–47.
[39] Lepow ML, Coyne N, Thompson LB, et al. A clinical epidemiologic and laboratory study of aseptic meningitis during the four-year period, 1955–1958: II. The clinical disease and its sequelae. N Engl J Med 1962;266:1188–93.
[40] Lennette EH, Magoffin RL, Knouf EG. Viral central nervous system disease: an etiologic study conducted at the Los Angeles County General Hospital. JAMA 1962;179:687–95.
[41] Buescher EL, Artenstein MS, Olson LC. Central nervous system infections of viral etiology: the changing pattern. Res Pub Assoc Nerv Ment Dis 1968;44:147–63.
[42] Deibel R, Barron A, Millian S, et al. Central nervous system infections in New York State: etiologic and epidemiologic observations, 1972. N Y State J Med 1974;74:1929–35.
[43] Deibel R, Flanagan TD, Smith V. Central nervous system infections in New York State: etiologic and epidemiologic observations, 1974. N Y State J Med 1975;75:2337–42.
[44] Deibel R, Flanagan TD, Smith V. Central nervous system infections in New York State: etiologic and epidemiologic observations, 1975. N Y State J Med 1977;75:1398–404.
[45] Deibel R, Flanagan TD. Central nervous system infections: etiologic and epidemiologic observations in New York State, 1976–1977. N Y State J Med 1979;79:689–95.
[46] Flanagan TD, Deibel R. Central nervous system infections: etiologic and epidemiologic observations in New York State, 1978–1979. N Y State J Med 1981;81:346–53.
[47] Berlin LE, Rorabaugh ML, Heldrich F, et al. Aseptic meningitis in infants <2 years of age: diagnosis and etiology. J Infect Dis 1993;168:888–92.
[48] Rotbart HA. Enteroviruses. In: Murray PR, Baron EJ, Pfaller MA, editors. Manual of clinical microbiology. Washington, DC: ASM Press; 1999. p. 990–8.
[49] Sawyer MH, Holland D, Aintablian N. Diagnosis of enteroviral central nervous system infection by polymerase chain reaction during a large community outbreak. Pediatr Infect Dis J 1994;3:177–82.
[50] Lee BE, Chawla R, Langley JM, et al. Pediatric Investigators Collaborative Network on Infections in Canada (PICNIC) study of aseptic meningitis. BMC Infect Dis 2006;6:68–75.
[51] Michos AG, Syriopoulou VP, Hadjichristodoulou C, et al. Aseptic meningitis in children: analysis of 506 cases. PLoS ONE 2007;2(7):e674.
[52] Abzug MJ, Kevserling HL, Lee ML, et al. Neonatal enterovirus infection: virology, serology, and effects of intravenous immune globulin. Clin Infect Dis 1995;20:1201–6.
[53] Webster AD, Rotbart HA, Warner T, et al. Diagnosis of enterovirus brain disease in hypogammaglobulinemic patients by polymerase chain reaction. Clin Infect Dis 1993;17: 657–61.
[54] Abzug MJ, Cloud G, Bradley J, et al, National Institute of Allergy and Infectious Diseases Collaborative Antiviral Study Group. Double blind placebo-controlled trial of pleconaril in infants with enteroviral meningitis. Pediatr Infect Dis J 2003;22:335–41.
[55] Desmond RA, Accortt NA, Talley L, et al. Enteroviral meningitis: natural history and outcome of pleconaril therapy. Antimicrob Agents Chemother 2006;50:2409–14.

[56] Gilden DH, Kleinschmidt-DeMasters BK, LaGuardia JJ, et al. Neurologic complications of the reactivation of varicella-zoster virus. N Engl J Med 2000;342:635–45.

[57] Kincaid O, Lipton HL. Viral myelitis: an update. Curr Neurol Neurosci Rep 2006;6:469–75.

[58] Fux CA, Pfister S, Nohl F, et al. Cytomegalovirus-associated acute transverse myelitis in immunocompetent adults. Clin Microbiol Infect 2003;9:1187–90.

[59] World Health Organization. Performance of acute flaccid paralysis (AFP) surveillance and incidence of poliomyelitis 2005–2006. Wkly Epidemiol Rec 2007;82:89–92.

[60] Transverse Myelitis Consortium Working Group. Proposed diagnostic criteria and nosology of acute transverse myelitis. Neurology 2002;59:499–505.

[61] Saad M, Youssef S, Kirschke D, et al. Acute flaccid paralysis: the spectrum of a newly recognized complication of West Nile infection. J Infect 2005;51:120–7.

[62] Sejvar JJ, Bode AV, Marfin AA, et al. West Nile virus-associated flaccid paralysis. Emerg Infect Dis 2005;11:1021–7.

[63] Glass JD, Samuels O, Rich MM. Poliomyelitis due to West Nile virus. N Engl J Med 2002;347:1280–1.

[64] Leis A, Stokic D, Webb R, et al. Clinical spectrum of muscle weakness in human West Nile virus infection. Muscle Nerve 2003;28:302–8.

[65] Irani DN, Kerr DA. 14-3-3 protein in the cerebrospinal fluid of patients with acute transverse myelitis. Lancet 2000;355:901.

[66] Tyler KL. Acute viral myelitis. In: Scheld WM, Whitley RJ, Marra CM, editors. Infections of the central nervous system. 3rd edition. Philadelphia: Lippincott, Williams & Wilkins; 2004. p. 305–22.

[67] Galanakis E, Bikouvarakis S, Mamoulakis D, et al. Transverse myelitis associated with herpes virus infections. J Child Neurol 2001;16:866–7.

[68] Nakajima H, Furutama D, Kimura F, et al. Herpes simplex virus myelitis: clinical manifestations and diagnosis by the polymerase chain reaction method. Eur Neurol 1998;39:163–7.

[69] So YT, Olney RK. Acute lumbosacral polyradiculopathy in acquired immunodeficiency syndrome: experience in 23 patients. Ann Neurol 1994;35:53–8.

[70] Grose C, Feorino PM. Epstein-Barr virus and transverse myelitis. Lancet 1973;1:892.

[71] Bodian D, Horstmann DM. Polioviruses. In: Horsefall FL, Tamm I, editors. Viral and rickettsial infections of man. 4th edition. Philadelphia: Lippincott; 1965. p. 430–73.

[72] Davies NW, Brown LJ, Irish D, et al. Factors influencing PCR detection of viruses in cerebrospinal fluid of patients with suspected CNS infections. J Neurol Neurosurg Psychiatry 2005;76:82–7.

[73] Rotbart HA, Webster AD, Pleconaril Treatment Registry Group. Treatment of potentially life-threatening enterovirus infections with pleconaril. Clin Infect Dis 2001;32:228–35.

[74] Irani DN, Prow NA. Neuroprotective interventions targeting detrimental host immune responses protect mice from fatal alphavirus encephalitis. J Neuropathol Exp Neurol 2007;66:533–44.

[75] Prow NA, Irani DN. The opioid receptor antagonist, naloxone, protects spinal motor neurons in a murine model of alphavirus encephalomyelitis. Exp Neurol 2007;205:461–70.

[76] Nolan MA, Craig ME, Lahra MM, et al. Survival after pulmonary edema due to enterovirus 71. Neurology 2006;60:1651–6.

ELSEVIER
SAUNDERS

Neurol Clin 26 (2008) 657–674

NEUROLOGIC
CLINICS

Herpes Simplex Infections of the Nervous System

J. Richard Baringer, MD

Department of Neurology, University of Utah Health Sciences Center, 50 North Medical Drive, Salt Lake City, UT 84132, USA

Herpes simplex viruses types 1 and 2 are members of a group of eight herpes viruses, consisting of the following

Herpes simplex virus type 1 (HSV-1)
Herpes simplex virus type 2 (HSV-2)
Varicella-Zoster virus
Cytomegalovirus
Epstein-Barr virus
Human herpes virus type 6 (HHV-6)
Human herpes virus type 7 (HHV-7)
Human herpes virus type 8 (HHV-8)

These viruses share a capacity for establishing a state of latency [1] in humans, and under a variety of conditions, for causing nervous system disease. This article concentrates on the biology and disease-producing ability of the two herpes simplex viruses, mentioning the others as they are relevant.

Herpes simplex infections are best known by their ability to produce recurrent infections of epithelial surfaces. The word herpes derives from the Greek term to creep, and the descriptions of the skin lesions were allegedly first contributed by Herodotus.

Over the years, it has become recognized that herpes viruses types 1 and 2 may cause recurrent oral, labial, genital, cutaneous, and corneal lesions in presumably immunocompetent individuals, based on the ability of the viruses, after initial primary infection, to cause latent infections in sensory ganglia.

Herpes simplex viruses types 1 and 2 are closely related double-stranded DNA viruses. The DNA core is surrounded by 262 capsomeres in a pattern of icosahedral symmetry. This in turn is surrounded in mature virions by

E-mail address: richard.baringer@hsc.utah.edu

doi:10.1016/j.ncl.2008.03.005 ***neurologic.theclinics.com***

a tegument and an envelope. The process of envelopment takes place as the virus is extruded from nuclei of infected cells. The virus in cells is readily recognized by its distinct arrangement of the capsomeres forming a hexagonal pattern, and the dense DNA core (Fig. 1). HSV-1 and 2 are not distinguishable from one another under the electron microscope.

The two viruses share DNA homology of approximately 50%, and are antigenically distinguishable. Roizman and Sears have over the past several decades worked out the sequential order of viral gene products essential for the replication and invasiveness of the viruses [2].

The ability of HSV-1 and HSV-2 to cause recurrent, often highly localized infections of epithelial surfaces is based on the ability of the viruses to establish latent infections in sensory ganglia, which in turn serve as reservoirs for reactivated virus that can travel back down sensory fibers to the cutaneous surface and result in localized vesicular eruptions. An early evidence of this was provided by the neurosurgeon Cushing [3], who observed that resection of the trigeminal ganglion for tic douloureux resulted in the appearance of vesicles on the face. Later neurosurgical experience [4] documented the frequent occurrence of herpetic lesions in the mouth or on the lips in patients undergoing trigeminal root section for tic.

Marinesco and Dragonesco in Paris [5] and later Goodpasture [6,7] showed that after inoculation of HSV 1 (taken from the lips of patients) on the cornea of rabbits, herpetic lesions could be demonstrated in the trigeminal ganglia, demonstrating convincingly that after inoculation at the periphery, the virus could ascend along sensory fibers to the ganglia. Cook and his colleagues [8,9] at UCLA showed in a series of investigations that latent HSV could be demonstrated in the trigeminal ganglia of mice.

Studies in humans have shown that latent HSV-1 exists in the trigeminal ganglia of unselected individuals at the time of postmortem examination [10,11], and that in an analogous fashion, HSV-2 resides in the sacral ganglia [12]. It is presumed that the residence of HSV-1 in trigeminal ganglia

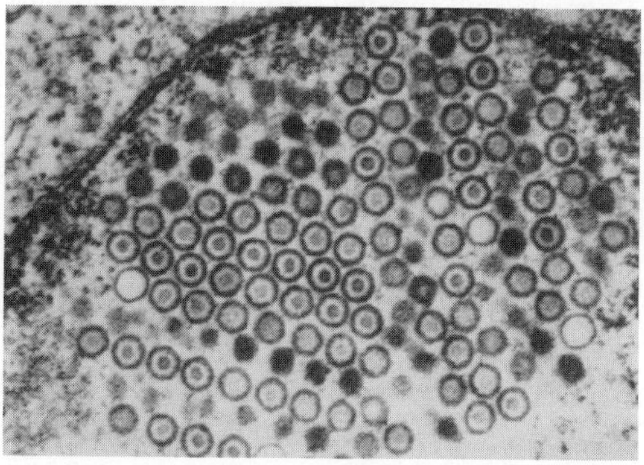

Fig. 1. HSV virions within the nucleus of an infected cell. Some contain DNA cores, whereas others appear empty. The nuclear membrane is at the top left.

arises from oral infection with the virus, a common occurrence in infancy or childhood, and that the residence of HSV-2 arises from infection of the genitalia during sexual contact.

The possibility that HSV established latency in brain was supported by studies of brain specimens obtained at autopsy from individuals who have no known neurologic disease [13]. HSV-1 genomes were demonstrated by polymerase chain reaction (PCR) predominantly in medulla, pons, and olfactory bulb, but were absent from other regions of the cerebrum and cerebellum.

Herpes simplex virus type 1 infections

Ganglionic herpes simplex virus type 1 infections

As indicated above, the most frequent infections by HSV-1 are the recurrent eruptions on the lips experienced by many individuals. It is presumed that these arise from virus latent in trigeminal ganglia, and that in response to a variety of stimuli, the virus has the ability to traverse the sensory fibers to cause a cutaneous lesion. Mentioned above is the frequent occurrence of recurrent lesions after any surgical manipulation of the trigeminal ganglia or root. Common provoking factors in many are fever ("fever blisters"), exposure to sunlight, menstruation, or facial trauma.

There is considerable evidence that sporadic facial paralysis (Bell's palsy) may be the result of latent HSV-1 in the geniculate ganglia [14,15].

Herpes simplex virus encephalitis

The most serious consequence of infection by HSV-1 is that of encephalitis. It is considered that HSV-1 is the most frequent cause of sporadic necrotizing encephalitis. The clinical picture of herpes simplex virus encephalitis (HSVE) is the spontaneous and rapid development of fever, headache, confusion, and seizures in a child or adult. Occasionally, the process may be ushered in by the appearance of olfactory or gustatory hallucinations, but these need not be present or recognized. The tempo of development of symptoms is variable, often developing over a few days, but they may be strokelike in evolution. As the process develops, progressive obtundation takes place, and focal disturbances in the form of aphasia or hemiparesis may develop. Untreated, the mortality approaches 70%. The clinical description and features of a large number of cases were well-described in the National Institute of Allergy and Infectious Diseases Collaborative Antiviral Study Group (NIAID/CASG) study under the direction of Whitley and colleagues [16–18].

Pathology of herpes simplex virus encephalitis infections

The pathology is relatively stereotyped in immunocompetent hosts, consisting of a necrotizing process in the medial and inferior temporal lobes

(Fig. 2), spreading later into the subfrontal and insular regions. Often lesions are seen in the cingulate cortex as well, reflecting the propensity for the process to involve limbic structures. All cellular elements are affected, particularly neurons and astrocytes and oligodendrocytes. Cowdry type A inclusions, characteristic of herpes virus infections, can be seen in the nuclei of affected cells (Fig. 3), but in numerous cases, these may be missing. As the process continues, petechial hemorrhages are often seen, and later, cavitation develops.

The pathogenesis of HSVE is not clear. Work by Whitley and colleagues [19] comparing viral isolates from lip lesions and brain using restriction endonuclease analysis revealed in some cases that the virus from the periphery and the brain were identical, but in others were not identical. Work by Esiri [20] indicated that the distribution of HSV antigen concentrated in the olfactory and limbic system, suggesting a possible exogenous olfactory origin. Latent HSV genomes in the central nervous system (CNS) in normal individuals at autopsy were largely confined to olfactory bulbs, pons, and medulla [13].

Although HSVE does not seem to be more frequent in immunocompromised hosts, the process in these patients is often more widespread in brain, and may present with atypical features [21].

Occasionally, retinal necrosis has been reported coincident with or after episodes of HSVE [22,23].

Diagnosis of herpes simplex virus encephalitis

The diagnosis of the disease has been greatly advanced by the development of MRI and the application of PCR technology to demonstrate HSV genomes in cerebrospinal fluid (CSF). Whereas brain biopsy was formerly considered necessary to establish the diagnosis of HSVE, it is now

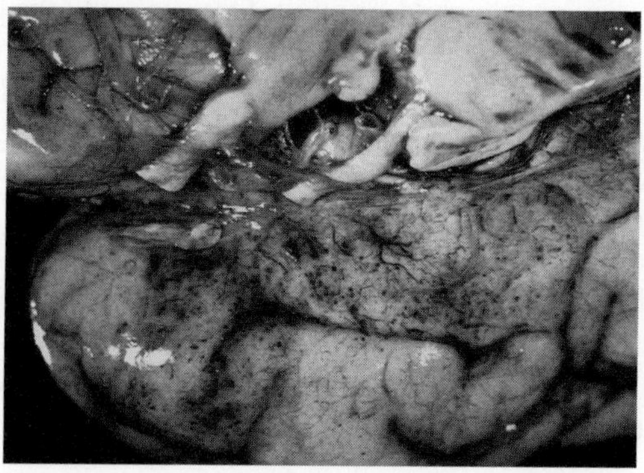

Fig. 2. View of the medial surface of the brain in a patient dying of herpes simplex encephalitis. Note petechial hemorrhages on the surface of the brain. (*Reprinted from* Baringer JR. Handbook of clinical neurology, vol. 34, Infections of the nervous system. Vinken PJ, Bruyn GW, editors. Amsterdam-New York-Oxford: North Holland Publishers; 1978. p. 149, Fig. 1; with permission.)

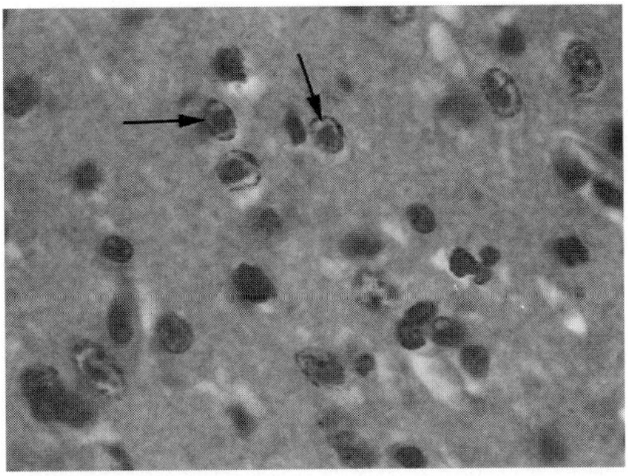

Fig. 3. Neurons in the brain of an animal that has HSV encephalitis. Note the dense eosinophilic inclusions within the nuclei of infected cells (*arrows*). Such inclusions are often missing in brains of humans with HSVE.

necessary only in rare instances when MR imaging and CSF studies do not clarify the diagnosis or an alternative diagnosis is seriously considered. If biopsy is necessary, fresh tissue should be used for fluorescent antibody staining for herpes antigen, culture for HSV, and PCR, as well as studies for other causes of the pathologic process.

MRI scans are abnormal in the great majority of cases when symptoms bring the patient to medical attention. Reflecting the pathologic process, the scans show abnormal signal earliest in fluid-attenuated inversion recovery (FLAIR) sequences (Figs. 4 and 5) and diffusion restriction images in the medial temporal lobes and in the insulas [24], as well as in contrast-enhanced images. In some, lesions in the cingulate cortex can also be seen. Computerized tomographic (CT) scans are of less use in the diagnosis because of

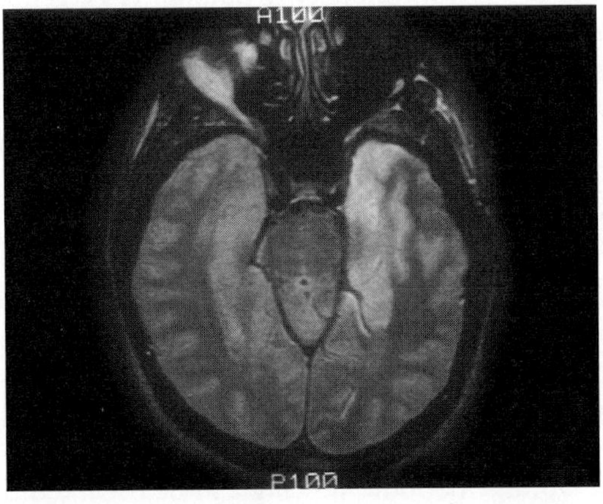

Fig. 4. An acute HSVE lesion in the medial temporal lobe of a patient with herpes encephalitis.

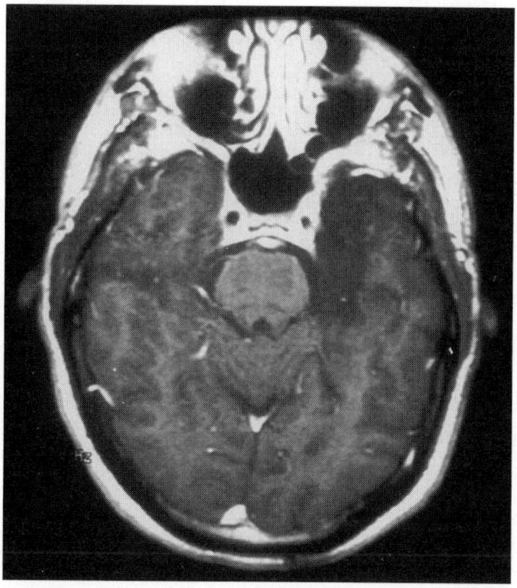

Fig. 5. The same area as displayed in Fig. 4 several years later.

reduced sensitivity and lesser resolution than MRI scans [25]. Often the diagnosis has been missed early in the disease because CT scans were interpreted as normal or were considered to show ischemic lesions.

The CSF is almost always abnormal, although normal CSF has been described in a small proportion of patients early in the disease process. Usually, the CSF contains up to 200 lymphocytes, with protein elevation to 200 mg per deciliter or more. Glucose values are usually normal. The presence of red blood cells in the CSF is not useful in distinguishing HSVE from encephalitides of other cause.

The application of PCR technology to amplify HSV genomes in CSF has provided a high degree of sensitivity (98%) and specificity (94%)for diagnosis of the disease, and in most cases has obviated the necessity for brain biopsy to establish a diagnosis. In rare very early cases, the PCR may be negative, and it becomes negative somewhere in the second to third week of the disease [26–29]. Also, the amplification procedure may be impaired in the presence of blood in the CSF sample, although current extraction methods make this less problematic. Later in the disease, HSV antigen can be detected, and even later, antibody to HSV. PCR has largely replaced antigen detection because of greater sensitivity. Antibody detection may be useful retrospectively in some instances.

The electroencephalogram (EEG) is universally abnormal in HSVE, and frequently shows slowing in the temporal and frontal regions. Classically, intermittent high voltage periodic slowing can be observed, but this pattern is often lacking, especially in younger patients. The EEG lacks specificity compared with MRI and CSF studies, but is useful in monitoring seizure activity when that is present or suspected.

Despite recent advances in imaging, the diagnosis of HSVE is often not made early enough in the process to enable the early application of treatment. The diagnosis rests on a high degree of suspicion in patients presenting with confusion, fever, headache, and seizures. The differential diagnosis [30] includes cerebrovascular disease (Fig. 6), bacterial or fungal diseases, and other viral infections of the brain, but most of these can be distinguished by their patterns on MRI examinations, and contrasted with the relatively stereotyped appearance of HSV lesions. Lesions that more closely mimic those of HSVE on MRI examinations are produced by mitochondrial encephalopathy (MELAS) [31,32], by HHV-6 infections [33], by paraneoplastic limbic encephalitis (Figs. 7 and 8), by listeriosis [34], and by the recently described voltage gated potassium channel antibody syndrome [35]. This latter disorder presents usually in middle age with memory loss, confusion, and seizures. Usually the tempo is slower than HSVE, progressing over weeks to months. MRI scans show bilateral hippocampal lesions (Fig. 9). The patients are frequently hyponatremic and serum contains high levels of antibody to voltage gated potassium channels. Patients who have limbic encephalitis related to underlying tumors may have MRI scans resembling those seen in HSVE, (see Figs. 7 and 8) but the tempo of the disease is usually much more gradual. Most of these differential possibilities are obviously excluded by PCR studies on CSF.

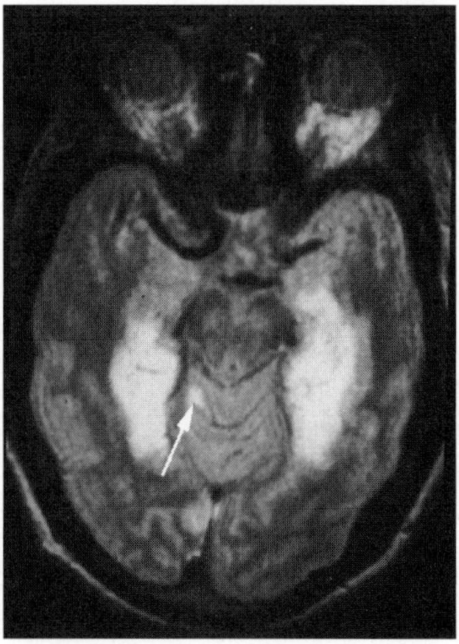

Fig. 6. Bilateral embolic lesions in temporal lobes mimicking those seen in HSVE. Note a small lesion in the cerebellar vermis (*arrow*), seen in other sections through the cerebellum, indicating the vascular nature of the process.

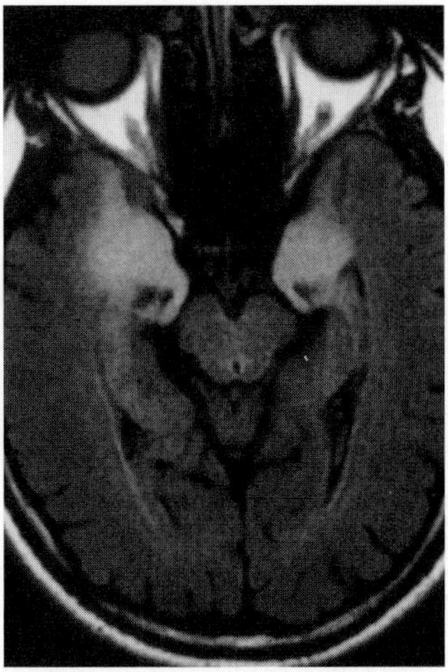

Fig. 7. An MRI of a patient with paraneoplastic limbic encephalitis showing medial temporal lobe involvement.

Recently, Seeley and coworkers [36] have called attention to an acute limbic encephalitis related to reactivation of HHV-6 seen in bone marrow transplant patients. This occurred in the setting of graft-versus-host disease, and presented with dense amnesia accompanied by a confusional state and seizures. The MRI scans (Fig. 10) closely resembled those seen in HSVE, prompting initial treatment with acyclovir until the PCR for HHV-6, present in six of the nine cases, was positive.

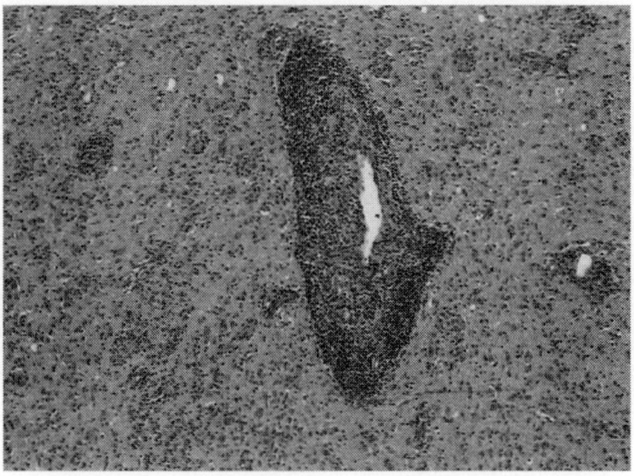

Fig. 8. Prominent perivascular and parenchymal inflammatory process in the same patient as shown in Fig. 7. (*Courtesy of* John E. Greenlee, MD and Jeannette J. Townsend, MD, Salt Lake City, UT.)

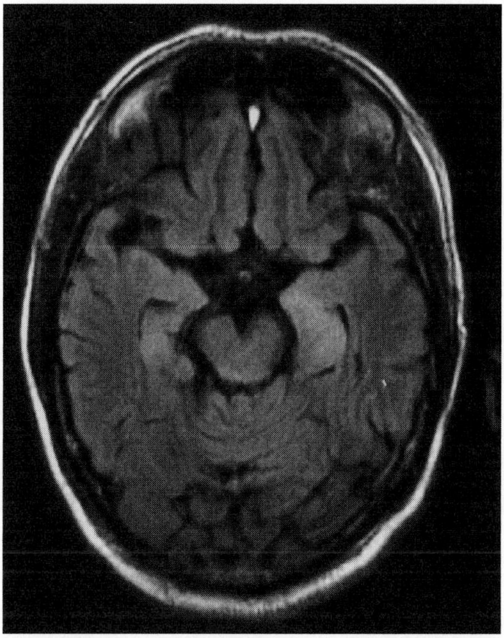

Fig. 9. Subtle abnormal signal in the medial temporal lobes of a patient with voltage gated potassium channel antibody syndrome. (*Courtesy of* John E. Greenlee, MD, Salt Lake City, UT.)

Treatment of herpes simplex virus encephalitis

The development of acyclovir as an effective antiviral drug has greatly advanced the ability to treat HSV encephalitis. The drug has a unique mode of action in that it inhibits HSV-encoded thymidine kinase in preference to cellular thymidine kinase. The triphosphate derivative of acyclovir inhibits viral DNA polymerase and terminates the growth of the viral DNA chain. The study sponsored by the National Institute of Allergy and Infectious Disease Collaborative Antiviral Studies Group (NIAID/CASG) and led by Dr. Richard Whitley at the University of Alabama, Birmingham, convincingly demonstrated the beneficial effect of acyclovir in reducing the morbidity and mortality of HSVE [16,37,38]. The study showed that acyclovir was superior to its predecessor vidarabine, and reduced mortality from 70% to 20% to 30%. Also the study demonstrated that a lower initial level of consciousness and age over 30 years had an adverse effect on morbidity and mortality.

Despite these studies, anecdotal experience of the author's group and others would indicate that there are exceptions to these data, and that patients over 30 years of age and in severe states of obtundation may obtain satisfactory recoveries with aggressive treatment. We have also observed spontaneous recovery in a very few instances. The recommended treatment

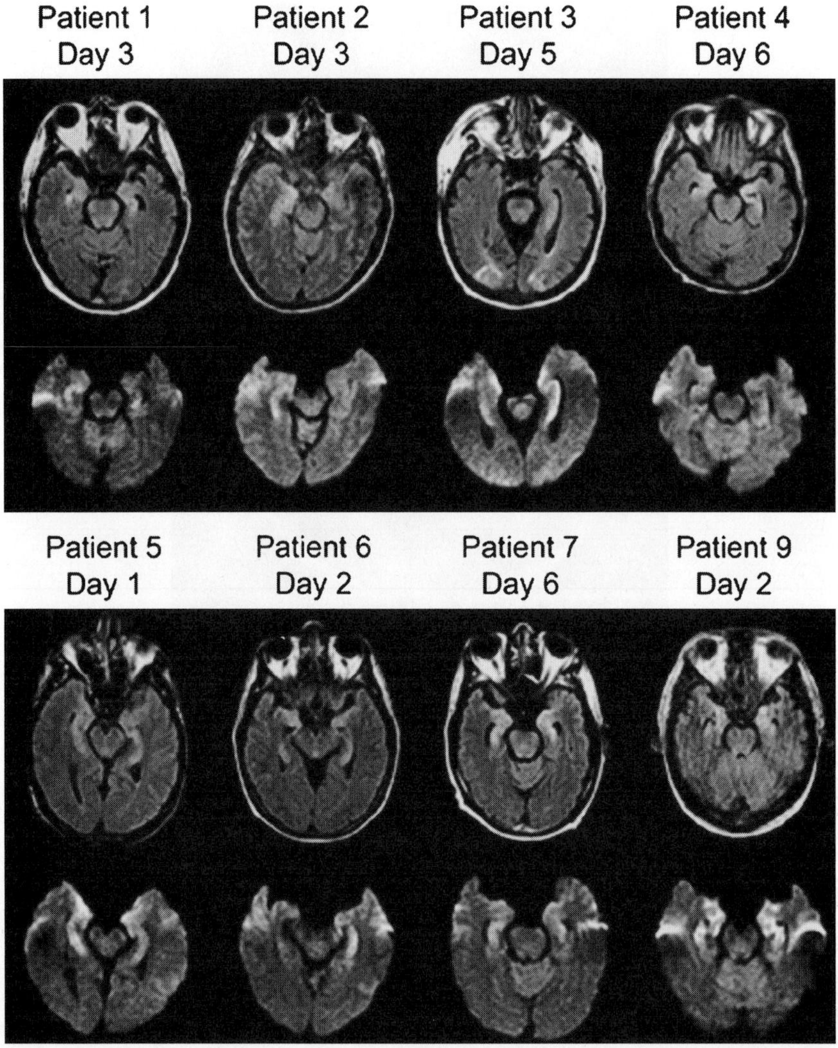

Fig. 10. MRI findings in acute HHV-6 encephalitis after bone marrow transplantation. Fluid inversion recovery images above and diffusion weighted images below. (*Reprinted from* Seeley WW, Marty FM, Holmes TM, et al. Post-transplant acute limbic encephalitis: clinical features and relationship to HHV6. Neurology 2007;69:159; with permission.)

for children and adults is 10 mg/kg intravenously every 8 hours for 14 to 21 days. In patients who have renal failure crystals may develop in the kidney, and a drug-induced encephalopathy can be seen; dosage adjustment is necessary in these patients. Some authorities suggest 20 mg/kg with the same schedule in very young children with HSVE, with close monitoring of renal function.

Because delay in treatment allows rapid replication of virus in brain, treatment should be considered an emergency, and in cases presenting with appropriate clinical, imaging and CSF findings, treatment should be instituted urgently even before PCR results are available. Whether to discontinue acyclovir if PCR for HSV is negative is a matter of judgment. If the clinical, CSF, and imaging studies all point strongly to the diagnosis of HSVE, it may be better to repeat the CSF PCR study before making the decision to stop treatment. Acyclovir treatment involves far less risk than the risk of untreated HSVE.

Whether or when to employ corticosteroids in the treatment of HSVE has been unclear. A recent nonrandomized retrospective study of steroids in HSVE administered to 22 of 45 patients suggested that the use of corticosteroids, along with younger age and higher Glasgow coma scale at entry was one of the favorable factors influencing outcome [39]. The author's own experience supports this. Our group has seen rapid positive responses to corticosteroids in acyclovir-treated HSVE patients who appeared to be undergoing clinical deterioration even without significant changes in brain edema. Mild forms of HSVE may also be observed, particularly in children [40].

Relapses of HSVE are uncommon but have been described in the literature. Some of the relapses have been unaccompanied by any evidence of recurrent viral presence [41,42], but others [43–45] have shown abundant evidence of recurrent infection. In most cases resuming treatment seems to be the prudent course.

Herpes simplex virus type 2 infections

Ganglionic latency

In a fashion analogous to that for HSV-1, HSV-2 infections arising from initial genital infections result in establishment of latency of HSV-2 in sacral ganglia. The initial genital infection resulting from sexual activity is often attended by fever, back pain, and local lymphadenopathy, and it is presumed that at this time, HSV travels to the sacral ganglia, where it establishes a latent infection. The initial genital infection is sometimes attended by CSF pleocytosis. Recurrent infections present with vesicular eruptions on the penis in men or on the labia in women.

Either at the time of recurrent infections or at other intervals, patients may experience radicular pain, often mimicking the pain of disc protrusion. In the author's experience, the recurrent radiculopathy is sometimes attended by the appearance of a small patch of vesicles over the buttock, similar to that described in a patient who has recurrent meningitis [46].

In recent years it has been recognized that in patients who have recurrent episodes of lymphocytic meningitis, so-called "Mollaret's meningitis," PCR studies have revealed the presence of HSV-2 genomes in CSF [27,47,48]. It appears that this process is self-limited in the majority of cases. Although treatment with acyclovir would seem to be indicated in hopes of shortening the course of the disease [49–52], in the author's experience, most of these meningitic processes subside spontaneously. It is not uncommon to encounter patients who have survived several untreated episodes without adverse sequelae. Whether to treat patients who have recurrent episodes of HSV-2 meningitis prophylactically depends on the frequency and severity of recurrences.

Rare cases of myelopathy have been ascribed to HSV-2 infections, most often in immunocompromised individuals.

Encphalitis caused by herpes simplex virus type 2

The most severe consequence of HSV-2 infections is that of neonatal encephalitis. This topic has been the subject of excellent and comprehensive recent reviews by Kimberlin and coworkers [53–55] and by Elbers and colleagues [40]. Neonatal encephalitis is the unfortunate result of maternal genital HSV-2 infection. Although the majority of cases result from HSV-2, a smaller proportion result from maternal genital infection with HSV-1, or postnatal contact with individuals who have cutaneous HSV-1 lesions. The majority of neonatal HSV infections result from infection of the infant in passage through an infected birth canal. It appears that the important considerations are the frequency and severity of the maternal infections, whether there is recent evidence of infection in the birth canal, the duration of rupture of membranes before delivery, and the status of the maternal antibody and hence transplacental derived antibody in the newborn [53]. Intrauterine infection constitutes a much smaller proportion of HSV infections, and may occur either by transmission of virus through the cervix or transplacentally. These infections often are associated with severe cerebral lesions with cystic encephalomalacia or hydranencephaly (Fig. 11), and may resemble the lesions seen with congenital varicella.

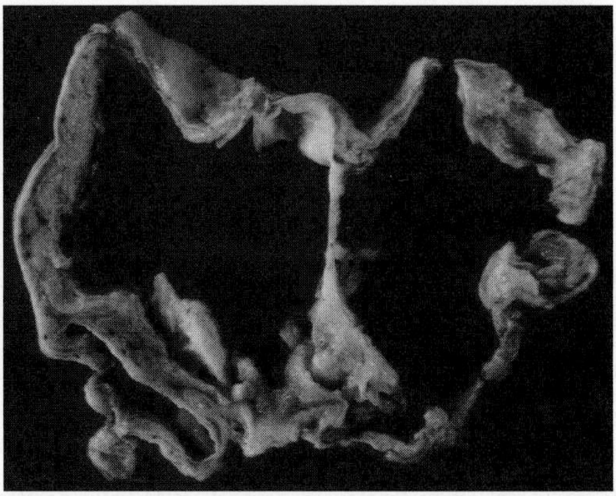

Fig. 11. Severe cavitary lesions and hydranencephaly in an infant who died 3 months after HSVE acquired during delivery. (*Courtesy of* Jeannette J. Townsend, MD, Salt Lake City, UT.)

Unfortunately, genital HSV-2 infections are not uncommon, and often do not produce symptoms, so that the possibility of HSV infection in the newborn may be overlooked, especially because the initial symptoms may be nonspecific and develop only after the newborn is discharged from the hospital [56].

A small proportion of cases of neonatal HSV encephalitis may result from contact with nursery personnel or relatives infecting the infant after delivery.

Although HSV infections of the newborn may be limited to disease of the skin, eyes, and mucous membranes, a proportion of these go on to infection of the brain. Often, however, the infection in the newborn is more generalized, involving systemic organs including lung, liver, adrenal, and brain, with a correspondingly worse prognosis [53,57,58].

The involvement of the central nervous system by HSV may initially involve the temporal and frontal cortices, as in the case with adults, but often the disease progresses to other portions of the brain [40].

Diagnosis

The diagnosis of HSV involvement in the newborn is made easier by the presence of vesicles on the skin. Biopsy or smear of the lesions and detection of virus antigen by fluorescent antibody or other tests can confirm the nature of the skin lesions. Suspicion is enhanced by the history of recent or even remote maternal genital infection, but this information is often lacking. The symptoms in the newborn consist of lethargy, failure to feed, tremors, irritability, or seizures, and in the absence of skin lesions, may be difficult to distinguish from a variety of other conditions affecting the newborn [59]. PCR studies for the presence of HSV genomes in the CSF are also helpful in establishing the diagnosis of HSV encephalitis, and distinguishing it from other newborn encephalitides caused by toxoplasmosis, cytomegalovirus, or other viruses.

When HSV involvement of the brain is suspected, CSF should be analyzed for presence of HSV by culture and PCR. Unlike the situation in adult HSVE, cultures of the CSF for HSV are more often positive in newborns. In contrast, it would appear that PCR is less sensitive for the detection of HSV genomes in neonatal encephalitis than in adult disease. Additionally in the experience at the Hospital for Sick Children in Toronto, Canada, PCR for HSV may be negative early in the illness [40]. In the absence of skin lesions, extensive sampling of other sites (throat, urine, skin, or rectal swabs) may be necessary to confirm the presence of the virus [53].

As in the case with encephalitis in children or adults, the encephalitis in newborns can be detected by MRI scanning. The pattern of involvement in newborns is more diffuse than that seen in children or adults, and may involve cavitary lesions and hemorrhages (Fig. 12) [60]. Leonard and colleagues [60] have described an intermediate form of distribution of lesions in infants and young children who have HSV-1 induced encephalitis. Diffusion weighted imaging may be more sensitive than routine MRI in the detection of early lesions in HSVE in neonates [61,62].

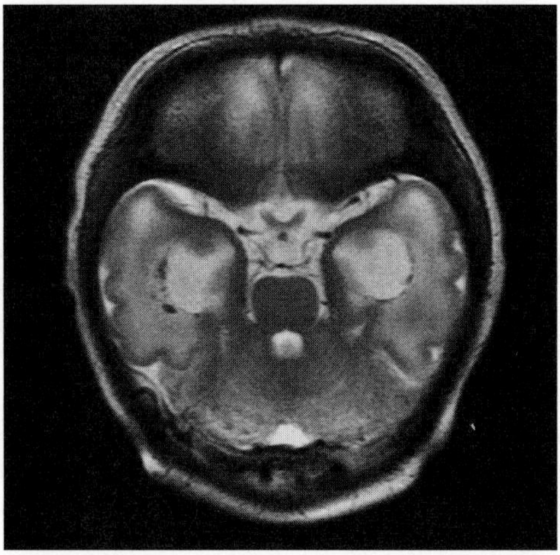

Fig. 12. Cystic enlargement of the temporal horns with surrounding hemorrhages (*black dots*) in an infant with intrauterine infection with HSV. (*Courtesy of* Richard S. Boyer, MD, Salt Lake City, UT.)

Treatment

The treatment of HSV infections in the newborn is complex, and has been recently extensively reviewed by Kimberlin [53,54]. For disseminated HSV infections or HSVE, the treatment with acyclovir is recommended at 20 mg/kg every 8 hours for 14 to 21 days [54]. A shorter course of 14 days is recommended for treatment of skin, eye, and mouth disease if CNS disease can be excluded. At these higher doses, neutrophil counts should be monitored. As in the case with disease in adults, early recognition of the disease and institution of treatment achieves the best outcomes. Some authorities suggest that a CSF PCR for HSV should be obtained at the completion of treatment to confirm the eradication of HSV. Unfortunately, the disease may not announce itself until the infant is at home, and diagnosis in this situation is often delayed. The overall prognosis for favorable recovery is correspondingly lessened [40]. The use of acyclovir in maternal genital infection at or before delivery appears to reduce the shedding of virus at delivery, and may obviate the necessity for Caesarean section to avoid HSV infection in the neonate [63]. These considerations have been comprehensively reviewed [53,54].

Summary

Dramatic progress has been made in the last 2 decades in the ability to noninvasively diagnose and treat HSVE in infants, children, and adults. The advent of PCR technology applied to detection of HSV DNA in spinal fluid and the advances in MRI have greatly enhanced our ability to diagnose the illness early in its course, and noninvasively. Acyclovir is of proven efficacy both experimentally and in clinical trials, and its administration is generally well-tolerated.

The major problem in terms of clinical management is that the pathologic process in the brain is usually well-advanced by the time the patient presents to medical attention, and the symptoms, particularly in newborns or infants, are often nonspecific at the outset. This, combined with a too frequent failure to recognize the nature and the seriousness of the process, results in further delay in diagnosis and treatment.

There needs to be an increased awareness by adult physicians and pediatricians of the early signs and symptoms of the presentation of HSVE, and an awareness of the imperative for early treatment.

For adults the combination of alteration in consciousness and fever should prompt immediate investigation. For pediatricians, the task may be more difficult, because of more subtle signs in newborns that are difficult to interpret. Certainly a history of maternal HSV infection should increase vigilance, although it is clear that many cases of HSVE in newborns occur in the absence of such a history.

Acknowledgment

The author would like to thank Dr. James F, Bale, Pediatric Neurology Division, Primary Children's Hospital and the University of Utah Health Science Center for advice in preparation of this article.

References

[1] Roizman B, Sears AE. An inquiry into the mechanisms of herpes simplex virus latency. Annu Rev Microbiol 1987;41:543–71.

[2] Whitley RJ, Kimberlin DW, Roizman B. Herpes simplex viruses. Clin Infect Dis 1998;26(3): 541–53, [quiz 554–545].

[3] Cushing H. Perineal zoster, with notes upon cutaneous segmentation postaxial to the lower limb. Am J Med Sci 1904;127(3):375–91.

[4] Carton CA, Kilbourne ED. Activation of latent herpes simplex by trigeminal sensory-root section. N Engl J Med 1952;246:172–6.

[5] Marinesco G, Draganesco S. Recherches expérimentales sur le neurotropisme du virus herpétique. Ann Inst Pasteur 1923;37:753–83 [in French].

[6] Goodpasture EW. The axis-cylinders of peripheral nerves as portals of entry to the central nervous system for the virus of herpes simplex in experimentally infected rabbits. Am J Pathol 1925;1:11–28.

[7] Goodpasture EW, Teague O. Transmission of the virus of herpes febrilis along nerves in experimentally infected rabbits. J Med Res 1923;44:139.

[8] Cook ML, Bastone VB, Stevens JG. Evidence that neurons harbor latent herpes simplex virus. Infect Immun 1974;9(5):946–51.

[9] Cook ML, Stevens JG. Pathogenesis of herpetic neuritis and ganglionitis in mice: evidence for intra-axonal transport of infection. Infect Immun 1973;7(2):272–88.

[10] Bastian FO, Rabson AS, Yee CL, et al. Herpesvirus hominis: isolation from human trigeminal ganglion. Science 1972;178(58):306–7.

[11] Baringer JR, Swoveland P. Recovery of herpes-simplex virus from human trigeminal ganglions. N Engl J Med 1973;288(13):648–50.
[12] Baringer JR. Recovery of herpes simplex virus from human sacral ganglions. N Engl J Med 1974;291(16):828–30.
[13] Baringer JR, Pisani P. Herpes simplex virus genomes in human nervous system tissue analyzed by polymerase chain reaction. Ann Neurol 1994;36(6):823–9.
[14] Murakami S, Mizobuchi M, Nakashiro Y, et al. Bell palsy and herpes simplex virus: identification of viral DNA in endoneurial fluid and muscle. Ann Intern Med 1996; 124(1 Pt 1):27–30.
[15] Baringer JR. Herpes simplex virus and Bell palsy. Ann Intern Med 1996;124(1 Pt 1):63–5.
[16] Whitley RJ, Soong SJ, Linneman C Jr, et al. Herpes simplex encephalitis. Clinical assessment. JAMA 1982;247(3):317–20.
[17] Whitley RJ, Gnann JW. Viral encephalitis: familiar infections and emerging pathogens. Lancet 2002;359(9305):507–13.
[18] Whitley RJ. Herpes simplex virus infections of the central nervous system. Encephalitis and neonatal herpes. Drugs 1991;42(3):406–27.
[19] Whitley R, Lakeman AD, Nahmias A, et al. DNA restriction-enzyme analysis of herpes simplex virus isolates obtained from patients with encephalitis. N Engl J Med 1982; 307(17):1060–2.
[20] Esiri MM. Herpes simplex encephalitis. An immunohistological study of the distribution of viral antigen within the brain. J Neurol Sci 1982;54(2):209–26.
[21] Grover D, Newsholme W, Brink N, et al. Herpes simplex virus infection of the central nervous system in human immunodeficiency virus-type 1-infected patients. Int J STD AIDS 2004;15(9):597–600.
[22] Kim C, Yoon YH. Unilateral acute retinal necrosis occurring 2 years after herpes simplex type 1 encephalitis. Ophthalmic Surg Lasers 2002;33(3):250–2.
[23] Numazaki K, Itoh N. Herpes simplex virus type 1-induced acute retinal necrosis. Ann Neurol 2001;50(2):273–4.
[24] McCabe K, Tyler K, Tanabe J. Diffusion-weighted MRI abnormalities as a clue to the diagnosis of herpes simplex encephalitis. Neurology 2003;61(7):1015–6.
[25] Schroth G, Gawehn J, Thron A, et al. Early diagnosis of herpes simplex encephalitis by MRI. Neurology 1987;37(2):179–83.
[26] Tyler KL. Update on herpes simplex encephalitis. Rev Neurol Dis 2004;1(4):169–78.
[27] Tyler KL. Herpes simplex virus infections of the central nervous system: encephalitis and meningitis, including Mollaret's. Herpes 2004;(11 Suppl 2):57A–64A.
[28] Lakeman FD, Whitley RJ. Diagnosis of herpes simplex encephalitis: application of polymerase chain reaction to cerebrospinal fluid from brain-biopsied patients and correlation with disease. National institute of allergy and infectious diseases collaborative antiviral study group. J Infect Dis 1995;171(4):857–63.
[29] Kennedy PG, Chaudhuri A. Herpes simplex encephalitis. J Neurol Neurosurg Psychiatry 2002;73(3):237–8.
[30] Whitley RJ, Cobbs CG, Alford CA Jr, et al. Diseases that mimic herpes simplex encephalitis. Diagnosis, presentation, and outcome. NIAD collaborative antiviral study group. JAMA 1989;262(2):234–9.
[31] Johns DR, Stein AG, Wityk R. MELAS syndrome masquerading as herpes simplex encephalitis. Neurology 1993;43(12):2471–3.
[32] Kimata KG, Gordan L, Ajax ET, et al. A case of late-onset MELAS. Arch Neurol 1998; 55(5):722–5.
[33] McCullers JA, Lakeman FD, Whitley RJ. Human herpesvirus 6 is associated with focal encephalitis. Clin Infect Dis 1995;21(3):571–6.
[34] Protopsaltis J, Kokkoris S, Brestas PS, et al. Neurolisteriosis mimicking herpes simplex encephalitis in an immunocompromized patient. Scand J Infect Dis 2006;38(9):825–8.

[35] Vincent A, Buckley C, Schott JM, et al. Potassium channel antibody-associated encephalopathy: a potentially immunotherapy-responsive form of limbic encephalitis. Brain 2004; 127(Pt 3):701–12.
[36] Seeley WW, Marty FM, Holmes TM, et al. Post-transplant acute limbic encephalitis: clinical features and relationship to HHV6. Neurology 2007;69:156–65.
[37] Whitley RJ, Lakeman F. Herpes simplex virus infections of the central nervous system: therapeutic and diagnostic considerations. Clin Infect Dis 1995;20(2):414–20.
[38] Whitley RJ, Roizman B. Herpes simplex virus infections. Lancet 2001;357(9267):1513–8.
[39] Kamei S, Sekizawa T, Shiota H, et al. Evaluation of combination therapy using aciclovir and corticosteroid in adult patients with herpes simplex virus encephalitis. J Neurol Neurosurg Psychiatry 2005;76(11):1544–9.
[40] Elbers JM, Bitnun A, Richardson SE, et al. A 12-year prospective study of childhood herpes simplex encephalitis: is there a broader spectrum of disease? Pediatrics 2007;119(2):e399–407.
[41] Skoldenberg B, Aurelius E, Hjalmarsson A, et al. Incidence and pathogenesis of clinical relapse after herpes simplex encephalitis in adults. J Neurol 2006;253(2):163–70.
[42] Dennett C, Klapper PE, Cleator GM. Polymerase chain reaction in the investigation of "relapse" following herpes simplex encephalitis. J Med Virol 1996;48(2):129–32.
[43] Yamada S, Kameyama T, Nagaya S, et al. Relapsing herpes simplex encephalitis: pathological confirmation of viral reactivation. J Neurol Neurosurg Psychiatry 2003;74(2):262–4.
[44] Valencia I, Miles DK, Melvin J, et al. Relapse of herpes encephalitis after acyclovir therapy: report of two new cases and review of the literature. Neuropediatrics 2004;35(6):371–6.
[45] De Tiege X, Rozenberg F, Burlot K, et al. Herpes simplex encephalitis: diagnostic problems and late relapse. Dev Med Child Neurol 2006;48(1):60–3.
[46] Gonzales N, Tyler KL, Gilden DH. Recurrent dermatomal vesicular skin lesions: a clue to diagnosis of herpes simplex virus 2 meningitis. Arch Neurol 2003;60(6):868–9.
[47] Tedder DG, Ashley R, Tyler KL, et al. Herpes simplex virus infection as a cause of benign recurrent lymphocytic meningitis. Ann Intern Med 1994;121(5):334–8.
[48] Tyler KL, Adler D. Twenty-eight years of benign recurring Mollaret meningitis. Arch Neurol 1983;40(1):42–3.
[49] Bachmeyer C, de la Blanchardiere A, Lepercq J, et al. Recurring episodes of meningitis (Mollaret's meningitis) with one showing an association with herpes simplex virus type 2. J Infect 1996;32(3):247–8.
[50] Bergstrom T, Alestig K. Treatment of primary and recurrent herpes simplex virus type 2 induced meningitis with acyclovir. Scand J Infect Dis 1990;22(2):239–40.
[51] Bergstrom T, Vahlne A, Alestig K, et al. Primary and recurrent herpes simplex virus type 2-induced meningitis. J Infect Dis 1990;162(2):322–30.
[52] Cohen BA, Rowley AH, Long CM. Herpes simplex type 2 in a patient with Mollaret's meningitis: demonstration by polymerase chain reaction. Ann Neurol 1994;35(1):112–6.
[53] Kimberlin D. Herpes simplex virus, meningitis and encephalitis in neonates. Herpes 2004;(11 Suppl 2):65A–76A.
[54] Kimberlin DW. Advances in the treatment of neonatal herpes simplex infections. Rev Med Virol 2001;11(3):157–63.
[55] Kimberlin DW, Lakeman FD, Arvin AM, et al. Application of the polymerase chain reaction to the diagnosis and management of neonatal herpes simplex virus disease. National institute of allergy and infectious diseases collaborative antiviral study group. J Infect Dis 1996;174(6):1162–7.
[56] Frenkel LM. Challenges in the diagnosis and management of neonatal herpes simplex virus encephalitis. Pediatrics 2005;115(3):795–7.
[57] Whitley R, Arvin A, Prober C, et al. Predictors of morbidity and mortality in neonates with herpes simplex virus infections. The national institute of allergy and infectious diseases collaborative antiviral study group. N Engl J Med 1991;324(7):450–4.
[58] Whitley RJ. Neonatal herpes simplex virus infections. J Med Virol 1993;(Suppl 1):13–21.

[59] Toth C, Harder S, Yager J. Neonatal herpes encephalitis: a case series and review of clinical presentation. Can J Neurol Sci 2003;30(1):36–40.

[60] Leonard JR, Moran CJ, Cross DT III, et al. MR imaging of herpes simplex type 1 encephalitis in infants and young children: a separate pattern of findings. AJR Am J Roentgenol 2000;174(6):1651–5.

[61] Kubota T, Ito M, Maruyama K, et al. Serial diffusion-weighted imaging of neonatal herpes encephalitis: a case report. Brain Dev 2007;29(3):171–3.

[62] Teixeira J, Zimmerman RA, Haselgrove JC, et al. Diffusion imaging in pediatric central nervous system infections. Neuroradiology 2001;43(12):1031–9.

[63] Scott LL, Sanchez PJ, Jackson GL, et al. Acyclovir suppression to prevent cesarean delivery after first-episode genital herpes. Obstet Gynecol 1996;87(1):69–73.

ELSEVIER
SAUNDERS

Neurol Clin 26 (2008) 675–697

NEUROLOGIC
CLINICS

Varicella Zoster Virus Infection: Clinical Features, Molecular Pathogenesis of Disease, and Latency

Niklaus H. Mueller, PhD[a], Donald H. Gilden, MD[a,b,*], Randall J. Cohrs, PhD[a], Ravi Mahalingam, PhD[a], Maria A. Nagel, MD[a]

[a]*Department of Neurology, University of Colorado School of Medicine, 4200 East 9th Avenue, Mail Stop B182, Denver, CO 80262, USA*
[b]*Department of Microbiology, University of Colorado School of Medicine, 4200 East 9th Avenue, Mail Stop B182, Denver, CO 80262, USA*

Varicella zoster virus (VZV) is an exclusively human neurotropic alpha-herpesvirus. Primary infection causes varicella (chickenpox), after which virus becomes latent in cranial nerve ganglia, dorsal root ganglia, and autonomic ganglia along the entire neuraxis. Years later, in association with a decline in cell-mediated immunity in elderly and immunocompromised individuals, VZV reactivates and causes a wide range of neurologic disease, including herpes zoster, postherpetic neuralgia, vasculopathy, myelopathy, retinal necrosis, cerebellitis and zoster sine herpete (Fig. 1). Importantly, many of these complications occur without rash. This article discusses the clinical manifestations, treatment, and prevention of VZV infection and reactivation; pathogenesis of VZV infection; and current research focusing on VZV latency, reactivation, and animal models.

This work was supported in part by Public Health Service grants NS32623 and AG06127 from the National Institutes of Health. Drs. Maria Nagel and Niklaus Mueller are supported by Public Health Service grant NS07321 from the National Institutes of Health.

* Corresponding author. Department of Neurology, University of Colorado School of Medicine, 4200 East 9th Avenue, Mail Stop B182, Denver, CO 80262.

E-mail address: don.gilden@uchsc.edu (D.H. Gilden).

0733-8619/08/$ - see front matter
doi:10.1016/j.ncl.2008.03.011

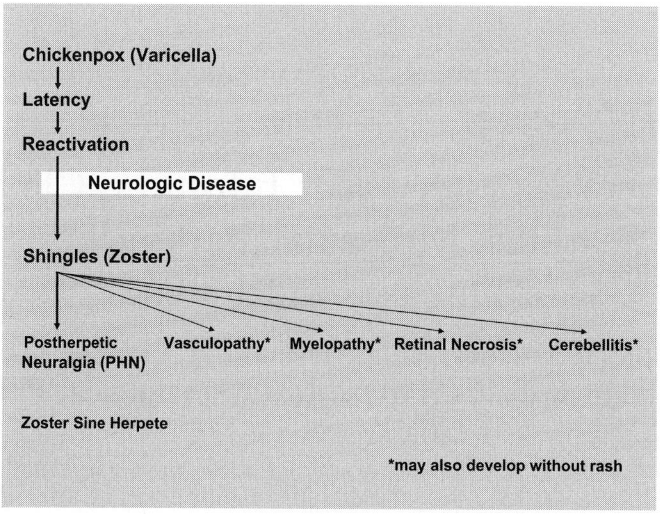

Fig. 1. The neurologic complications of varicella zoster virus reactivation.

Clinical manifestations of primary varicella zoster virus infection

Varicella

Initial infection with VZV results in chickenpox (varicella), which is typically seen in children 1 to 9 years of age [1]. Primary infection in adults is usually more severe and may be accompanied by interstitial pneumonia. Infection in immunocompromised individuals often causes severe, disseminated disease. Climate seems to affect the epidemiology of varicella. In most temperate climates, more than 90% of people are infected before adolescence [2–5] with an incidence of 13 to 16 cases per 1000 people per year [6–8]. In tropical climates, VZV infection occurs later in life and adults are more susceptible than children [9–11]. Varicella has a peak incidence in the late winter and spring [10,12–14], and epidemics tend to occur every 2 to 5 years [12–14].

Varicella is characterized by fever concurrent with a self-limiting rash on the skin and sometimes mucosa. Headache, malaise, and loss of appetite are also seen. The rash begins as macules, rapidly progresses to papules, followed by a vesicular stage and crusting of lesions. Crusts slough off after 1 to 2 weeks. VZV is highly infectious and transmission occurs by direct contact with skin lesions or by respiratory aerosols from infected individuals. Central nervous system complications include self-limiting cerebellar ataxia in 1 in 4000 cases [15], meningitis, meningoencephalitis, and vasculopathy [16]. Strokes may occur months after varicella secondary to VZV vasculopathy and are not always easy to diagnose (see section on VZV vasculopathy).

Diagnosis of varicella is based on the characteristic vesicular rash. Treatment is aimed at symptomatic relief. Acetaminophen is used to control fever, fluids are given for hydration, and topical medications are given for the pruritic rash. Treatment with intravenous acyclovir is mandatory in patients at risk for or with clinical evidence of disseminated disease, or in newborns who were exposed to VZV shortly after birth. In otherwise healthy children, antiviral treatment is not mandatory, but Dunkle and colleagues [17] have shown that treatment with oral acyclovir within 24 hours of illness results in a 1-day reduction in the duration of fever and a reduced severity of cutaneous and systemic symptoms and signs.

Clinical manifestations of varicella zoster virus reactivation

Herpes zoster

Zoster affects approximately 1 million individuals in the United States per year. Most patients are over age 60 [18] or immunocompromised [19]. The annual incidence of zoster is approximately 5 to 6.5 per 1000 individuals at age 60, increasing to 8 to 11 per 1000 at age 70 [19]. Unlike varicella, which occurs primarily in the spring, there is no seasonal predilection for zoster. The development of zoster may be viewed in the context of a continuum in immunodeficient individuals, ranging from a natural decline in VZV-specific cell-mediated immunity with age, to more serious immune deficits seen in cancer patients and transplant recipients, and ultimately in patients with AIDS [20]. Not surprisingly, zoster in otherwise young, healthy individuals may be the first manifestation of HIV infection [21]. Interestingly, varicella in infancy predisposes to zoster earlier in life [22].

Herpes zoster usually begins with a prodromal phase characterized by pain, itching, paresthesias (numbness or tingling), dysesthesias (unpleasant sensations), or sensitivity to touch (allodynia) in one to three dermatomes. A few days later, a unilateral maculopapular rash appears on the affected area, which then evolves into vesicles. These vesicles usually scab over in 10 days, after which the lesions are not contagious. Dissemination may occur in immunosuppressed patients, such as patients with a hematologic malignancy or iatrogenic immunosuppression. In most patients, the disappearance of skin lesions is accompanied by decreased pain and complete resolution of pain in 4 to 6 weeks. In zoster, MRI has shown enhancement of dorsal root ganglia and affected nerve roots [23].

Zoster affects any level of the neuraxis. The most common site is the chest, followed by lesions on the face, typically in the ophthalmic distribution of the trigeminal nerve. In immunocompromised patients, multidermatomal involvement is common and may be the first clue to the immunodeficient condition. Herpes zoster ophthalmicus is often accompanied by zoster keratitis, which can lead to blindness if unrecognized and not treated. If visual symptoms are present in these patients, an immediate slit-lamp examination by an ophthalmologist is imperative, especially if skin lesions extend to the medial side of the nose (Hutchinson's sign). Involvement of the optic nerves with subsequent optic neuritis and neuropathy has occurred rarely in association with herpes zoster ophthalmicus and other cutaneous zoster eruptions [24,25]. Ophthalmoplegia after zoster most frequently involves cranial nerves III and VI, and less frequently cranial nerve IV [24,26–28]; in addition, involvement of the maxillary and mandibular distribution of the trigeminal nerve can produce osteonecrosis and spontaneous tooth exfoliation [29–31].

Zoster affecting the seventh cranial nerve (geniculate) ganglion causes weakness or paralysis of ipsilateral facial muscles, with rash in the external auditory canal (zoster oticus) or tympanic membrane, or on the ipsilateral anterior two thirds of the tongue or hard palate. Lesions in these areas are often missed. The combination of peripheral facial weakness and zoster oticus constitutes the Ramsay Hunt syndrome [32]. Although the Ramsay Hunt syndrome is traditionally defined as lower motor neuron facial palsy with zoster oticus, many of these patients also have tinnitus, hearing loss, nausea, vomiting, vertigo, and nystagmus, indicating involvement not only of the geniculate ganglion, but also the eighth cranial nerve within the bony facial canal. Rarely, cranial nerves V, VI, IX, and X may also be involved [33]. Compared with Bell's palsy (peripheral facial paralysis without rash), individuals with the Ramsay Hunt syndrome often have more severe facial paralysis at onset and are less likely to recover completely [34]. In addition, peripheral facial paralysis caused by VZV may develop in the absence of rash as demonstrated by a four-fold rise in antibody to VZV or the presence of VZV DNA in auricular skin, mononuclear cells (MNCs), middle ear fluid, or saliva [35]. Some patients with idiopathic facial weakness actually represent another variant of zoster sine herpete (pain without rash).

Cervical or lumbar distribution zoster may be followed by lower motor neuron type weakness in the arm or leg, respectively [36,37]. Cervical zoster may rarely be followed by diaphragmatic weakness [38]. Rare cases of thoracic zoster have been associated with abdominal muscle weakness, which can result in abdominal herniation [39].

Zoster is presumed to develop by retrograde transport of virus from ganglia to skin in a host partially immune to VZV. VZV has also been isolated from the blood of immunocompromised patients with localized and disseminated zoster [40], suggesting a role for hematogenous spread. Cardinal pathologic features of zoster are inflammation and hemorrhagic necrosis with associated neuritis, localized leptomeningitis, unilateral segmental poliomyelitis, and degeneration of related motor and sensory roots [41,42]. Demyelination has also been observed in areas with MNC infiltration and microglial proliferation. Intranuclear inclusions, viral antigen, and herpesvirus particles have been detected in acutely infected ganglia [43–46].

Treatment for zoster should consider the patient's immune status and age. In immunocompetent individuals under age 50, analgesics are used to relieve pain. Antivirals (famciclovir, 500 mg orally three times daily, or valacyclovir, 1 g three times daily for 7–10 days) are not required, but speed healing of rash. In immunocompetent individuals age 50 and older, treatment with both analgesics and antivirals is recommended and is essential in patients with ophthalmic distribution zoster. Similarly, treatment of patients with the Ramsay Hunt syndrome within 7 days of onset reportedly improves recovery [47,48], although prospective randomized treatment trials remain to be conducted. The authors also use prednisone (1 mg/kg body weight once a day for 5 days) to reduce the inflammatory response, although double-blind placebo-controlled studies to prove additional efficacy are lacking. In immunocompromised patients, intravenous acyclovir (10 mg/kg three times per day for no less than 7 days) is recommended.

Postherpetic neuralgia

About 40% of zoster patients over age 60 experience postherpetic neuralgia (PHN) [49,50]. PHN is characterized by constant, severe, stabbing or burning, dysesthetic pain that persists for at least 3 months and sometimes years after resolution of rash. The cause and pathogenesis of PHN are unknown. Two nonmutually exclusive theories are that excitability of ganglionic or even spinal cord neurons is altered, and persistent or low-grade productive virus infection exists in ganglia. The concept that PHN is produced by low-level ganglionitis is supported by the detection of VZV DNA and proteins in blood MNCs of many patients with PHN [51–53], and by a favorable response of some PHN patients to antiviral treatment [54–56].

Although not life-threatening, PHN is difficult to manage. Treatment is supportive with use of neuroleptic drugs and various analgesics, including opiates to alleviate pain, but no universally accepted treatment exists. Gabapentin (300 mg daily with gradually increasing doses, up to a maximum of 3600 mg/day in three doses) is one of the most widely accepted treatments [57,58]. Lidocaine is administered as 5% patches with up to three patches applied topically at one time for up to 12 hours within a 24-hour period. Pregabalin is given initially at a dose of 75 mg orally twice a day or 50 mg orally three times a day, then gradually increased to a maximum dose of 300 mg per day based on efficacy and tolerability. If minimal relief is obtained at 300 mg per day for 2–4 weeks, the dose can be increased to 600 mg per day in two or three divided doses, although dosing needs to be adjusted based on side effects of the drug as well as the patient's renal function.

In addition, oxycodone (controlled release, 10–40 mg orally every 12 hours) or controlled-release morphine sulfate and tricyclic antidepressants are used [59]. Levorphanol produces morphine-like analgesia, at a dose of 2 mg orally every 6 to 8 hours as needed with maximum doses of 6 to 12 mg daily [60]. Combination treatment with morphine and gabapentin also decreases pain more than either drug alone or placebo [61]. Tricyclic antidepressants, including amitriptyline (10–25 mg orally at bedtime with a maximum dose of 150–200 mg/day), nortriptyline, mapotriline, and despramine, lessen the pain of PHN.

Numerous studies indicate that antiviral therapy with oral acyclovir, famcyclovir, or valacyclovir may reduce the duration and severity of pain after zoster [62–64]. A recent prospective, open-label phase I/II clinical trial treated 15 patients with moderate to severe PHN with intravenous acyclovir for 2 weeks, followed by oral valacyclovir for 1 month; 8 (53%) of 15 patients reported improvement of pain [56].

Varicella zoster virus vasculopathy

VZV vasculopathy results from productive virus infection in large or small cerebral arteries, or both. Patients present with headache; fever; mental status changes; transient ischemic attacks; and focal deficits (stroke). The clinical spectrum of VZV vasculopathy is protean. For example, one case of VZV vasculopathy was characterized by posterior ischemic optic neuropathy with a normal cerebral angiogram and MRI [65]. Cerebral aneurysms and hemorrhage can also develop from viral invasion of vessels [66,67]. VZV vasculopathy often occurs without rash [68,69].

The cerebrospinal fluid (CSF) usually, but not always, reveals a mononuclear pleocytosis and oligoclonal bands. The oligoclonal IgG has been shown to be antibody directed against VZV [70]. Brain imaging usually reveals ischemic or hemorrhagic infarcts, more deep-seated than cortical lesions and at gray-white matter junctions. Cerebral angiography also reveals areas of focal arterial stenosis or occlusion. Macroscopically, a predominance of gray-white matter junction lesions is seen. Microscopically, virus is present in affected cerebral arteries [71] but not in areas of infarction, although in chronic cases virus may be seen in brain parenchyma, usually close to arteries and veins. The primary site of VZV is in cerebral arteries, which contain multinucleated giant cells, Cowdry A inclusion bodies, and herpes virus particles. Postmortem virologic analysis has revealed not only VZV DNA, but also VZV antigen in cerebral vessels [71].

Confirmation of VZV vasculopathy requires virologic analysis to detect amplifiable VZV DNA or anti-VZV IgG antibodies or both in the CSF. PCR is the diagnostic test of choice for herpes simplex virus (HSV) encephalitis, with HSV DNA present early in the course of acute disease, whereas antiviral antibody is detected only in the second week [72]. In cases of VZV vasculopathy, the CSF does not always contain PCR-amplifiable VZV DNA, but does contain anti-VZV IgG [73]. The detection of anti-VZV IgG, but not VZV DNA, likely reflects the chronic, protracted course of disease. Testing for both VZV DNA and anti-VZV IgG must be done, and only when both are negative can the diagnosis of VZV vasculopathy be excluded. Also, because VZV vasculopathy can occur without rash, all vasculopathies of unknown etiology should be evaluated for VZV. Rapid diagnosis of VZV vasculopathy is important because the mortality without treatment is 25% [74], whereas treatment with intravenous acyclovir, even after neurologic disease has been present for months, can be curative [65].

Varicella zoster virus myelopathy

Two clinical presentations of VZV myelitis predominate. The first is a self-limiting, monophasic spastic paraparesis, with or without sensory features and sphincter problems. This so-called "postinfectious myelitis" usually occurs in immunocompetent patients, days to weeks after acute varicella or zoster. Its pathogenesis is unknown. The CSF usually reveals a mild mononuclear pleocytosis with a normal or slightly elevated protein. Steroids are used to treat these patients [75], although some improve spontaneously [76].

The second presentation is an insidious, progressive, and sometimes fatal myelitis, seen mostly in immunocompromised individuals. Because AIDS is so prevalent, this has become the most common condition associated with VZV myelitis. CSF examination reveals a mild, predominantly mononuclear pleocytosis with elevated protein. MRI reveals longitudinal serpiginous enhancing lesions. Diagnosis is confirmed by the presence of VZV DNA or anti-VZV IgG or both in CSF [77]. Pathologic and virologic analyses of the spinal cord from fatal cases have shown frank invasion of VZV in the parenchyma [78] and, in some instances, spread of virus to adjacent nerve roots [79]. Not surprisingly, several patients have responded favorably to antiviral therapy [80–82]. Importantly, VZV myelitis may develop with rash. Early diagnosis and aggressive treatment with intravenous acyclovir has been helpful, even in immunocompromised patients [80]. The benefit of steroids in addition to antiviral agents is unknown.

Most recently, a case of VZV spinal cord infarction was identified by diffusion-weighted MRI and confirmed virologically [83]. This indicates that VZV vasculopathy can cause stroke in the spinal cord and the brain, and that abnormalities on diffusion-weighted MRI is crucial for diagnosis.

Varicella zoster virus retinal necrosis

VZV-induced necrotizing retinitis manifests as two clinical syndromes: acute retinal necrosis (ARN) and progressive outer retinal necrosis (PORN).

Acute retinal necrosis

ARN is seen in both immunocompetent and immunocompromised hosts. Patients present with periorbital pain and floaters with hazy vision and loss of peripheral vision. ARN is a full-thickness retinal necrosis characterized by focal, well-demarcated areas of necrosis in the retina located beyond the major temporal vascular arcades. Distinguishing features of this occlusive vasculopathy are prominent intraocular inflammation in the anterior chamber and vitreous [84]. In addition to VZV as the causal agent [85], both HSV-1 and -2 can induce ARN [86,87]. Patients are typically treated with intravenous acyclovir, steroids, and aspirin followed by oral acyclovir [88]. Intravitreal injections of foscarnet and oral acyclovir have been used in early, milder cases. In ARN caused by VZV, brivudine (not available in the United States) and valgancyclovir have demonstrated good results [89].

Progressive outer retinal necrosis

PORN is caused almost exclusively by VZV. After cytomegalovirus, VZV-associated PORN is the second most common opportunistic retinal infection among AIDS patients in North America [90]. PORN occurs primarily in AIDS patients with CD4 counts typically less than 10 cells/mm^3 of blood [91], and in other immunosuppressed individuals [92]. PORN may be preceded by retrobulbar optic neuritis and aseptic meningitis [93], central retinal artery occlusion, or ophthalmic distribution zoster [94], and may occur together with multifocal vasculopathy or myelitis. Patients present with sudden painless loss of vision, floaters, and constricted visual fields with resultant retinal detachment. Unlike ARN, there is little or no inflammation in the anterior chamber or vitreous and no occlusive vasculitis. Multifocal, discrete opacified lesions begin in the outer retinal layers peripherally or posterior pole; only late in disease are inner retinal layers involved. Diffuse retinal hemorrhages and whitening with macular involvement bilaterally are characteristic findings.

VZV was shown to be the causative agent of PORN based on the detection of VZV DNA, VZV antigen, and virus particles in aqueous-vitreous biopsies [95], and in vitreous-retinal cultures [96], and by histologic examination of necropsy specimens from eyes and brains combined with in situ hybridization [97]. There have also been rare reports of cytomegalovirus and HSV-1 antigens detected in the retina of patients with PORN [98], and cytomegalovirus DNA has been amplified in the vitreous of patients with PORN [99]. Nevertheless, multiple studies have shown that VZV is the most common cause.

The immediate recognition and treatment of PORN is essential because of its destructive nature and high likelihood of retinal detachment. Unfortunately, multiple combinations of antiviral medications and other experimental treatments have not successfully treated all cases of PORN. Treatment with intravenous acyclovir has given poor or inconsistent results [100], and even when acyclovir helped, VZV retinopathy recurred when drug was tapered or stopped. PORN patients treated with a combination of ganciclovir and foscarnet or with ganciclovir alone had a better final visual acuity than those treated with acyclovir or foscarnet [101]. In one instance, oral bromovinyldeoxyuridine treatment was successful when acyclovir failed [102]. Aggressive combined antiviral treatment over a prolonged period with repair of retinal detachment may save the patient's vision. The best treatment for PORN in AIDS patients may be prevention with highly active antiretroviral therapy, which seems to have decreased the incidence of this syndrome [103].

Zoster sine herpete

Zoster sine herpete (pain without rash) is caused by reactivation of VZV [104], a concept first supported by the description of dermatomal distribution radicular pain in areas distinct from pain with rash in zoster patients [105]. Currently, most clinicians regard zoster sine herpete exclusively as the rare occurrence of chronic radicular pain without rash with virologic confirmation of VZV reactivation. In recent years, the detection of VZV DNA and anti-VZV antibody in patients with meningoencephalitis, vasculopathy, myelitis, cerebellar ataxia, and polyneuritis cranialis, all without rash, has expanded the spectrum of VZV infection without rash.

The first verification of zoster sine herpete was in a physician who developed acute trigeminal distribution pain without rash, associated with a fourfold rise in serum antibody specific to VZV [106]. Schott [107] reported four patients who developed trigeminal distribution zoster followed years later by zoster sine herpete in the same distribution of the trigeminal nerve as their previous zoster; unfortunately, none were studied virologically. Further virologic verification of zoster sine herpete came from PCR analysis of CSF from two men with prolonged thoracic distribution radicular pain without rash; amplifiable VZV DNA, but not HSV DNA, was found in their CSF and blood MNCs, and pain resolved after treatment with intravenous acyclovir [104]. An additional virologically confirmed case demonstrated electromyographic fibrillation potentials restricted to chronically painful thoracic roots [108]. MRI of another patient with virologically verified active VZV infection revealed inflammation in ganglia and nerve roots corresponding to persistent pain [23].

Virologic analyses have demonstrated the association of VZV with meningoencephalitis, vasculopathy, myelitis, cerebellar ataxia, and polyneuritis cranialis, all without rash. Powell and coworkers [109] reported a patient with meningoencephalitis without rash in whom VZV DNA was detected in the CSF, and Mancardi and coworkers [110] described a patient with encephalomyelitis without rash in whom anti-VZV antibody was found in the CSF. Kleinschmidt-DeMasters and coworkers [111] reported an HIV-positive patient with a fatal encephalomyelitis and necrotizing vasculitis, pathologically verified to be caused by VZV without rash. Cases of unifocal and multifocal VZV vasculopathy in the absence of zoster rash resulting in stroke have been verified virologically [68,69]. Two patients with myelopathy in the absence of rash have been described: one developed myelopathy 5 months after zoster rash, at which time amplifiable VZV DNA was detected in the CSF; the second patient developed myelopathy concurrent with zoster and the myelopathy recurred 6 months later in the absence of rash, at which time both VZV DNA and VZV antibody were detected in the CSF [104].

Although it is well recognized that cerebellar ataxia may complicate childhood varicella [112], there is one report of a child who became ataxic 5 days before chickenpox developed [113]. Most recently, acute cerebellar ataxia without rash has been reported in adults whose CSF revealed VZV DNA and VZV antibody [114,115]. Polyneuritis cranialis without rash caused by VZV infection has also been described in a patient with involvement of cranial nerves IX, X, and XI, and upper cervical nerve roots without rash, and with anti-VZV antibody in the CSF [116]. There is also a growing body of literature on ocular abnormalities associated with zoster sine herpete. Goon and coworkers [117] reported a case of severe, unremitting eye pain without rash proved to be caused by VZV infection by the detection of VZV DNA in nasal and conjunctival samples. In addition, cases of third cranial nerve palsies [118], retinal periphlebitis [119], uveitis [118,120], iridocyclitis [121], and disciform keratitis [122], all without rash and confirmed virologically to be caused by VZV, have been reported.

Two remarkable cases of VZV infection without rash deserve mention. The first was a 77-year-old man with T-cell lymphoma and no history of zoster rash who developed an acute fatal meningoradiculitis of cranial nerve roots and cauda equina, pathologically and virologically confirmed to be caused by VZV [123]. The second case was an immunocompetent adult who had experienced relentless trigeminal distribution pain for more than a year with no history of zoster rash; pathologic and virologic analysis of a trigeminal ganglionic mass confirmed chronic active VZV ganglionitis [124].

Prevalence estimates of VZV-induced pathology without rash await virologic analysis of additional patients with prolonged radicular pain or other neurologic symptoms and signs. Analysis should include both a test for anti-VZV IgG and PCR to amplify VZV DNA in CSF, and examination of blood MNCs for VZV DNA. Finally, the nosologic entity of zoster sine herpete has considerable implications for analysis and treatment of patients with PHN. Overall, VZV reactivation from latency in ganglia produces a variety of neurologic disorders all caused by the same pathogen and in the absence of zoster rash.

Vaccination

Widespread, aggressive VZV vaccination has reduced the total number of varicella cases by approximately 85% and the number of moderate-to-severe cases by 95% to 100% [125]. Now, like the live childhood varicella vaccine, there is a live zoster vaccine that seems to be safe and effective clinically. The results of a prospective, double-blind, placebo-controlled trial of attenuated VZV vaccine designed to prevent zoster and PHN in men and women over the age of 60 were recently reported [126]. Otherwise healthy adults age 60 years or older (median 69 years) were vaccinated with placebo or an attenuated Oka/Merck-VZV vaccine containing 18,700 to 60,000 plaque-forming units of virus, considerably greater than the approximately 1350 plaque-forming units in the Oka/Merck-VZV vaccine administered to American children since 1995. More than 38,000 recipients of the "zoster vaccine" were followed closely for 3 years. The incidence of zoster in the placebo group was 11.1 per 1000-person years, approximating the results of an epidemiologic survey performed a decade ago, which revealed zoster exceeding 10 cases per 1000-person years in individuals older than 75 years [19]. The effect of zoster vaccine was impressive; compared with placebo, vaccination reduced the incidence of shingles by 51%, the incidence of PHN by 66%, and the burden of illness by 61%.

Overall, serious adverse effects and deaths occurred in 1.4% of both vaccine and placebo recipients. In more than 6000 subjects who kept daily diaries of minor adverse effects for 42 days, 48% of vaccine recipients reported injection site erythema, pain or tenderness, swelling, and pruritis, compared with 16% of placebo recipients. In the same 6000 subjects, serious adverse effects were significantly more frequent ($P = 0.03$) in vaccine recipients (1.9%) compared with placebo recipients (1.3%), although no specific serious effects emerged. The relative impact of these side effects on the elderly (age $\geq$70) compared with younger patients was not examined but might be important in future analyses, because the at-risk population over age 70 years is projected to increase substantially in the coming decades. Although the Oka/Merck VZV vaccine on rare occasions unmasks a childhood immunodeficiency disorder, no cases of disseminated zoster that might have been attributed to zoster vaccine in a person with undiagnosed lymphoma, leukemia, or the like were reported.

In 2006, zoster vaccine received FDA approval for healthy VZV-seropositive adults over age 60. Zoster vaccine increases cell-mediated immunity to VZV in such individuals, and the boost is likely to last for decades. Because zoster and its attendant neurologic complication of PHN are common and serious, it seems prudent to recommend zoster vaccine. The Census Bureau projects that by the year 2050, there will be more than 21 million Americans 85 years of age or older (http://www.census.gov/ipc/www/usinterimproj/natprojtab02a.pdf).

Despite the development of a vaccine to prevent zoster, even if every healthy adult in the United States over age 60 years is vaccinated, there would still be approximately 500,000 zoster cases annually, about 200,000 of whom will experience PHN, and stroke, blindness, and myelopathy caused by VZV reactivation. Furthermore, because zoster vaccine is not approved for immunocompromised individuals, neurologic disease produced by VZV reactivation in this population is a continuing problem.

Physical and molecular properties of varicella zoster virus

VZV is one of the eight human herpesviruses and is morphologically indistinguishable from HSV-1, the prototype alphaherpesvirus. The VZV genome is a linear, double-stranded DNA molecule 124,884 nucleotides in length [127] with close homology to the HSV genome. The lipid envelope encloses the icosahedral nucleocapsid, which consists of 162 capsomeres. Virions are pleomorphic with a 150- to 200-nm diameter. In tissue culture,

VZV produces a cytopathic effect in approximately 3 days, characterized by the formation of large multinucleated syncytia without the release of significant quantities of stable infectious virions. VZV is highly cell-associated making it difficult to raise sufficient quantities of cell-free virus for molecular analysis.

The entire virus genome from 13 independent isolates has been sequenced. The VZV prototype consists of a long and short region, each bounded by inverted and terminal repeat sequences [127]. The long segment contains 104,836 base pairs (bp) of unique DNA flanked by 88-bp terminal repeat sequences. The short segment consists of 5232 bp of unique DNA flanked by 7320-bp repeat sequences. Analysis of the 124,884-bp VZV genome has identified 71 predicted open reading frames (ORF) numbered consecutively from the leftward end of the virus genome. ORFs 62, 63, and 64 map within the internal repeat region of the short segment of the VZV genome and are duplicated (although in opposite orientation) as ORFs 71, 70, and 69, respectively, within the terminal repeat region. ORFs 42 and 45 may be exons from the same approximately 5.7-kbp primary transcript. There are 68 predicted unique VZV genes. Additionally, two novel VZV genes (ORFs 9A and 33.5) have been identified experimentally, indicating a potential coding capacity of 70 unique genes.

Alphaherpesvirus gene transcription is classified into three distinct kinetic groups: (1) immediate-early, (2) early, and (3) late [128–130]. Immediate-early genes are transcribed in the absence of de novo protein synthesis and regulate transcription of early and late virus genes. The onset of early gene transcription precedes virus DNA replication. Transcription of early genes is induced by immediate-early proteins and early proteins, which are predominately involved in virus DNA replication and accumulate in the presence of inhibitors of DNA synthesis. Late proteins, which include the major virus structural proteins, are transcribed from progeny viral DNA, and their transcription is blocked by inhibitors of virus DNA synthesis.

During productive infection of cells in culture, transcripts mapping to all predicted VZV genes have been detected. PCR-based macroarrays developed to detect VZV transcription showed that major regions of gene transcription are not clustered, and instead are located throughout the virus genome [131]. These results were confirmed in a study using oligonucleotide-based microarrays [132]. Slight differences that were noted between the two array analyses are most likely caused by differences in virus strain and host cells.

Latency

The hallmark of herpesviruses is their ability to establish a life-long latent infection punctuated by periods of virus recrudescence. Alphaherpesvirus latency is characterized by the ability to reactivate infectious virus, the presence of virus DNA in ganglionic neurons, and limited virus gene transcription.

Virologic features in latently infected human ganglia

The hypothesis that VZV can establish a latent infection in sensory ganglia was first proposed by Head and Campbell [41], who noticed dermatomal distribution, varicella-like lesions in zoster cases. Serologic data indicated that varicella and zoster were caused by the same virus [133], but it was not until restriction endonuclease analysis of DNA from varicella and zoster lesions in the same individual that VZV was confirmed to cause both diseases [134].

VZV becomes latent in neurons [135–138] in cranial nerve, dorsal root, and autonomic ganglia along the entire human neuraxis. Latent VZV DNA assumes a circular or concatameric (end-to-end) state [139] and is present at a frequency of two to nine copies in 1% to 7% of individual neurons, which correlates to a virus burden of 30 to 3500 VZV DNA copies per 100 ng of total ganglionic DNA [137,138,140–144]. The wide range of VZV copy number during latent infection may reflect the degree of primary infection. For example, during varicella, the VZV DNA burden in blood ranges from 200 to 500 copies per 15,000 peripheral blood MNCs, 100 to 1000 copies per milliliter of whole blood, and 100 to more than 10,000 copies per milliliter of serum [145,146]. Furthermore, during the many decades from the time of infection until death, the amount of virus in latently infected ganglia is likely to be affected by exposure of adults to children with varicella or to other adults with zoster, or by spontaneous subclinical reactivation of VZV [147–149].

Varicella zoster virus gene expression in latently infected human ganglia

Transcripts corresponding to VZV genes 21, 29, 62, 63, and 66 have been identified in latently infected human ganglia [150–152]. The VZV gene 63 transcript is the most prevalent and abundant detected [153]. IE63, the immediate-early protein encoded by ORF 63, was the first VZV protein to be detected in latently infected human ganglia [154]. Subsequently, immunohistochemistry detected proteins encoded by VZV genes 62 and 66 [155] and by VZV genes 4, 21, and 29 [156,157]. The detection of protein encoded by VZV gene 4 requires confirmation because VZV gene 4 transcripts have not yet been found in latently infected human ganglia.

Animal models of varicella zoster virus infection

Development of an experimental animal model that recapitulates the pathogenesis of VZV seen in humans has been a goal ever since the realization that VZV infects only humans. Important criteria for any animal model of VZV latency include: (1) ability to reactivate the virus; (2) presence of virus nucleic acids in ganglia, but not in nonganglionic tissues; (3) presence of virus exclusively in neurons; and (4) limited transcription of virus genes.

Experimental inoculation of VZV into small animals including rabbits, mice, and rats leads to seroconversion in the absence of clinical signs [158–163]. After corneal inoculation of VZV in mice, viral DNA is detected in ganglionic neurons and nonneuronal cells, and in nonganglionic tissues 1 month after infection [164]. Intramuscular inoculation of guinea pigs with VZV produces a papular exanthem without vesicles [165], and VZV DNA has been detected by PCR in ganglia of guinea pigs 80 days after subcutaneous inoculation [166]. VZV RNA has also been detected in ganglia of guinea pigs by in situ hybridization 5 weeks after ocular inoculation [167]. It is difficult, however, to evaluate the usefulness of the guinea pig model because the absence of virus nucleic acids in nonganglionic tissues has not been shown. In addition, reactivation of latent VZV has not been demonstrated in guinea pigs.

Sadzot-Delvaux and coworkers [168] inoculated VZV subcutaneously into adult rats. Although no clinical signs developed, virus nucleic acids and proteins were detected in dissociated ganglionic neurons up to 9 months after experimental infection. Because the ganglia were cultured for 3 to 12 days, in vitro reactivation could not be excluded. Latent VZV infection has been reported in rats inoculated by footpad and sacrificed 1 month later [169,170]. The detection of VZV DNA in both neurons and nonneuronal cells [171] of dorsal root ganglia harvested 1 to 3 months after footpad-inoculation, however, questions the validity of the rat model. In addition, in vivo VZV reactivation in rats has not been reported.

Direct inoculation of VZV into human thymus and liver implants under the kidney capsule of severe combined immunodeficient mice results in virus infection as evidenced by the detection of virus proteins for 3 weeks after infection in $CD4^+$ and $CD8^+$ T cells [172]. Similar studies using human ganglionic implants have been used to demonstrate virus infection [173]. The absence of an intact immune system in these animals makes it difficult, however, to study latency or reactivation. Overall, although VZV reaches ganglia after experimental infection of small animals, nonganglionic tissues have not been studied and reactivation has not been demonstrated.

Oral-nasal-conjunctival application of the attenuated vaccine strain of VZV in marmosets results in mild pneumonia and an immune response without clinical disease [174]. Subcutaneous inoculation of the Oka VZV (vaccine strain) into the breast of chimpanzees produces viremia and a mild rash restricted to the site of inoculation [175]. VZV latency or reactivation has not been studied, however, in chimpanzees.

Simian varicella virus as a model for human varicella zoster virus disease

Immunologic, virologic, and pathologic features of simian varicella virus (SVV) infection of nonhuman primates closely resemble those of human VZV infection. Like VZV in humans, primary infection of primates with SVV leads to varicella followed by virus latency and spontaneous reactivation [176]. Intratracheal inoculation of SVV into nonhuman primates results in persistence of virus DNA in multiple organs, including blood MNCs for 2 years [177,178]. SVV DNA can be detected in ganglia 6 to 7 days after intratracheal or intravenous inoculation before varicella rash, pointing to hematogenous spread of the virus [179]. In monkeys intratracheally inoculated with SVV, virus DNA has been detected in both neurons and nonneuronal cells 9 to 10 months after infection, and exclusively in neurons 2 years after infection [180].

The authors developed a model of natural SVV infection in monkeys by exposing SVV-seronegative monkeys to other monkeys that had been inoculated intratracheally with SVV [181]. These naturally infected monkeys harbor latent SVV DNA in ganglionic neurons [182] at multiple levels of the neuraxis [181], and both clinical and subclinical reactivation of SVV were induced by immunosuppression and stress [183]. Subclinical reactivation of latent SVV also occurs after irradiation in rhesus monkeys [184].

The model of SVV infection of nonhuman primates is well-suited for studies to dissect the relative role of immunosuppression in varicella reactivation. Such studies assume importance in light of the association between VZV reactivation frequency and extent of immunosuppression, with a higher incidence in patients receiving chemotherapy and radiotherapy than in those receiving either alone [185]. Moreover, varicella reactivation results in serious neurologic complications in immunosuppressed individuals.

Summary

VZV is an exclusively human, highly neurotropic alphaherpesvirus. Primary infection causes chickenpox (varicella), after which virus becomes latent in cranial nerve ganglia, dorsal root ganglia, and autonomic ganglia along the entire neuraxis. Decades later, VZV may reactivate to cause herpes zoster (shingles), pain, and rash in one to three dermatomes. Multiple neurologic complications after VZV reactivation include PHN; vasculopathy; myelitis; necrotizing retinitis; and zoster sine herpete (pain without rash). Many may occur without rash and are difficult to recognize. Virologic confirmation requires testing the CSF for VZV DNA and anti-VZV IgG. Immediate treatment with antiviral agents may be warranted. The relative role of immunosuppression in the frequency and consequences of VZV reactivation awaits elucidation in the animal model of infection of monkeys with SVV, the only system to date that closely recapitulates the human disease. Successful mass vaccination against herpes zoster will have a profound impact on the health and quality of life of a steadily growing elderly population. Even if every healthy adult in the United States over age 60 years is vaccinated, however, there would still be approximately 500,000 zoster cases annually, about 200,000 of whom will experience PHN, and stroke, loss of vision, and myelopathy caused by VZV reactivation.

Acknowledgments

The authors thank Marina Hoffman for editorial review and Cathy Allen for preparing the manuscript.

References

[1] Finger R, Hughes JP, Meade BJ, et al. Age-specific incidence of chickenpox. Public Health Rep 1994;109:750–5.
[2] Heininger U, Braun-Fahrländer C, Desgrandchamps D, et al. Seroprevalence of varicella-zoster virus immunoglobulin G antibodies in Swiss adolescents and risk factor analysis for seronegativity. Pediatr Infect Dis J 2001;20:775–8.
[3] Wutzler P, Farber I, Wagenpfeil S, et al. Seroprevalence of varicella-zoster virus in the German population. Vaccine 2001;20:121–4.
[4] Kudesia G, Partridge S, Farrington CP, et al. Changes in age related seroprevalence of antibody to varicella zoster virus: impact on vaccine strategy. J Clin Pathol 2002;55:154–5.
[5] Kilgore PE, Kruszon-Moran D, Seward JF, et al. Varicella in Americans from NHANES III: implications for control through routine immunization. J Med Virol 2003;70(Suppl 1): S111–8.
[6] Choo PW, Donahue JG, Manson JE, et al. The epidemiology of varicella and its complications. J Infect Dis 1995;172:706–12.
[7] Boelle PY, Hanslik T. Varicella in non-immune persons: incidence, hospitalization and mortality rates. Epidemiol Infect 2002;129:599–606.
[8] Gershon AA, Hambleton S. Varicella vaccine for susceptible adults: do it today. Clin Infect Dis 2004;39:1640–1.
[9] Ooi PL, Goh KT, Doraisingham S, et al. Prevalence of varicella-zoster virus infection in Singapore. Southeast Asian J Trop Med Public Health 1992;23:22–5.
[10] Lee BW. Review of varicella zoster seroepidemiology in India and Southeast Asia. Trop Med Int Health 1998;3:886–90.
[11] Mandal BK, Mukherjee PP, Murphy C, et al. Adult susceptibility to varicella in the tropics is a rural phenomenon due to the lack of previous exposure. Infect Dis 1998;178(Suppl 1): S52–4.
[12] Deguen S, Chau NP, Flahault A. Epidemiology of chickenpox in France (1991–1995). J Epidemiol Community Health 1998;52(Suppl 1):46–9.
[13] Tobias M, Reid S, Lennon D, et al. Chickenpox immunisation in New Zealand. N Z Med J 1998;111:274–81.
[14] Bramley JC, Jones IG. Epidemiology of chickenpox in Scotland:1981 to 1998. Commun Dis Public Health 2000;3:282–7.
[15] Guess HA, Broughton DD, Melton LJ. Population-based studies of varicella complications. Pediatrics 1986;78(Suppl 4):723–7.
[16] Askalan R, Laughlin S, Mayank S, et al. Chickenpox and stroke in childhood: a study of frequency and causation. Stroke 2001;32:1257–62.
[17] Dunkle LM, Arvin AM, Whitley RJ, et al. A controlled trial of acyclovir for chickenpox in normal children. N Engl J Med 1991;325:1539–44.
[18] Harnisch JP. Zoster in the elderly: clinical, immunologic and therapeutic considerations. J Am Geriatr Soc 1984;32:789–93.
[19] Donahue JG, Choo PW, Manson JE, et al. The incidence of herpes zoster. Arch Intern Med 1995;155:1605–9.
[20] Gilden DH, Cohrs RJ, Mahalingam R. Clinical and molecular pathogenesis of varicella virus infection. Viral Immunol 2003;16:243–58.
[21] Leppard B, Naburi AE. Herpes zoster: an early manifestation of HIV infection. Afr Health 1998;21:5–6.

[22] Kakourou T, Theodoridou M, Mostrou G, et al. Herpes zoster in children. J Am Acad Dermatol 1998;39:207–10.
[23] Blumenthal DT, Salzman KL, Baringer JR, et al. MRI abnormalities in chronic active varicella zoster infection. Neurology 2004;63:1538–9.
[24] Carroll WM, Mastaglia FL. Optic neuropathy and ophthalmoplegia in herpes zoster oticus. Neurology 1979;29:726–9.
[25] Meenken C, van den Horn GJ, de Smet MD, et al. Optic neuritis heralding varicella zoster virus retinitis in a patient with acquired immunodeficiency syndrome. Ann Neurol 1998;43: 534–6.
[26] Archambault P, Wise JS, Rosen J, et al. Herpes zoster ophthalmoplegia: report of six cases. J Clin Neuroophthalmol 1988;8:185–93.
[27] Sodhi PK, Goel JL. Presentations of cranial nerve involvement in two patients with herpes zoster ophthalmicus. J Commun Dis 2001;33:130–5.
[28] Karmon Y, Gadath N. Delayed oculomotor nerve palsy after bilateral cervical zoster in an immunocompetent patient. Neurology 2005;65:170.
[29] Garty B-Z, Dinari G, Sarnat H, et al. Tooth exfoliation and osteonecrosis of the maxilla after trigeminal herpes zoster. J Pediatr 1985;106:71–3.
[30] Manz HJ, Canter HG, Melton J. Trigeminal herpes zoster causing mandibular osteonecrosis and spontaneous tooth exfoliation. South Med J 1986;79:1026–8.
[31] Volvoikar P, Patil S, Dinkar A. Tooth exfoliation, osteonecrosis and neuralgia following herpes zoster of trigeminal nerve. Indian J Dent Res 2002;13:11–4.
[32] Sweeney CJ, Gilden DH. Ramsay Hunt syndrome. J Neurol Neurosurg Psychiatry 2001;71: 149–54.
[33] Asnis DS, Micic L, Giaccio D. Ramsay Hunt syndrome presenting as a cranial polyneuropathy. Cutis 1996;57:421–4.
[34] Robillard RB, Hilsinger RL Jr, Adour KK. Ramsay Hunt facial paralysis: clinical analyses of 185 patients. Otolaryngol Head Neck Surg 1986;95:292–7.
[35] Furuta Y, Ohtani F, Aizawa H, et al. Varicella-zoster virus reactivation is an important cause of acute peripheral facial paralysis in children. Pediatr Infect Dis J 2005;24:97–101.
[36] Merchet MP, Gruener G. Segmental zoster paresis of limbs. Electromyogr Clin Neurophysiol 1996;36:369–75.
[37] Yoleri O, Olmez N, Oztura I, et al. Segmental zoster paresis of the upper extremity: a case report. Arch Phys Med Rehabil 2005;86:1492–4.
[38] Anderson JP, Keal EE. Cervical herpes zoster and diaphragmatic paralysis. Br J Dis Chest 1969;63:222–6.
[39] Tjandra J, Mansel RE. Segmental abdominal herpes zoster paresis. Aust N Z J Surg 1986; 56:807–8.
[40] Gold E. Serologic and virus-isolation studies of patients with varicella or herpes-zoster infection. N Engl J Med 1966;274:181–5.
[41] Head H, Campbell AW. The pathology of herpes zoster and its bearing on sensory localization. Brain 1900;23:353–523.
[42] Denny-Brown D, Adams RD, Fitzgerald PJ. Pathologic features of herpes zoster: a note on geniculate herpes. Arch Neurol Psychiatry 1944;51:216–31.
[43] Cheatham WJ, Weller TH, Dolan TF, et al. Varicella: report on two fatal cases with necropsy, virus isolation, and serologic studies. Am J Pathol 1956;32:1015–35.
[44] Esiri MM, Tomlinson AH. Herpes zoster: demonstration of virus in trigeminal nerve and ganglion by immunofluorescence and electron microscopy. J Neurol Sci 1972;15:35–48.
[45] Ghatak NR, Zimmerman HM. Spinal ganglion in herpes zoster. Arch Pathol 1973;95: 411–5.
[46] Nagashima K, Nakazawa M, Endo H. Pathology of the human spinal ganglia in varicella-zoster virus infection. Acta Neuropathol 1975;33:105–17.
[47] Murakami S, Hato N, Horiuchi J, et al. Treatment of Ramsay Hunt syndrome with acyclovir-prednisone: significance of early diagnosis and treatment. Ann Neurol 1997;41:353–7.

[48] Furuta Y, Ohtani F, Mesuda Y, et al. Early diagnosis of zoster sine herpete and antiviral therapy for the treatment of facial palsy. Neurology 2000;55:708–10.
[49] Rogers RS III, Tindall JP. Herpes zoster in the elderly. Postgrad Med 1971;50:153–7.
[50] Bowsher D. The effects of pre-emptive treatment of postherpetic neuralgia with amitriptyline: a randomized, double-blind, placebo-controlled trial. J Pain Symptom Manage 1997; 13:327–31.
[51] Vafai A, Wellish M, Gilden DH. Expression of varicella-zoster virus in blood mononuclear cells of patients with postherpetic neuralgia. Proc Natl Acad Sci USA 1988;85: 2767–70.
[52] Devlin ME, Gilden DH, Mahalingam R, et al. Peripheral blood mononuclear cells of the elderly contain varicella-zoster virus DNA. J Infect Dis 1992;165:619–22.
[53] Mahalingam R, Wellish M, Brucklier J, et al. Persistence of varicella-zoster virus DNA in elderly patients with postherpetic neuralgia. J Neurovirol 1995;1:130–3.
[54] Terada K, Niizuma T, Kawano S, et al. Detection of varicella-zoster virus DNA in peripheral mononuclear cells from patients with Ramsay Hunt syndrome or zoster sine herpete. J Med Virol 1998;56:359–63.
[55] Gilden DH, Cohrs RJ, Hayward AR, et al. Chronic varicella zoster virus ganglionitis: a possible cause of postherpetic neuralgia. J Neurovirol 2003;9:404–7.
[56] Quan D, Hammack BN, Kittelson J, et al. Improvement of postherpetic neuralgia after treatment with intravenous acyclovir followed by oral valacyclovir. Arch Neurol 2006; 63:940–2.
[57] Rowbotham M, Harden N, Stacey B, et al. Gabapentin for the treatment of postherpetic neuralgia: a randomized controlled trial. JAMA 1998;280:1837–42.
[58] Rice AS, Maton S. Postperpetic Neuralgia Study Group. Gabapentin in postherpetic neuralgia: a randomised, double blind, placebo controlled study. Pain 2001;94:215–24.
[59] Dubinsky RM, Kabbani H, El-Chami Z, et al. Practice parameter: treatment of postherpetic neuralgia: an evidence-based report of the Quality Standards Subcommittee of the American Academy of Neurology. Neurology 2004;63:959–65.
[60] Rowbotham MC, Manville NS, Ren J. Pilot tolerability and effectiveness study of levetiracetam for postherpetic neuralgia. Neurology 2003;61:866–7.
[61] Gilron I, Bailey JM, Tu D, et al. Morphine, gabapentin, or their combination for neuropathic pain. N Engl J Med 2005;352:1324–34.
[62] Huff JC. Antiviral treatment in chickenpox and herpes zoster. J Am Acad Dermatol 1988; 18:204–6.
[63] Beutner KR, Friedman DJ, Forszpaniak C, et al. Valaciclovir compared with acyclovir for improved therapy for herpes zoster in immunocompetent adults. Antimicrob Agents Chemother 1995;39:1546–53.
[64] Tyring S, Barbarash RA, Nahlik JE, et al. Famciclovir for the treatment of acute herpes zoster: effects on acute disease and postherpetic neuralgia: a randomized, double-blind, placebo-controlled trial. Collaborative Famciclovir Herpes Zoster Study Group. Ann Intern Med 1995;123:89–96.
[65] Gilden DH, Lipton HL, Wolf JS, et al. Two patients with unusual forms of varicella-zoster virus vasculopathy. N Engl J Med 2002;347:1500–3.
[66] Fukumoto S, Kinjo M, Hokamura K, et al. Subarachnoid hemorrhage and granulomatous angiitis of the basilar artery: demonstration of the varicella-zoster-virus in the basilar artery lesions. Stroke 1986;17:1024–8.
[67] O'Donohue JM, Enzmann DR. Mycotic aneurysm in angiitis associated with herpes zoster ophthalmicus. AJNR Am J Neuroradiol 1987;8:615–9.
[68] Gilden DH, Bennett JL, Kleinschmidt-DeMasters BK, et al. The value of cerebrospinal fluid antiviral antibody in the diagnosis of neurologic disease produced by varicella zoster virus. J Neurol Sci 1998;159:140–4.
[69] Nau R, Lantsch M, Stiefel M, et al. Varicella zoster virus-associated focal vasculitis without herpes zoster: recovery after treatment with acyclovir. Neurology 1998;51:914–5.

[70] Burgoon MP, Hammack BN, Owens GP, et al. Oligoclonal immunoglobulins in cerebrospinal fluid during varicella zoster virus (VZV) vasculopathy are directed against VZV. Ann Neurol 2003;54:459–63.
[71] Gilden DH, Kleinschmidt-DeMasters BK, Wellish M, et al. Varicella zoster virus, a cause of waxing and waning vasculitis: the New England Journal of Medicine case 5-1995 revisited. Neurology 1996;47:1441–6.
[72] Aurelius E, Johansson B, Skoldenberg B, et al. Rapid diagnosis of herpes simplex encephalitis by nested polymerase chain reaction assay of cerebrospinal fluid. Lancet 1991;337: 189–92.
[73] Nagel MA, Forghani B, Mahalingam R, et al. The value of detecting anti-VZV IgG antibody in CSF to diagnose VZV vasculopathy. Neurology 2007;68:1069–73.
[74] Hilt DC, Buchholz D, Krumholz A, et al. Herpes zoster ophthalmicus and delayed contralateral hemiparesis caused by cerebral angiitis: diagnosis and management approaches. Ann Neurol 1983;14:543–53.
[75] Pina MA, Ara JR, Capablo JL, et al. Myelitis and optic neuritis caused by varicella. Rev Neurol 1997;25:1575–6.
[76] Celik Y, Tabak F, Mert A, et al. Transverse myelitis caused by varicella. Clin Neurol Neurosurg 2001;103:260–1.
[77] Gilden DH, Beinlich BR, Rubinstien EM, et al. Varicella-zoster virus myelitis: an expanding spectrum. Neurology 1994;44:1818–23.
[78] Kleinschmidt-DeMasters BK, Gilden DH. Varicella-zoster virus infections of the nervous system: clinical and pathologic correlates. Arch Pathol Lab Med 2001;125: 770–80.
[79] Devinsky O, Cho ES, Petito CK, et al. Herpes zoster myelitis. Brain 1991;114:1181–96.
[80] de Silva SM, Mark AS, Gilden DH, et al. Zoster myelitis: improvement with antiviral therapy in two cases. Neurology 1996;47:929–31.
[81] Chua HC, Tjia H, Sitoh YY. Concurrent myelitis and Guillain-Barré syndrome after varicella infection. Int J Clin Pract 2001;55:643–4.
[82] Schvoerer E, Frechin V, Warter A, et al. Persistent multiple pulmonary nodules in a nonimmunocompromised woman after varicella-related myelitis treated with acyclovir. J Clin Microbiol 2003;41:4904–5.
[83] Orme HT, Smith G, Nagel MA, et al. VZV spinal cord infarction identified by diffusion-weighted magnetic resonance imaging (DWI). Neurology 2007;69:398–400.
[84] Holland GN. The progressive outer retinal necrosis syndrome. Int Ophthalmol 1994;18: 163–5.
[85] Soushi S, Ozawa H, Matsuhashi M, et al. Demonstration of varicella-zoster virus antigens in the vitreous aspirates of patients with acute retinal necrosis syndrome. Ophthalmology 1988;95:1394–8.
[86] Thompson WS, Culbertson WW, Smiddy WE, et al. Acute retinal necrosis caused by reactivation of herpes simplex virus type 2. Am J Ophthalmol 1994;118:205–11.
[87] Tan JCH, Byles D, Stanford MR, et al. Acute retinal necrosis in children caused by herpes simplex virus. Retina 2001;21:344–7.
[88] Bonfioli AA, Eller AW. Acute retinal necrosis. Semin Ophthalmol 2005;20:155–60.
[89] Savant V, Saeed T, Denniston A, et al. Oral valganciclovir treatment of varicella zoster virus acute retinal necrosis. Eye 2004;18:544–5.
[90] Engstrom RE Jr, Holland GN, Margolis TP, et al. The progressive outer retinal necrosis syndrome: a variant of necrotizing herpetic retinopathy in patients with AIDS. Ophthalmology 1994;101:1488–502.
[91] Guex-Crosier Y, Rochat C, Herbort CP. Necrotizing herpetic retinopathies: a spectrum of herpes virus-induced diseases determined by the immune state of the host. Ocul Immunol Inflamm 1997;5:259–65.
[92] Lewis JM, Nagae Y, Tano Y. Progressive outer retinal necrosis after bone marrow transplantation. Am J Ophthalmol 1996;122:892–5.

[93] Franco-Paredes C, Bellehemeur T, Merchant A, et al. Aseptic meningitis and optic neuritis preceding varicella-zoster progressive outer retinal necrosis in a patient with AIDS. AIDS 2002;16:1045–9.
[94] Menerath JM, Gerard M, Laurichesse H, et al. Bilateral acute retinal necrosis in a patient with acquired immunodeficiency syndrome. J Fr Ophthalmol 1995;18:625–33.
[95] Tran TH, Rozenberg F, Cassoux N, et al. Polymerase chain reaction analysis of aqueous humour samples in necrotising retinitis. Br J Ophthalmol 2003;87:79–83.
[96] Margolis TP, Lowder CY, Holland GN, et al. Varicella-zoster virus retinitis in patients with the acquired immunodeficiency syndrome. Am J Ophthalmol 1991;112:119–31.
[97] van den Horn GJ, Meenken C, Troost D. Association of progressive outer retinal necrosis and varicella zoster encephalitis in a patient with AIDS. Br J Ophthalmol 1996;80:982–5.
[98] Kashiwase M, Sata T, Yamauchi Y, et al. Progressive outer retinal necrosis caused by herpes simplex virus type 1 in a patient with acquired immunodeficiency syndrome. Ophthalmology 2000;107:790–4.
[99] Biswas J, Choudhry S, Priya K, et al. Detection of cytomegalovirus from vitreous humor in a patient with progressive outer retinal necrosis. Indian J Ophthalmol 2002;50:319–21.
[100] Johnston WH, Holland GN, Engstrom RE Jr, et al. Recurrence of presumed varicella-zoster virus retinopathy in patients with acquired immunodeficiency syndrome. Am J Ophthalmol 1993;116:42–50.
[101] Moorthy RS, Weinberg DV, Teich SA, et al. Management of varicella zoster virus retinitis in AIDS. Br J Ophthalmol 1997;81:189–94.
[102] Dullaert H, Maudgal PC, Leys A, et al. Bromovinyldeoxyurdine treatment of outer retinal necrosis due to varicella-zoster virus: a case-report. Bull Soc Belge Ophtalmol 1996;262: 107–13.
[103] Austin RB. Progressive outer retinal necrosis syndrome: a comprehensive review of its clinical presentation, relationship to immune system status, and management. Clin Eye Vis Care 2000;12:119–29.
[104] Gilden DH, Wright RR, Schneck SA, et al. Zoster sine herpete, a clinical variant. Ann Neurol 1994;35:530–3.
[105] Lewis GW. Zoster sine herpete. Br Med J 1958;34:418–21.
[106] Easton HG. Zoster sine herpete causing acute trigeminal neuralgia. Lancet 1970;2:1065–6.
[107] Schott GD. Triggering of delayed-onset postherpetic neuralgia. Lancet 1998;351:419–20.
[108] Amlie-Lefond C, Mackin GA, Ferguson M, et al. Another case of virologically confirmed zoster sine herpete, with electrophysiologic correlation. J Neurovirol 1996;2:136–8.
[109] Powell KF, Wilson HG, Corxson MO, et al. Herpes zoster meningoencephalitis without rash: varicella zoster virus DNA in CSF. J Neurol Neurosurg Psychiatry 1995;59:198–9.
[110] Mancardi GL, Melioli G, Traverso F, et al. Zoster sine herpete causing encephalomyelitis. Ital J Neurol Sci 1987;8:67–70.
[111] Kleinschmidt-DeMasters BK, Mahalingam R, Shimek C, et al. Profound cerebrospinal fluid pleocytosis and Froin's syndrome secondary to widespread necrotizing vasculitis in an HIV-positive patient with varicella zoster virus encephalomyelitis. J Neurol Sci 1998; 159:213–8.
[112] Connolly AM, Dodson WE, Prensky AL, et al. Course and outcome of acute cerebellar ataxia. Ann Neurol 1994;35:673–9.
[113] Dangond F, Engle E, Yessayan L, et al. Pre-eruptive varicella cerebellitis confirmed by PCR. Pediatr Neurol 1993;9:491–3.
[114] Moses H, Nagel MA, Gilden DH. Acute cerebellar ataxia in a 41-year-old woman. Lancet Neurol 2006;5:984–8.
[115] Ratzka P, Schlachetzki JC, Bahr M, et al. Varicella zoster virus cerebellitis in a 66-year-old patient without herpes zoster. Lancet 2006;367:182.
[116] Funakawa I, Terao A, Koga M. A case of zoster sine herpete with involvement of the unilateral IX, X and XI cranial and upper cervical nerves. Rinsho Shinkeigaku 1999;39: 958–60.

[117] Goon P, Wright M, Fink C. Ophthalmic zoster sine herpete. J R Soc Med 2000;93:191–2.
[118] Hon C, Au WY, Cheng VC. Ophthalmic zoster sine herpete presenting as oculomotor palsy after marrow transplantation for acute myeloid leukemia. Haematologica 2005;90: EIM04.
[119] Noda Y, Nakazawa M, Takahashi D, et al. Retinal periphlebitis as zoster sine herpete. Arch Ophthalmol 2001;119:1550–2.
[120] Akpek EK, Gottsch JD. Herpes zoster sine herpete presenting with hyphema. Ocul Immunol Inflamm 2000;8:115–8.
[121] Yamamoto S, Tada R, Shimomura Y, et al. Detecting varicella-zoster virus DNA in iridocyclitis using polymerase chain reaction: a case of zoster sine herpete. Arch Ophthalmol 1995;113:1358–9.
[122] Silverstein BE, Chandler D, Neger R, et al. Disciform keratitis: a case of herpes zoster sine herpete. Am J Ophthalmol 1997;123:254–5.
[123] Dueland AN, Devlin M, Martin JR, et al. Fatal varicella-zoster virus meningoradiculitis without skin involvement. Ann Neurol 1991;29:569–72.
[124] Hevner R, Vilela M, Rostomily R, et al. An unusual cause of trigeminal-distribution pain and tumour. Lancet Neurol 2003;2:567–71.
[125] Takahashi M. Effectiveness of live varicella vaccine. Expert Opin Biol Ther 2004;4:199–216.
[126] Oxman MN, Levin MJ, Johnson GR, et al. A vaccine to prevent herpes zoster and postherpetic neuralgia in older adults. N Engl J Med 2005;352:2271–84.
[127] Davison AJ, Scott JE. The complete DNA sequence of varicella-zoster virus. J Gen Virol 1986;67:1759–816.
[128] Honess RW, Roizman B. Proteins specified by herpes simplex virus. XIII. Glycosylation of viral polypeptides. J Virol 1975;16:1308–26.
[129] Zhang YF, Wagner EK. The kinetics of expression of individual herpes simplex virus type 1 transcripts. Virus Genes 1987;1:49–60.
[130] Zhang YF, Devi-Rao GB, Rice M, et al. The effect of elevated levels of herpes simplex virus alpha-gene products on the expression of model early and late genes in vivo. Virology 1987; 157:99–106.
[131] Cohrs RJ, Gilden DH, Kinchington PR, et al. Varicella-zoster virus gene 66 transcription and translation in latently infected human ganglia. J Virol 2003;77:6660–5.
[132] Kennedy PG, Grinfeld E, Craigon M, et al. Transcriptional analysis of varicella-zoster virus infection using long oligonucleotide-based microarrays. J Gen Virol 2005;86:2673–84.
[133] Hope-Simpson RE. The nature of herpes zoster: a long-term study and a new hypothesis. Proc R Soc Med 1965;58:9–20.
[134] Straus SE, Hay J, Smith H, et al. Genome differences among varicella-zoster isolates. J Gen Virol 1983;64:1031–41.
[135] Hyman RW, Ecker JR, Tenser RB. Varicella-zoster virus RNA in human trigeminal ganglia. Lancet 1983;2:814–6.
[136] Gilden DH, Rozemann Y, Murray R, et al. Detection of varicella-zoster virus nucleic acid in neurons of normal human thoracic ganglia. Ann Neurol 1987;22:377–80.
[137] LaGuardia JJ, Cohrs RJ, Gilden DH. Prevalence of varicella-zoster virus DNA in dissociated human trigeminal ganglion neurons and nonneuronal cells. J Virol 1999;73: 8571–7.
[138] Levin MJ, Cai GY, Manchak MD, et al. Varicella-zoster virus DNA in cells isolated from human trigeminal ganglia. J Virol 2003;77:6979–87.
[139] Clarke P, Beer T, Cohrs R, et al. Configuration of latent varicella-zoster virus DNA. J Virol 1995;69:8151–4.
[140] Kennedy PG, Grinfeld E, Gow JW. Latent varicella-zoster virus is located predominantly in neurons in human trigeminal ganglia. Proc Natl Acad Sci USA 1998;95:4658–62.
[141] Pevenstein SR, Williams RK, McChesney D, et al. Quantitation of latent varicella-zoster virus and herpes simplex virus genomes in human trigeminal ganglia. J Virol 1999;73: 10514–8.

[142] Cohrs RJ, Randall J, Smith J, et al. Analysis of individual human trigeminal ganglia for latent herpes simplex virus type 1 and varicella-zoster virus nucleic acids using real-time PCR. J Virol 2000;74:11464–71.

[143] Cohrs RJ, LaGuardia JJ, Gilden D. Distribution of latent herpes simplex virus type-1 and varicella zoster virus DNA in human trigeminal ganglia. Virus Genes 2005;31:223–7.

[144] Wang K, Lau TY, Morales M, et al. Laser-capture microdissection: refining estimates of the quantity and distribution of latent herpes simplex virus 1 and varicella-zoster virus DNA in human trigeminal ganglia at the single-cell level. J Virol 2005;79:14079–87.

[145] De Jong MD, Weel JF, Schuurman T, et al. Quantitation of varicella-zoster virus DNA in whole blood, plasma and serum by PCR and electrochemiluminescence. J Clin Microbiol 2000;38:2568–73.

[146] Mainka C, Fuss B, Geiger H, et al. Characterization of viremia at different stages of varicella-zoster virus infection. J Med Virol 1998;56:91–8.

[147] Ljungman P, Lonnqvist B, Gahrton G, et al. Clinical and subclinical reactivations of varicella-zoster virus in immunocompromised patients. J Infect Dis 1986;153:840–7.

[148] Schunemann S, Mainka C, Wolff MH. Subclinical reactivation of varicella-zoster virus in immunocompromised and immunocompetent individuals. Intervirology 1998;41: 98–102.

[149] Mehta SK, Cohrs RJ, Forghani B, et al. Stress-induced subclinical reactivation of varicella zoster virus in astronauts. J Med Virol 2004;72:174–9.

[150] Cohrs RJ, Srock K, Barbour MB, et al. Varicella-zoster virus (VZV) transcription during latency in human ganglia: construction of a cDNA library from latently infected human trigeminal ganglia and detection of a VZV transcript. J Virol 1994;68:7900–8.

[151] Cohrs RJ, Barbour MB, Mahalingam R, et al. Varicella-zoster virus (VZV) transcription during latency in human ganglia: prevalence of VZV gene 21 transcripts in latently infected human ganglia. J Virol 1995;69:2674–8.

[152] Cohrs RJ, Gilden DH. Varicella zoster virus transcription in latently infected ganglia. Anticancer Res 2003;23:2063–70.

[153] Cohrs RJ, Gilden DH. Prevalence and abundance of latently transcribed varicella-zoster virus genes in human ganglia. J Virol 2007;81:2950–6.

[154] Mahalingam R, Wellish M, Cohrs R, et al. Expression of protein encoded by varicella-zoster virus open reading frame 63 in latently infected human ganglionic neurons. Proc Natl Acad Sci USA 1996;93:2122–4.

[155] Theil D, Derfuss T, Paripovic I, et al. Latent herpesvirus infection in human trigeminal ganglia causes chronic immune response. Am J Pathol 2003;163:2179–84.

[156] Lungu O, Panagiotidis CA, Annunziato PW, et al. Aberrant intracellular localization of varicella-zoster virus regulatory proteins during latency. Proc Natl Acad Sci USA 1998; 95:7080–5.

[157] Grinfeld E, Kennedy PG. Translation of varicella-zoster virus genes during human ganglionic latency. Virus Genes 2004;29:317–9.

[158] Takahashi M, Okuno Y, Otsuka T, et al. Development of a live attenuated varicella vaccine. Biken J 1975;18:25–33.

[159] Myers MG, Duer HL, Hausler CK. Experimental infection of guinea pigs with varicella-zoster virus. J Infect Dis 1980;142:414–20.

[160] Myers MG, Stanberry LR, Edmond BJ. Varicella-zoster virus infection of strain 2 guinea pigs. J Infect Dis 1985;151:106–13.

[161] Matsunaga Y, Yamanishi K, Takahashi M. Experimental infection and immune response of guinea pigs with varicella-zoster virus. Infect Immun 1982;37:407–12.

[162] Wroblewska Z, Devlin M, Reilly K, et al. The production of varicella zoster virus antiserum in laboratory animals. Arch Virol 1982;74:233–8.

[163] Walz-Cicconi MA, Rose RM, Dammin GJ, et al. Inoculation of guinea pigs with varicella-zoster virus via the respiratory route. Arch Virol 1986;88:265–77.

[164] Wroblewska Z, Valyi-Nagy T, Otte J, et al. A mouse model for varicella-zoster virus latency. Microb Pathog 1993;15:141–51.
[165] Myers MG, Connelly BL, Stanberry LR. Varicella in hairless guinea pigs. J Infect Dis 1991; 163:746–51.
[166] Lowry PW, Sabella C, Koropchak CM, et al. Investigation of the pathogenesis of varicella-zoster virus infection in guinea pigs by using polymerase chain reaction. J Infect Dis 1993; 167:78–83.
[167] Tenser RB, Hyman RW. Latent herpesvirus infections of neurons in guinea pigs and humans. Yale J Biol Med 1987;60:159–67.
[168] Sadzot-Delvaux C, Merville-Louis MP, Delree P, et al. An in vivo model of varicella-zoster virus latent infection of dorsal root ganglia. J Neurosci Res 1990;26:83–9.
[169] Debrus S, Sadzot-Delvaux C, Nikkels AF, et al. Varicella-zoster virus gene 63 encodes an immediate-early protein that is abundantly expressed during latency. J Virol 1995;69: 3240–5.
[170] Kennedy PG, Grinfeld E, Bontems S, et al. Varicella-zoster virus gene expression in latently infected rat dorsal root ganglia. Virology 2001;289:218–23.
[171] Annunziato P, LaRussa P, Lee P, et al. Evidence of latent varicella-zoster virus in rat dorsal root ganglia. J Infect Dis 1988;178(Suppl 1):S48–51.
[172] Moffat JF, Stein MD, Kaneshima H, et al. Tropism of varicella-zoster virus for human $CD4^+$ and $CD8^+$ T lymphocytes and epidermal cells in SCID-hu mice. J Virol 1995;69: 5236–42.
[173] Zerboni L, Hinchliffe S, Sommer MH, et al. Analysis of varicella zoster virus attenuation by evaluation of chimeric parent Oka/vaccine Oka recombinant viruses in skin xenografts in the SCIDhu mouse model. Virology 2005;332:337–46.
[174] Provost PJ, Keller PM, Banker FS, et al. Successful infection of the common marmoset (Callithrix jacchus) with human varicella-zoster virus. J Virol 1987;61:2951–5.
[175] Cohen JI, Moskal T, Shapiro M, et al. Varicella in chimpanzees. J Med Virol 1996;50: 289–92.
[176] Gray WL. Simian varicella: a model for human varicella-zoster virus infections. Rev Med Virol 2004;14:363–81.
[177] White TM, Mahalingam R, Traina-Dorge V, et al. Simian varicella virus DNA is present and transcribed months after experimental infection of adult African green monkeys. J Neurovirol 2002;8:191–203.
[178] White TM, Mahalingam R, Traina-Dorge V, et al. Persistence of simian varicella virus DNA in $CD4^+$ and $CD8^+$ blood mononuclear cells for years after intratracheal inoculation of African green monkeys. Virology 2002;303:192–8.
[179] Mahalingam R, Wellish M, Soike K, et al. Simian varicella virus infects ganglia before rash in experimentally infected monkeys. Virology 2001;279:339–42.
[180] Grinfeld E, Kennedy PG. The pattern of viral persistence in monkeys intra-tracheally infected with simian varicella virus. Virus Genes 2007;35:289–92.
[181] Mahalingam R, Traina-Dorge V, Wellish M, et al. Naturally acquired simian varicella virus infection in African green monkeys. J Virol 2002;76:8548–50.
[182] Kennedy PG, Grinfeld E, Traina-Dorge V, et al. Neuronal localization of simian varicella virus DNA in ganglia of naturally infected African green monkeys. Virus Genes 2004;28:273–6.
[183] Mahalingam R, Traina-Dorge V, Wellish M, et al. Simian varicella reactivation in cynomologous monkeys. Virology 2007;368(1):50–9.
[184] Kolappaswamy K, Mahalingam R, Traina-Dorge V, et al. Disseminated simian varicella virus infection in an irradiated Rhesus macaque (Macaca mulatta). J Virol 2007;81:411–5.
[185] Mandal BK. Herpes zoster and the immunocompromised. J Infect 1987;14:1–5.

ELSEVIER
SAUNDERS

Neurol Clin 26 (2008) 699–715

NEUROLOGIC
CLINICS

Could Epstein-Barr Virus or Canine Distemper Virus Cause Multiple Sclerosis?

John A. Lincoln, MD, PhD[a],
Karolina Hankiewicz, MD[b], Stuart D. Cook, MD[a,*]

[a]*Department of Neurology and Neuroscience, Multiple Sclerosis Center at UMDNJ-New Jersey Medical School, Doctors Office Center, Suite 8121, 90 Bergen Street, Newark, NJ 07103, USA*

[b]*Hospital Clinico y Provincial, C/ Villarroel 170, Barcelona, 08036, Spain*

Multiple sclerosis (MS) is thought to be an acquired immune-mediated disease with resultant demyelination and axonal injury in the central nervous system (CNS). It has been proposed that exposure to various environmental factors may trigger the development of MS in genetically susceptible individuals. Many environmental factors may play a role in MS pathogenesis including cigarette smoking, vitamin D deficiency, and improved hygiene. Additionally, viruses have long been thought to be prime candidate agents to cause neural injury either directly or indirectly by initiating immune responses to as yet unidentified antigens within the CNS. Although many viral candidates have been proposed, there is strong and growing evidence for Epstein-Barr virus (EBV), a member of the human herpesvirus family, and lesser evidence for canine distemper virus (CDV), a member of the paramyxoviridae family, as causative agents for MS. This article reviews the available data on the possible role of EBV and CDV in MS.

Epidemiology of multiple sclerosis

MS is the most common demyelinating disease of the CNS. The prevalence of MS in the United States is approximately 1:1000 and it is estimated that as many as 2 million or more people worldwide have this disease. MS usually begins between the ages of 15 and 50 in most patients, with a female

* Corresponding author.
E-mail address: cooksd@umdnj.edu (S.D. Cook).

doi:10.1016/j.ncl.2008.03.004

to male ratio of 2 to 3:1 [1]. Although MS can be benign in as many as 20% to 30% of individuals [2], it is one of the primary causes of neurologic disability in those younger than 50 years of age. The often devastating personal, social, and economical impact on individuals and their families justifies the enormous efforts that are being undertaken to study the disease to uncover its cause. To date, however, the etiology of MS remains a mystery.

Population-based studies are a basic approach to understanding the relative roles that genes or environment might play in complex diseases, such as MS. For example, the worldwide distribution of MS is not uniform with a crude north-south gradient described in the northern hemisphere and a similar but inverse pattern seen in the southern hemisphere. MS is generally more common in higher latitudes and its prevalence usually diminishes in populations closer to the equator [3]. MS also seems to be less common in South America and Asia and is rare in Africa.

An increased genetic susceptibility to MS is thought to be related to the presence of various HLAs, such as HLA DRB1*1501, and other gene polymorphisms [4,5]. To date, however, the prevalence of MS cannot be explained by genetic linkages alone. Some studies have shown concordance rates of only 28% in monozygotic twins versus 3.7% in dizygotic twins, although other studies have found lower rates in both twin groups [6]. Overall, this suggests that in addition to genetic factors, there is also a strong influence of the environment in the development of MS. A few ambient factors that link MS to the environment have already been suggested to play a role in pathogenesis, including sunlight exposure, low vitamin D plasma levels, cigarette smoking, or viral infection [7–10].

The role that environmental factors might play is most striking when evaluating migrational data. The effect of migration on altering MS risk seems to be related to age at migration. Persons who move from regions with a higher prevalence of MS to regions with a lower prevalence of disease may diminish their risk for MS if migration occurs early in life, although evidence exists that this may be changing [11]. Although no increase in MS is seen in adult migrants from areas of low to high MS prevalence, some studies have described such an increase in their offspring [12,13].

Two overlapping theories have been proposed to explain these observations. The first one, termed the "poliomyelitis hypothesis," postulates that exposure to an as yet unknown virus early in life can confer protective immunity, whereas initial exposure during late childhood or adulthood results in an increased risk of developing disease [14]. Based on his epidemiologic studies of the Faroe Islands, Kurtzke [15] later proposed his "theory of prevalence" postulating that MS is an infectious disease caused by an as yet unknown pathogen more common in areas of greater prevalence. This agent could cause an asymptomatic but persistent infection in most of the population but might precipitate neurologic symptoms in only a small number of cases, years after the initial infection.

Central to both of these theories is the role that one or more infectious agents may play in triggering immune-mediated diseases, such as MS. The "hygiene hypothesis," a derivative of the poliomyelitis hypothesis, incorporates observations on the relationship between sanitation and MS that were made in Israel nearly 30 years earlier [16] with the ideas previously presented without specifying a candidate etiologic agent. Exposure to various infectious agents at an early age would be protective, whereas initial infection later in life would result in a greater risk of developing MS. This hypothesis accounts for both the reduced prevalence of MS seen in underdeveloped countries, where less effective sanitation allows for greater exposure to infective agents very early in life, and the vanishing north-south gradient now evident in the United States, as sanitary conditions improve in the southern regions of the country.

Although it may be difficult to isolate or culture a specific pathogen, conclusive data that identify a single candidate agent in tissue from MS patients has been unavailable, unconfirmed, and indirect. Evidence in support of a candidate agent as an instigator of disease includes the low concordance rates in twins, unusual worldwide distribution of disease, effects of migration on susceptibility, and increased risk in persons who develop late childhood infections. The likelihood of an infectious trigger becomes even more attractive if one considers that other conditions once thought to be idiopathic, such as Lyme (*Borrelia burgdorferi*), cat-scratch (*Rochalimea hensleae*), Whipple's (*Tropheryma whippelii*), and peptic ulcer disease (*Helicobacter pylori*), now have an infectious etiology. The identification of a viral or bacterial agent might expand the therapeutic options available to prevent or treat MS including vaccines and antiviral or antibacterial drugs.

Epstein-Barr virus

Epidemiology

EBV is a lymphotrophic DNA virus belonging to the herpesvirus family. EBV binds to the CD21 (C3d) receptor on the surface of the B cell resulting in viral infection with subsequent immortalization of both the cell and virus [17]. EBV infects greater than 90% of humans, typically following exposure to oral secretions. Infection in childhood is often asymptomatic or manifested by mild nonspecific symptoms indistinguishable from other illnesses. Infection in adolescents or adults causes a symptomatic EBV illness (ie, infectious mononucleosis), however, in nearly 50% of cases [18,19]. Following infection, EBV remains dormant in memory B cells in secondary lymph tissue but can undergo spontaneous periodic reactivation, which is usually asymptomatic.

When searching for a candidate agent in MS, EBV seems especially attractive, because infection has been found to occur at an earlier age in areas where MS is less common and at a later age in areas of greater prevalence of

MS [11,18,19]. Consistent with migrational data on MS risk and the hygiene hypothesis previously outlined, Elian and coworkers [12] and Kurtzke [15] found that migrants from the West Indies, Africa, or Asia, areas possibly with less effective sanitation and early EBV exposure, to Great Britain, an area with better sanitation but higher prevalence of MS, were protected from developing MS. The offspring of these migrants in Great Britain, however, had the same risk for developing MS as the native population [12].

There have been many small case-control studies that have shown an increased risk of MS in patients who previously developed infectious mononucleosis [19–21]. Possible recall-bias, however, limits the validity of the conclusions that can be drawn from such data. This limitation can be overcome by using large population-based studies that identify patients with infectious mononucleosis and prospectively determine risk of MS. One such study by Marrie and colleagues [3], using the United Kingdom General Practice Research Database, found that previous infectious mononucleosis was associated with a greater than fivefold increase in the risk for developing MS. Additionally, studies that evaluated heterophile-positive infectious mononucleosis patients from Sweden and Denmark found an increased risk of developing MS with reported relative risks of 2.8 and 3.7 respectively [22,23]. For the group seronegative for EBV, however, the risk of developing MS was 20-fold lower, despite sharing the same environment and hygiene conditions [24].

Finally, if MS was resultant from late EBV infection, clusters of MS cases associated with EBV outbreaks might also be expected. Such a mini cluster, consisting of eight MS patients, was reported to have occurred in Denmark. These patients were found to have the same subtype of EBV, a subtype not found in controls, which included healthy classmates [15,25].

Animal and human models

Human autoimmune diseases caused by Epstein-Barr virus

EBV has been a leading candidate to trigger autoimmune states. Evans [26] initially identified increased EBV-specific antibody titers in patients with systemic lupus erythematosis. Recently, evidence for increased antibody titers against various EBV antigens has been described in association with other autoimmune diseases. Antibody against EBV nuclear antigen-1 has been described in Hashimoto's thyroiditis and polymyositis. Antibody against EBV early antigen has been associated with numerous autoimmune diseases, including systemic lupus erythematosis, antiphospholipid syndrome, Sjögren's syndrome, systemic sclerosis, rheumatoid arthritis, and Churg-Strauss vasculitis [27]. Finally, antibody against EBV viral capsid antigen has been seen in polyarteritis nodosa and Wegener's granulomatosis in addition to several autoimmune states mentioned previously.

Further, in case control serologic studies, acute infection with EBV has been implicated in the causation or exacerbation of demyelinating disorders, such as Acute Disseminated Encephalomyelitis (ADEM) and Acute Inflammatory Demyelinating Polyneuropathy (AIDP), (Guillain-Barré syndrome) [24,28–31]. For example, Bray and colleagues [28] prospectively followed five patients with heterophile-positive EBV infection and found evidence of both acute and chronic demyelinating disease in each. In a more recent study, acute EBV infection, as determined by increased EBV-specific antibody titers, was found to induce ADEM in 4 of 10 cases. Additionally, this study found a chronic EBV infection resulting in one case of encephalitis and one case of relapsing ADEM, although no increase in antibody against viral antigens was detected in serum from the latter case [29].

Based on the wide range of autoimmune disorders associated with EBV-positive serology, the suggestion has been made that EBV might be involved not only in triggering MS but may also participate in the pathogenesis of multiple autoimmune disorders [32]. Unlike systemic lupus erythematosis, however, where up to a 40-fold increase in EBV genome can be found in blood lymphocytes, little or no increase has been shown in MS patients [33].

Animal models

Although numerous animal models have been used to study aspects of EBV biology, there is no robust animal model of CNS demyelination associated with EBV infection.

Laboratory evidence

Serology

Many studies have found EBV-specific antibodies in the serum and CSF of patients with MS [33–36]. Indeed, studies have consistently shown that adult patients with MS, and systemic lupus erythematosis and rheumatoid arthritis, are nearly 100% seropositive for EBV compared with between 84% and 95% for controls [37]. Recent studies from Canada and Germany and a large multinational study have also detected EBV-specific antibodies, but not antibodies directed against herpes simplex virus, cytomegalovirus, parvovirus B-19, or varicella-zoster virus, in the serum of children who developed MS [36,38,39]. Overall, these studies found EBV-specific antibodies in greater numbers of children with MS (83%–98.6%) compared with matched controls (42%–72.1%) [30,36,38,39]. Because EBV infects 90% to 95% of the normal population, if EBV causes MS, it would be an uncommon complication of a common infection.

In addition to seroprevalence, many studies have found increased absolute titers of EBV-specific antibodies in the serum of MS patients [36,39]. Elevated titers of EBV-specific antibodies have even been found years before onset of MS. For example, Levin and colleagues reported that the level of antibodies against EBV nuclear antigen proteins (EBNA1) rise dramatically after 20 years of age and were associated with an approximately 10-fold increased risk of developing MS [40]. A more recent study found that elevated titers to

EBNA1 might occur as early as 15 to 20 years before the onset of MS symptoms [41]. These elevations in antibody titers may not necessarily be specific to EBV, however, because others have found similar increases in antibodies directed against HHV-6 and measles antigens among some MS patients [42].

Most MS patients have increased intrathecal synthesis of immunoglobulins and CSF oligoclonal bands. Elevated antibody titers against EBV-specific antigens have also been seen in the CSF of patients with MS. In one study, comigration of EBNA1-specific immunoglobulins with CSF oligoclonal bands were found in 5 of 15 MS patients but 0 of 12 controls [43]. The same study showed that CSF from 85% of MS patients but only 13% of controls contained antibody against EBNA. Additionally, an independent study by Cepok and colleagues [44] confirmed that oligoclonal bands from MS patients specifically bound the purified EBV proteins, EBNA1 and BRRF2 (a structural EBV protein).

Finally, two studies found that reactivation of latent EBV infection might be associated with disease activity. Wandinger and colleagues [45] followed 19 MS patients monthly for 1 year and found an increase in antibodies to early antigens, indicative of active viral replication, in 72.7% of patients with active disease and none with stable disease. Their study also found positive serum DNA in MS patients with active disease. In another study, Buljevac and colleagues [46] prospectively followed Relapsing-Remitting Multiple Sclerosis (RRMS) patients and showed that viral reactivation (increased antibody to EBV early antigen) was associated with gadolinium enhancement, although this assertion has not been confirmed.

Although not conclusive, the data presented suggest an association between EBV infection and the occurrence of MS. Additionally, there are some data in support of a possible temporal relationship of infection to disease onset [40,42,43,47].

Infection in brain tissue

Little evidence is available that convincingly demonstrates the presence of EBV in brain tissue. A recent unconfirmed study by Serafi and colleagues [48] found markers of EBV infection in brain tissue from MS patients. This study noted the presence of B cells within ectopic follicles surrounding areas of demyelination and axonal loss. EBV-encoded small nuclear mRNA was isolated from B cells in 21 of 22 MS cases but was not detected in other neuroinflammatory diseases including one patient with primary CNS vasculitis, one with herpes simplex virus encephalitis, and one with arbovirus encephalitis. In addition, ectopic follicles were more prevalent in MS cases with active inflammation and expressed EBV proteins consistent with abortive replication. Their data suggest that seeding with EBV may be an early event in MS and taken together with serologic data from Wandinger and coworkers [45] and Buljevac and coworkers [46] provide evidence that reactivation may be associated with ongoing disease activity.

Other studies, however, have failed to find similar evidence of viral proteins or nuclear materials in MS brain tissue. Opsahl and Kennedy [49] were unable to find evidence of EBV latent or lytic proteins that indicated viral expression in preserved brain tissue from MS patients. Finally, others have been unsuccessful at isolating EBV-specific DNA in autopsied brain and CSF specimens [50]. Differences in isolation technique and absence of inflammatory activity in previously studied brain specimens, however, resulting possibly in minimal if any reactivation of latent virus, might have limited the identification of markers of EBV infection.

Overall, there is little direct evidence at this time that viral DNA is present in MS brain. Additionally, if EBV is involved in MS pathogenesis, it remains unclear that viral persistence is involved with disease activity.

Pathogenesis of Epstein-Barr virus in multiple sclerosis

There are several hypotheses proposed to explain the mechanism by which EBV might induce autoimmune diseases, such as MS. During persistent infection, EBV stimulates both strong T- and B-cell responses. Lünemann and colleagues [51] found that EBNA-specific $CD4^+$ T cells were crucial in EBV-specific control in both MS patients and healthy controls. MS patients were found to have an increased number of EBNA-specific T cells, however, with greater ability to recognize diverse antigens and produce more interferon-γ. This cross-recognition of antigens might further enhance the activation of autoreactive T-cell clones possibly perpetuating inflammation.

EBV proteins may also stimulate cross-reactive autoimmune B-cell clones either directly or by molecular mimicry. Work by Swanson-Mungerson and Longnecker [52] have established that EBV manipulates B-cell activation using viral proteins, such as latent membrane protein (LMP) 2A. Infection of autoreactive B-cell clones can result in immortalization by LMP2A-mediated increase in Bcl-2, an antiapoptotic factor. Additionally, LMP2A can provide a constitutive B-cell receptor signal that could enhance the production of antibody with reactivation of latent infection, furthering a perpetual autoimmune state. Alternatively, expansion of B-cell clones could occur by molecular mimicry with shared antigenic determinants resulting in autoimmune B-cell activation. This autoimmune response might also be transient or self-perpetuating.

The mechanisms described previously have been attributed to direct EBV-mediated effects on both T and B cells. Yet another possible mechanism of immune-mediated injury is through a bystander effect [53]. Repeated activation of B cells by latent EBV could result in the release of various factors that might subsequently stimulate previously primed autoimmune cells.

Finally, EBV may induce T- and B-cell activity without maintaining persistent infection. A transient "hit and run" theory of infection might still incite autoimmunity by molecular mimicry, epitope spread, or bystander effect [53].

Treatment of Epstein-Barr virus in multiple sclerosis

Acyclovir has been used for many years to treat various herpes viral infections. Acyclovir is a nucleoside analog that selectively inhibits viral replication and has been shown to have low toxicity to other tissues. If EBV infection induces MS or if viral reactivation is indeed associated with relapse, then it is reasonable to assume that patients with MS could respond to an antiviral therapy, such as acyclovir. Lycke and colleagues conducted a randomized, placebo-controlled, double-blinded trial with 60 patients with relapsing forms of MS [54]. These patients were randomized to receive either placebo or 800 mg acyclovir three times a day for 2 years. Although overall there was no significant difference ($P = .083$) between treatment groups, when patients were grouped according to exacerbation frequency they found significant differences between treatments.

Bech and colleagues performed additional randomized studies using valacyclovir, a valine ester of acyclovir with increased intestinal absorption 6- to 10-fold greater than the parent drug [55]. In this short-term treatment study, patients received either 3000 mg/d of valacyclovir or placebo for 24 weeks and were evaluated by monthly MRI. Their study found that overall valacyclovir did not reduce the number of active MS lesions. A planned subgroup analysis stratified patients according to baseline level of activity on MRI, however, and found a significant decrease ($P = .025$) in the number of newly active lesions between patients on valacyclovir (median 2.0) and placebo (6.5).

Not all studies, however, have reached the same conclusions. A study by Friedman and colleagues [56] that evaluated both clinical and MRI parameters of disease activity failed to find significant differences between treatment groups in 58 patients with MS randomized to receive either placebo or 3000 mg/d of valacyclovir for 2 years. The authors noted trends in favor of the valacyclovir treatment group, however, and also noted that, unlike previous trials, patients with both early and more advanced forms of MS were recruited for their study population.

Summary of evidence for Epstein-Barr virus

EBV is considered by many to be a leading candidate agent causing MS. Epidemiologic data from large cohorts have demonstrated that patients diagnosed with infectious mononucleosis have up to a fivefold increased risk of developing MS, whereas their seronegative counterparts maintain nearly a 20-fold lower risk of MS. Serologic evidence favoring this association include the presence of EBV-specific antibodies in nearly 100% of patients with MS, the reactivity of CSF oligoclonal bands from MS patients to EBV antigens, and the occurrence of EBV-specific antibodies years before the onset of MS. EBV can also directly infect B cells inducing immortalization. There is evidence that aberrant B- and T-cell functions, resultant from EBV infection, can potentiate an immune-mediated state, such as that seen in MS. Finally, a recent report of EBV protein and mRNA in MS brain tissue adds to the growing pool of evidence supporting a role for EBV in MS pathogenesis.

Canine distemper virus

Epidemiology

Members of the morbillivirus genus of the family *Paramyxoviridae* include CDV, measles virus, and rinderpest virus among others. These viruses are single-stranded, negative-sense RNA viruses that infect both humans and terrestrial and aquatic animals [57,63,74,75]. Not only can CDV infect a wide range of animals, but interspecies spread has also been well documented [58,59,60]. CDV is a virulent neurotropic virus transmitted by aerosol to the lymphatic tissue of the upper respiratory tract where it effectively spreads to various epithelial tissues of the body and the CNS. CNS infection can occur through either hematogenous spread or by anterograde spread through the olfactory nerve [61]. Animals infected with CDV typically undergo an acute and subsequent chronic phase of neurologic disease with sequelae including optic neuritis, myelopathy, or encephalopathy. Whereas measles, prevaccine, produced demyelination in 0.1% of infected individuals, demyelinating disease can be seen in up to 90% of dogs infected with some strains of CDV [62]. Neurologic illness is commonly acute and monophasic but can be relapsing or progressive with lesions predominating in white matter [63,64,65].

Although a vaccine for CDV has been available for many years and has resulted in a decline of disease seen in dogs, CDV infection has occurred even in previously vaccinated dogs and numerous outbreaks have been documented within the past 10 years [66,67]. If one assumes that spread of CDV from dog to human can in some instances cause MS, then one would expect greater exposure of MS patients to dogs before onset of symptoms when compared with controls. Several, but not all studies have shown greater exposure to dogs in the years before onset of MS [68,69]. Further, clusters of patients developing MS would also be expected in areas where CDV was common in the canine population. Such an association may have occurred in the Faeroe Islands, where an "outbreak" of MS was postulated to have resulted from British troops bringing infected dogs onto the islands during World War II [70], although others have rejected this hypothesis. In addition, one would expect an increased incidence of CDV or greater exposure to dogs in areas where MS prevalence is highest. Norman and colleagues [69] showed that in such areas as the northern United States, where MS is more prevalent, exposure to infected dogs is more likely because dogs are kept indoors in the colder climate. In addition, the proportion of dogs residing indoors is higher in North America and Europe where MS is more prevalent than in India and, probably, China and Japan where MS prevalence is low [63,64,71]. Finally, CDV epidemics are more common in cold, damp climates and it is posited that CDV, like influenza, is more stable at colder temperatures [70].

The altitudinal and cultural gradient seen with dog ownership (ie, the number of animals and whether kept indoors or outdoors) and the potential ease of infection, because the disease is spread by respiratory route even with minimal exposure, make CDV an intriguing candidate for jumping species and causing MS. Indeed, there have been two documented zoonotic transmissions of morbillivirus from horse to humans with death in one case [72]. There have also been three reports of increased exposure of MS patients to dogs with a CDV-like illness, with one reporting that exposure was greater in MS patients than controls in the 5 years before onset of MS [68,73]. The problem with epidemiologic studies of dog exposure is the large number of controls who also own dogs and potential bias and difficulties in recounting possible canine infection.

Animal and human models

Members of the morbillivirus genus include, among others, measles virus and CDV. Measles virus has been implicated in various immune-mediated human diseases, such as acute disseminated encephalomyelitis and forms of postinfectious encephalitidies including subacute sclerosing panencephalitis. This virus is also speculated to be involved in the pathogenesis of Paget's disease of the bone, systemic lupus erythematosis, Crohn's disease, and ulcerative colitis.

In contrast to measles virus, the natural hosts for CDV are terrestrial and aquatic mammals [57,74,75]. Gordon and colleagues [76], however, found an association between CDV infection and Paget's disease of humans. Indeed, CDV has been considered a naturally occurring model for MS [62,65] but, despite documented zoonotic infection of humans by morbillivirus, no strong evidence exists that clearly links CDV to human disease.

There are both large and smaller animal models for CDV infection. Dogs are a natural host for CDV and have been used to evaluate the pathogenesis of the acute and chronic phase of infection. In addition, von Messling and colleagues [77] recently developed a ferret model of CDV infection. The continued development of smaller animal models for CDV infection would more readily allow experimentation into the role of morbillivirus in demyelination.

Laboratory evidence

Many early studies found higher titers of serum antibodies to CDV in MS patients compared with controls [63,64,78]. One study found the highest antibody titers to a virulent rather than vaccine strain of CDV with no increase in antibody to six other dog viruses [79]. The ability of measles virus–specific antibodies to cross-react with CDV, however, made it difficult definitively to demonstrate selectivity of CDV infection in MS patients.

More definitive evidence was offered by Rohowsky-Kochan and colleagues [80] when they were able to synthesize three CDV hemagglutinin (CDV-H) peptides that were structurally distinct from each other and from measles virus. They showed that serum antibodies from animals immunized with CDV reacted with all three CDV-H peptides, whereas none of the measles antibody–positive sera reacted. Further, analysis of sera from a large number of MS patients and age- and sex-matched normal individuals and patients with other neurologic (including Guillain-Barré, subacute sclerosing panencephalitis, stroke) and immunologic (including systemic lupus erythematosis, acute measles infection) diseases showed increased antibody titers to all three CDV-H peptides in MS patients compared with controls [80]. The study also found a strong relationship between elevated CDV-H antibody levels and the diagnosis of MS (odds ratio of 5.0). In contrast, no increase in viral antibody titer was found to varicella zoster or polio virus in this study.

Another study by Geeraedts and colleagues [81] looked for the presence of morbillivirus antigens in brain tissue from MS patients. They found that eight of nine MS and two of five herpes simplex encephalitis brain samples stained positive with a monoclonal antibody recognizing a conserved epitope on the fusion protein of morbilliviruses. A similar staining pattern was not found, however, for samples from six controls or four patients with ischemic stroke [81].

Although these studies support the view that humans might be infected by CDV, the evidence fails conclusively to tie CDV infection with an increased risk of developing MS. Indeed, high titers of CDV antibody have been found in some individuals without MS and low titers found in many patients with MS. In addition, no studies have been performed to assess whether CSF oligoclonal bands react with CDV proteins and, to date, no viral RNA has been amplified from samples of brain tissue from MS patients. Finally, unlike the decline in incidence rates of subacute sclerosing panencephalitis associated with the use of the measles vaccine, there has been no similar decrease in MS despite the widespread use of a CDV vaccine in dogs.

Pathogenesis of canine distemper virus in multiple sclerosis

CDV is transmitted as an aerosol infection to the lymphoid tissue in the upper respiratory tract. CDV is a neurotropic virus that causes an acute and chronic demyelination [82–84]. Acute demyelination is thought to be resultant from direct infection and subsequent cell lysis of neurons, astrocytes, and to a lesser extent oligodendrocytes by virus particles [85]. Zurbriggen and colleagues found small numbers of oligodendrocytes containing CDV mRNA at the edge of lesions in animals infected with CDV [84]. Further, infection of cultures of canine oligodendrocytes overlayed on astrocytes showed loss of organelles and metabolic dysfunction with a decrease in myelin transcription [84,86,87]. The acute phase of CDV infection is also marked by an up-regulation of major histocompatibility complex II and a relative increase in the $CD8^+$ T-cell population [88].

The chronic phase of CDV infection in a dog model is associated with a strong inflammatory reaction, with proinflammatory cytokines up-regulated and anti-inflammatory cytokines unchanged, and with intrathecal antibody synthesis [85,89,90]. In addition, Wünschmann and colleagues [91] found an altered $CD8^+/CD4^+$ ratio in the brain parenchyma of dogs with spontaneous chronic CDV infection. The precise role, however, that these T-cell populations play in maintaining a chronic inflammation has not yet been determined.

The mechanism by which CDV induces chronic demyelination in the dog has not been elucidated. One theory incorporates a potential role for bystander-mediated injury. Griot and colleagues [92] found that antibodies against CDV antigens could induce generation of reactive oxygen species from macrophages, found at the inflammation site, resulting in cell activation. Activated macrophages could cause bystander-mediated destruction of nearby oligodendrocytes resulting in chronic demyelination.

Persistent viral infection was proposed as a mechanism of chronic demyelination by Vandevelde and Zurbriggen [65]. They proposed that persistent infection might be caused by a defective virus that spreads by noncytolytic restricted infection thereby escaping immune surveillance and inducing a recurrent immune reaction.

Finally, a role for CDV-infected astrocytes was investigated by De Keyser and colleagues [93]. They examined paraffin-embedded brain sections from 11 dogs spontaneously infected with CDV and found that astrocytes in chronic demyelinating lesions failed to express β_2-adrenergic receptors, a finding also seen in MS plaques and in normal-appearing white matter but not in other inflammatory conditions. They also found, from other in vitro studies, support for a direct CDV-mediated effect of β_2-adrenergic receptor expression on astrocytes, a neural cell recently proposed to play a role in autoimmune reactions, such as MS [94].

Summary of evidence for canine distemper virus

CDV in its demyelinating form in dogs may be a naturally occurring model for MS [62,65]. Based on clinical, laboratory, and pathologic findings, this highly neurotropic virus, notorious for species spread, can cause both an acute virally mediated demyelination and a chronic relapsing disease in dogs. Mechanisms used to cause neurologic sequelae include bystander-mediated damage, direct viral persistence in glial cells, and possibly by enhancing the numbers of autoimmune $CD8^+$ T cells. In addition, a role for demyelinating antibodies cannot be excluded. Although the presence of viral RNA has not been convincingly found in human tissue, it is possible that CDV could trigger immune-mediated demyelination as seen in MS. The similarities between the demyelinating encephalitis seen in dogs infected with CDV and that seen in humans with MS are remarkable and warrant further study.

Summary

This article reviews the data in favor of an infectious cause of MS and presents evidence for either EBV or CDV as possible candidate agents. Additionally, the most relevant epidemiology, serology, and pathogenesis of both viral candidates specifically in relation to a potential role in triggering immune-mediated diseases, such as MS, have been outlined. At the present time, neither agent has been conclusively proved to cause MS. Evidence is growing, however, that EBV may be an important contributor to MS pathogenesis. Future research may provide more conclusive proof for the causative role of these ubiquitous viruses in MS. The exciting possibility exists that, if EBV, CDV, or another as yet undetermined pathogen causes most cases of MS, it might be possible to focus research so as not only to treat infection better, but also to develop vaccines to prevent disease.

References

[1] Orton SM, Herrera BM, Yee IM, et al. Sex ratio of multiple sclerosis in Canada: a longitudinal study. Lancet Neurol 2006;5:932–6.

[2] Weinshenker BG. The natural history of multiple sclerosis. Neurol Clin 1995;13:119–46.

[3] Marrie RA, Wolfson C, Stukenboom MC, et al. Multiple sclerosis and antecedent infections: a case-control study. Neurology 2000;54:2307–10.

[4] Bielekova B, Richert N, Howard T, et al. Humanized anti-CD25 (daclizumab) inhibits disease activity in multiple sclerosis patients failing to respond to interferon beta. Proc Natl Acad Sci U S A 2004;101:8705–8.

[5] Lundmark F, Duvefelt K, Iacobaeus E, et al. Variation in interleukin 7 receptor alpha chain (IL7R) influences risk of multiple sclerosis. Nat Genet 2007;39:1108–13.

[6] Sadovnick AD, Armstrong H, Rice GPA, et al. A population based study of multiple sclerosis in twins: update. Ann Neurol 1993;33:281–5.

[7] van der Mei IA, Ponsonby AL, Blizzard L, et al. Regional variation in multiple sclerosis prevalence in Australia and its association with ambient ultraviolet radiation. Neuroepidemiology 2001;20:168–74.

[8] Goldberg P. Multiple sclerosis: vitamin D and calcium as environmental determinants of prevalence (a viewpoint). Part 1: sunlight, dietary factors and epidemiology. Int J Environ Stud 1974;6:19–27.

[9] Hernán MA, Olek MJ, Ascherio A. Cigarette smoking and incidence of multiple sclerosis. Am J Epidemiol 2001;154:69–74.

[10] Hernán MA, Jick SS, Logroscino G, et al. Cigarette smoking and the progression of multiple sclerosis. Brain 2005;128:1461–5.

[11] Gale CR, Martyn CN. Migrant studies in multiple sclerosis. Prog Neurobiol 1995;47:425–48.
[12] Elian M, Nightingale S, Dean G. Multiple sclerosis among United Kingdom-born children of immigrants from the Indian subcontinent, Africa and the West Indies. J Neurol Neurosurg Psychiatry 1990;53:906–11.
[13] Dean G, Elian M. Age at immigration to England of Asian and Caribbean immigrants and the risk of developing multiple sclerosis. J Neurol Neurosurg Psychiatry 1997;63:565–8.
[14] Poskanzer DC, Walker AM, Yonkondy J, et al. Studies in the epidemiology of multiple sclerosis in the Orkney and Shetland Islands. Neurology 1976;26:14–7.
[15] Kurtzke JF. MS epidemiology worldwide: one view of current status. Acta Neurol Scand 1995;161:23–33.
[16] Leibowitz U, Antonovsky A, Medalie JM, et al. Epidemiological study of multiple sclerosis in Israel. II. Multiple sclerosis and level of sanitation. J Neurol Neurosurg Psychiatry 1966; 29:60–8.
[17] Fingeroth JD, Weis JJ, Tedder TF, et al. Epstein-Barr virus receptor of human B lymphocytes is the C3d receptor CR2. Proc Natl Acad Sci U S A 1984;81:4510–4.
[18] Cohen JI. Epstein-Barr virus infection. N Engl J Med 2000;343:481–92.
[19] Wolfson C. Multiple sclerosis and antecedent infections. Epidemiology 2001;12:298–9.
[20] Hernan MA, Zhang SM, Lipworth L, et al. Multiple sclerosis and age at infection with common viruses. Epidemiology 2001;12:301–6.
[21] Operskalski EA, Visscher BR, Malmgren RM, et al. A case-control study of multiple sclerosis. Neurology 1989;39:825–9.
[22] Lindberg C, Andersen O, Vahlne A, et al. Epidemiological investigation of the association between infectious mononucleosis and multiple sclerosis. Neuroepidemiology 1991;10: 62–5.
[23] Haahr S, Koch-Henriksen N, Moller-Larsen A, et al. Increased risk of multiple sclerosis after late Epstein Barr virus infection. Mult Scler 1995;1:73–7.
[24] Ascherio A, Munger KL. Environmental risk factors for multiple sclerosis. Part I: the role of infection. Ann Neurol 2007;61:288–99.
[25] MacGregor HS, Latiwonk QI. Is multiple sclerosis and autoimmune disease or a chronic gammaherpes virus infection? J Clin Lab Immunol 1996;48:45–74.
[26] Evans AS. E.B. virus antibody in systemic lupus erythematosus. Lancet 1971;1:1023–4.
[27] Barzilai O, Sherer Y, Ram M, et al. Epstein-Barr virus and cytomegalovirus in autoimmune diseases are they truly notorious? A preliminary report. Ann N Y Acad Sci 2007;1108: 567–77.
[28] Bray PF, Culp KW, McFarlin DE, et al. Demyelinating disease alter neurologically complicated primary Epstein-Barr virus infection. Neurology 1992;42:278–82.
[29] Fujimoto H, Asaoka K, Imaizumi T, et al. Epstein-Barr virus infections of the central nervous system. Intern Med 2003;42:33–40.
[30] Imai S, Usui N, Sugiura M, et al. Epstein-Barr virus genomic sequences and specific antibodies in cerebrospinal fluid in children with neurologic complications of acute and reactivated EBV infection. J Med Virol 1993;40:278–84.
[31] Takahashia K, Kunishigea M, Shinoharaa M, et al. Guillain-Barré syndrome and hemophagocytic lymphohistiocytosis in a patient with severe chronic active Epstein-Barr virus infection syndrome. Clin Neurol Neurosurg 2005;108:80–3.
[32] Pender MP. Infection of autoreactive B lymphocytes with EBV, causing chronic autoimmune diseases. Trends Immunol 2003;24:584–8.
[33] Cook SD. Does Epstein-Barr virus cause multiple sclerosis? Rev Neurol Dis 2004;1:115–23.
[34] Sumaya CV, Myers LW, Ellison GW, et al. Increased prevalence and titer of Epstein-Barr virus antibodies in patients with multiple sclerosis. Ann Neurol 1985;17:371–7.
[35] Larsen PD, Bloomer LC, Bray PF. Epstein-Barr nuclear antigen and viral capsid antigen antibody titers in multiple sclerosis. Neurology 1985;35:435–8.
[36] Alotaibi S, Kennedy J, Tellier R, et al. Epstein-Barr virus in pediatric multiple sclerosis. JAMA 2004;291:1875–9.

[37] Lünemann JD, Münz C. Epstein-Barr virus and multiple sclerosis. Curr Neurol Neurosci Rep 2007;7:253–8.
[38] Pohl D, Krone B, Rostasy K, et al. High seroprevalence of Epstein-Barr virus in children with multiple sclerosis. Neurology 2006;67:2063–5.
[39] Banwell B, Krupp L, Kennedy J, et al. Clinical features and viral serologies in children with multiple sclerosis: a multinational observational study. Lancet Neurol 2007;6:773–81.
[40] Levin LI, Munger KL, Rubertone MV, et al. Temporal relationship between elevation of Epstein Barr virus antibody titers and initial onset of neurological symptoms in multiple sclerosis. JAMA 2005;239:2496–500.
[41] DeLorenze GN, Munger KL, Lennette ET, et al. Epstein-Barr virus and multiple sclerosis: evidence of association from a prospective study with long-term follow-up. Arch Neurol 2006;63:839–44.
[42] Sundström P, Juto P, Wadell G, et al. An altered immune response to Epstein-Barr virus in multiple sclerosis: a prospective study. Neurology 2004;62:2277–82.
[43] Rand KH, Houck H, Denslow ND, et al. Epstein-Barr virus nuclear antigen-1 (EBNA-1) associated oligoclonal bands in patients with multiple sclerosis. J Neurol Sci 2000;173: 32–9.
[44] Cepok S, Zhou D, Srivastava R, et al. Identification of Epstein-Barr virus proteins as putative targets of the immune response in multiple sclerosis. J Clin Invest 2005;115:1352–60.
[45] Wandinger KP, Jabs W, Siekhaus A, et al. Association between clinical disease activity and Epstein–Barr virus reactivation in MS. Neurology 2000;55:178–84.
[46] Buljevac D, van Doornum GJJ, Flach HZ, et al. Epstein-Barr virus and disease activity in multiple sclerosis. J Neurol Neurosurg Psychiatry 2005;76:1377–81.
[47] Ascherio A, Munger KL, Lennette ET, et al. Epstein-Barr virus antibodies and risk of multiple sclerosis: a prospective study. JAMA 2001;286:3083–8.
[48] Serafi B, Rosicarelli B, Franciotta D, et al. Dysregulated Epstein-Barr virus infection in the multiple sclerosis brain. J Exp Med 2007;204:2899–912.
[49] Opsahl ML, Kennedy PG. An attempt to investigate the presence of Epstein Barr virus in multiple sclerosis and normal control brain tissue. J Neurol 2007;254:425–30.
[50] Sanders VJ, Felisan S, Waddell A, et al. Detection of herpesviridae in postmortem multiple sclerosis brain tissue and controls by polymerase chain reaction. J Neurovirol 1996;2:249–58.
[51] Lünemann JD, Edwards N, Muraro PA, et al. Increased frequency and broadened specificity of latent EBV nuclear antigen-1-specific T cells in multiple sclerosis. Brain 2006;129: 1493–506.
[52] Swanson-Mungerson M, Longnecker R. Epstein–Barr virus latent membrane protein 2A and autoimmunity. Trends Immunol 2007;28:213–8.
[53] Lünemann JD, Kamradt T, Martin R, et al. Epstein-Barr virus: environmental trigger of multiple sclerosis? J Virol 2007;81:6777–84.
[54] Lycke J, Svennerholm B, Hjelmquist E, et al. Acyclovir treatment of relapsing-remitting multiple sclerosis: a randomized, placebo-controlled, double-blind study. J Neurol 1996; 243:214–24.
[55] Bech E, Lycke J, Gadeberg P, et al. A randomized, double-blind, placebo-controlled MRI study of anti-herpes virus therapy in MS. Neurology 2002;58:31–6.
[56] Friedman JE, Zabriskie JB, Plank C, et al. A randomized clinical trial of valacyclovir in multiple sclerosis. Mult Scler 2005;11:286–95.
[57] Harder TC, Osterhaus AD. Canine distemper virus: a morbillivirus in search of new hosts? Trends Microbiol 1997;5:120–4.
[58] Yoshikawa Y, Ochikubo F, Matsubara Y, et al. Natural infection with canine distemper virus in a Japanese monkey (Macaca fuscata). Vet Microbiol 1989;20:193–205.
[59] Appel MJ, Reggiardo C, Summers BA, et al. Canine distemper virus infection and encephalitis in javelinas (collared peccaries). Arch Virol 1991;119:147–52.
[60] Appel MJ, Yates RA, Foley GL, et al. Canine distemper epizootic in lions, tigers, and leopards in North America. J Vet Diagn Invest 1994;6:277–88.

[61] Rudd PA, Cattaneo R, Von Messling V. Canine distemper virus: review of structural and functional modulations in lymphoid tissues. J Virol 2006;80:9361–70.
[62] Vandevelde M, Higgins RJ, Kristensen B, et al. Demyelination in experimental canine distemper virus infection; immunological, pathologic and immunohistological studies. Acta Neuropathol 1982;56:285–93.
[63] Cook SD, Rohowsky-Kochan C, Bansil S, et al. Evidence for multiple sclerosis as an infectious disease. Acta Neurol Scand Suppl 1995;161:34–42.
[64] Cook SD. Epidemiology of multiple sclerosis: clues to etiology of a mysterious disease. Neuroscientist 1996;2:172–80.
[65] Vandevelde M, Zubriggen A. Demyelination in canine distemper virus infection: a review. Acta Neuropathol 2005;109:56–68.
[66] Blixenkrone-Møllera M, Svanssona V, Have P, et al. Studies on manifestations of canine distemper virus infection in an urban dog population. Vet Microbiol 1993;37:163–73.
[67] Lednicky JA, Dubach J, Kinsel MJ, et al. Genetically distant American canine distemper virus lineages have recently caused epizootics with somewhat different characteristics in raccoons living around a large suburban zoo in the USA. Virol J 2004;1:1–14.
[68] Cook SD, Natelson BH, Levin BE, et al. Further evidence of a possible association between house dogs and multiple sclerosis. Ann Neurol 1978;3:141–3.
[69] Norman J, Cook SD, Dowling PC. Pilot survey of household pets among veterans with multiple sclerosis and age-matched controls. Arch Neurol 1983;40:213–4.
[70] Chan WW. Multiple sclerosis and dogs. Lancet 1977;1:487–8.
[71] Bansil S, Singhal BS, Ahuja GK, et al. Multiple sclerosis in India: a case control study of environmental exposures. Acta Neurol Scand 1997;95:208–10.
[72] Murray K, Selleck P, Hooper P, et al. A morbillivirus that caused fatal disease in horses and humans. Science 1995;268:94–7.
[73] Warren SA, Warren KG, Greenhill S, et al. How multiple sclerosis is related to animal illness, stress and diabetes. Can Med Assoc J 1982;55:359–67.
[74] Hall AJ. Morbilliviruses in marine mammals. Trends Microbiol 1995;31:4–9.
[75] Kennedy S. Morbillivirus infections in aquatic mammals. J Comp Pathol 1998;119:201–25.
[76] Gordon MT, Mee AP, Anderson DC, et al. Canine distemper virus transcripts sequenced from pagetic bone. Bone Miner 1992;19:159–74.
[77] von Messling V, Milosevic D, Cattaneo R. Tropism illuminated: lymphocyte-based pathways blazed by lethal morbillivirus through the host immune system. Proc Natl Acad Sci U S A 2004;101:14216–21.
[78] Cook SD, Dowling PC, Russell WC. Neutralizing antibodies to canine distemper and measles virus in multiple sclerosis. J Neurol Sci 1979;41:61–70.
[79] Appel MJ, Glickman LI, Raine CS, et al. Viruses and multiple sclerosis. Neurology 1981;31:944–9.
[80] Rohowsky-Kochan C, Dowling PC, Cook SD. Canine distemper virus-specific antibodies in multiple sclerosis. Neurology 1995;45:1554–60.
[81] Geeraedts F, Wilczak N, van Binnendijk R, et al. Search for morbillivirus proteins in multiple sclerosis brain tissue. Neuroreprot 2004;15:27–32.
[82] Higgins RJ, Krakowka SG, Metzler AE, et al. Primary demyelination in experimental canine distemper virus induced encephalomyelitis in gnotobiotic dogs: sequential immunologic and morphologic findings. Acta Neuropathol 1982;58:1–8.
[83] Vandevelde M, Zurbriggen A, Higgins RJ, et al. Spread and distribution of viral antigen in nervous canine distemper. Acta Neuropathol 1985;67:211–8.
[84] Zurbriggen A, Schmid I, Graber H, et al. Oligodendroglial pathology in canine distemper. Acta Neuropathol 1998;95:71–7.
[85] Vandevelde M, Zurbriggen A, Steck A, et al. Studies on the intrathecal humoral immune response in canine distemper encephalitis. J Neuroimmunol 1986;11:41–51.
[86] Glaus TC, Griot C, Richard A, et al. Ultrastructural and biochemical findings in brain cell cultures infected with canine distemper virus. Acta Neuropathol 1990;80:59–67.

[87] Graber H, Muller C, Vandevelde A, et al. Restricted infection with canine distemper virus leads to down-regulation of myelin gene transcription in cultured oligodendrocytes. Acta Neuropathol 1999;97:179–87.

[88] Tipold A, Vandevelde M, Wittek R, et al. Partial protection and intrathecal invasion of CD8(+) T cells in acute canine distemper virus infection. Vet Microbiol 2001;83:189–203.

[89] Vandevelde M, Fankhauser R, Kristensen F, et al. Immunoglobulins in demyelinating lesions in canine distemper encephalitis: an immunohistological study. Acta Neuropathol 1981;54:31–41.

[90] Markus S, Failing K, Baumgartner W. Increased expression of pro-inflammatory cytokines and lack of up-regulation of anti-inflammatory cytokines in early distemper lesions. J Neuroimmunol 2002;125:30–41.

[91] Wünschmann A, Kremmer E, Baumgärtner W. Phenotypical characterization of T and B cell areas in lymphoid tissues of dogs with spontaneous distemper. Vet Immunol Immunopathol 2000;73:83–98.

[92] Griot C, Burge T, Vandevelde M, et al. Antibody-induced generation of reactive oxygen radicals by brain macrophages in canine distemper encephalitis: a mechanism for bystander demyelination. Acta Neuropathol 1989;78:396–403.

[93] De Keyser J, Wilczak N, Walter JH, et al. Disappearance of beta2-adrenergic receptors on astrocytes in canine distemper encephalitis: possible implications for the pathogenesis of multiple sclerosis. Neuroreport 2001;12(2):191–4.

[94] De Keyser J, Zeinstra E, Frohman E. Are astrocytes central players in the pathophysiology of multiple sclerosis? Arch Neurol 2003;60:132–6.

ELSEVIER
SAUNDERS

NEUROLOGIC
CLINICS

Neurol Clin 26 (2008) 717–726

Rabies

Alan C. Jackson, MD*

Departments of Internal Medicine (Neurology) and of Medical Microbiology, University of Manitoba, Winnipeg, Manitoba, Canada

Rabies remains an important public health problem in developing countries, and the indigenous threat of rabies continues in developed countries because of wildlife reservoirs. A diagnosis of rabies is often not considered by physicians until late in the clinical course or after death in North America and Europe, even with typical clinical presentations. Transmission of rabies virus has occurred in association with transplantation of tissues and also recently with organs. In 2004 a young patient survived rabies in Wisconsin, but the reasons for this favorable outcome remain elusive. This article reviews current information and developments on a variety of neurologic aspects of rabies.

Pathogenesis

Rabies virus is usually transmitted in the saliva from the bite of a rabid animal; however, rare human cases have been transmitted by aerosols in laboratory accidents [1], and in a cave containing millions of bats [2,3], and also by transplantation of tissues and organs [4]. After a bite exposure, saliva containing infectious rabies virus is deposited in muscle and subcutaneous tissues. The virus remains close to the site of exposure for the majority of the long incubation period, which usually lasts from 20 to 90 days (Fig. 1). Rarely, the incubation period may last for over 1 year [5]. Rabies virus binds to the nicotinic acetylcholine receptor in muscle [6], which is expressed on the postsynaptic membrane of the neuromuscular junction. Differences in the susceptibility of different species for rabies may be related, in part, to the quantity of nicotinic acetylcholine receptors in muscle, and a marked difference has been reported in muscle of foxes (high content

* Health Sciences Centre, GF543, 820 Sherbrook Street, Winnipeg, Manitoba R3A 1R9, Canada.

E-mail address: ajackson2@hsc.mb.ca

doi:10.1016/j.ncl.2008.03.010 ***neurologic.theclinics.com***

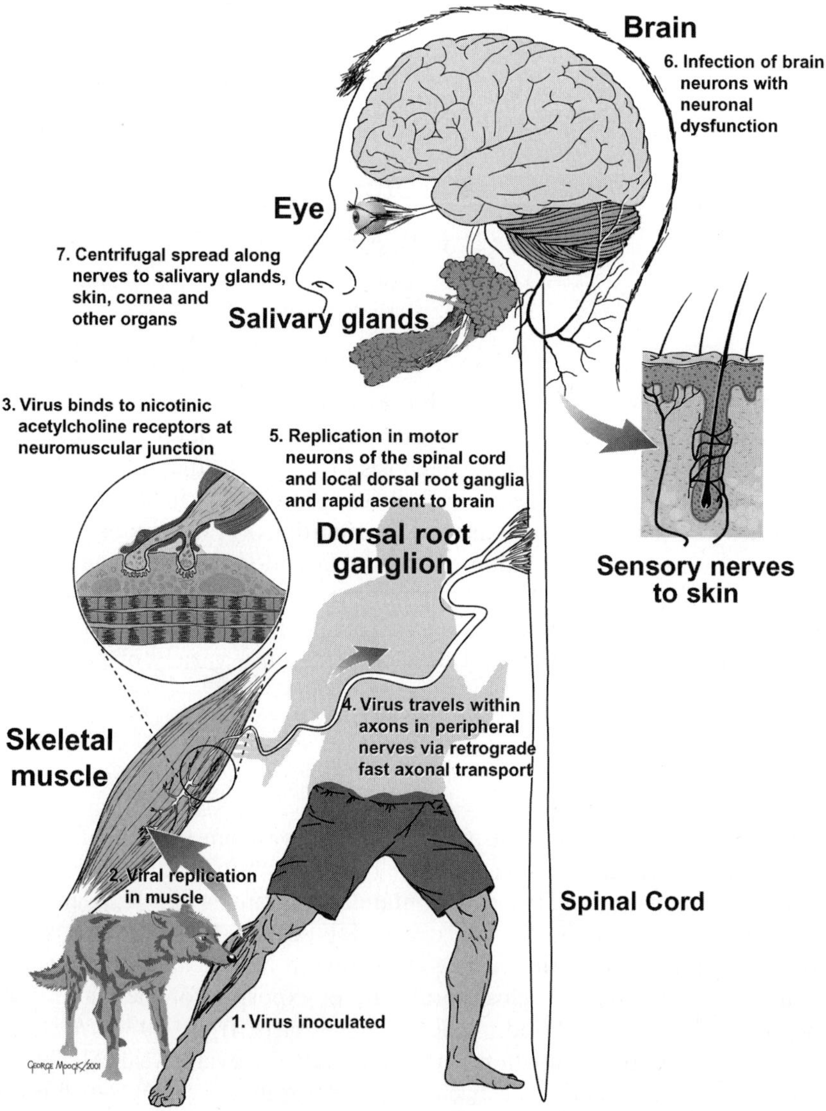

Fig. 1. Schematic diagram showing the sequential steps in the pathogenesis of rabies after an animal bite. (*Reproduced from* Jackson AC. Pathogenesis. In: Jackson AC, Wunner WH, editors. Rabies. San Diego: Academic Press; 2002; with permission.)

and high susceptibility) and opossums (low content and low susceptibility) [7]. After crossing the neuromuscular junction, the virus spreads within axons in peripheral nerves by retrograde fast axonal transport [8]. Once the virus reaches the central nervous system (CNS), widespread dissemination occurs by axonal transport along neuroanatomical connections [9]. The bases for behavioral changes in rabies vectors is unclear, but studies in a skunk model of rabies indicate that impaired serotonin neurotransmission caused by infection of raphe nuclei in the brainstem may play a role [10,11]. There are inflammatory changes in the brain, but neurodegenerative changes are quite mild, which gives rise to the concept that neuronal dysfunction, rather than neuronal death, explains the severe clinical disease with a fatal outcome [11,12]. The virus spreads from the CNS to multiple organs along autonomic and sensory neurons, which results the salivary gland infection in rabies vectors and, consequently, infectious virus is secreted in the saliva. There is involvement of nerves in close proximity to hair follicles, which is the basis for a skin biopsy as a diagnostic test for rabies in humans [13], and cardiac involvement may sometimes be associated with a myocarditis. There is infection of nerve plexuses in other organs and of the adrenal medulla [14].

Epidemiology

Globally, there are about 55,000 fatal human cases of rabies each year [15]. Because rabies often occurs in children, when the years of life lost are factored into an analysis, rabies is the seventh most important infectious disease [16]. The main problem resulting in the majority of human cases worldwide is endemic dog rabies in developing countries with dog-to-dog rabies virus transmission. In developed countries, rabies in wildlife is the main threat for indigenous cases. In the United States rabies is present in bats, raccoons, skunks, and foxes, and bat rabies is present in all states except Hawaii [17]. Bat rabies viruses are responsible for most human cases in the United States and Canada. Transmission can occur without awareness of either a bat bite or exposure to a bat. Bat rabies virus variants associated with silver-haired and eastern pipistrelle bats, and, less commonly, with Brazilian (Mexican) free-tailed bats are responsible for the majority of human deaths in the United States. Raccoon rabies is present throughout the eastern United States, but only a single known case of human rabies was associated with the raccoon rabies virus variant [18]. Imported cases may also occur, typically caused by dog exposures in developing countries.

Clinical features

Most cases of rabies can be diagnosed, or at least strongly suspected, on clinical grounds. Human rabies is often misdiagnosed or only recognized relatively late in the clinical course in developed countries because physicians are not familiar with the disease. Delayed diagnosis often leads to many exposures in health care workers, with a requirement for postexposure prophylaxis (see below). Prodromal symptoms include malaise, headache, fever, anxiety, and agitation. Pain, paresthesias, or pruritus may occur at the site of the exposure, at which time the wound has often healed, likely because of involvement of local sensory ganglia (eg, dorsal root ganglia).

There are two clinical forms of rabies: encephalitic (furious) rabies in about 80% of cases, and paralytic (dumb) rabies in about 20%. In encephalitic rabies there are episodes of generalized arousal or hyperexcitability that occur separated by lucid intervals, autonomic dysfunction, including hypersalivation, gooseflesh, cardiac arrhythmias, priapism, and hydrophobia. Hydrophobia, from the Latin meaning "fear of water," is characterized by spasms of inspiratory muscles, including the diaphragm, on attempts to drink [19]. This may become a conditioned reflex so that even the sight of water precipitates symptoms. It is thought that infection of brainstem neurons near nucleus ambiguus is responsible for this feature of the disease that is unique to rabies, and there is exaggeration of defensive reflexes that protect the respiratory tract [19,20]. Paralytic rabies is characterized by early motor weakness, often beginning in the bitten extremity, with progression to quadriparesis and often bilateral facial weakness. Sensory involvement is not usually prominent and sphincter involvement is common. Patients who have paralytic rabies usually survive longer than those who have encephalitis rabies. Patients who have both clinical forms of disease eventually progress to coma and, subsequently, often develop multiple organ failure, and cardiac and respiratory complications are quite common [21]. Patients survive longer with aggressive care in a critical care unit. Only six patients have survived rabies [4], and five of these patients received some rabies vaccine before the onset of their clinical disease (Table 1).

Diagnostic aspects

In most cases of rabies, there should be a strong clinical suspicion of the diagnosis based on the clinical presentation, and laboratory tests are important for diagnostic confirmation. MRI may be normal [22,23] or may show increased signals in gray matter areas [24,25]. Increased signals have been observed on T_2-weighted images in the medulla and pons [26]. Cerebrospinal fluid (CSF) analysis is often abnormal in human rabies. A CSF pleocytosis was found in 59% of cases in the first week of illness and in 87% after the first week [27]. The white cell count is usually less than 100 cells/μL, and the leukocytes are predominantly mononuclear cells. The CSF protein concentration may be mildly elevated, and glucose is usually in the normal range. The presence of neutralizing anti-rabies virus antibodies in serum is useful in unvaccinated patients, but antibodies may not appear until after the first week of clinical illness. Detection of rabies virus antigen, often using the fluorescent antibody technique, is a rapid diagnostic test that can be applied to punch skin biopsies, which are usually taken from the nape of the neck, because nerves adjacent to hair follicles may contain the viral antigen [13]. Brain tissue also contains viral antigen, but brain biopsies are not usually performed for rabies diagnosis. A recent advance in rabies diagnosis is detection of rabies virus RNA using reverse transcription–polymerase chain

Table 1
Cases of human rabies with recovery

Location	Year	Age of patient	Transmission	Immunization	Outcome	Reference
United States	1970	6	Bat bite	Duck embryo vaccine	Complete recovery	Baer et al [46]
Argentina	1972	45	Dog bites	Suckling mouse brain vaccine	Mild sequelae	Porras et al [47]
United States	1977	32	Laboratory (vaccine strain)	Pre-exposure vaccination	Sequelae	Tillotson et al [48]
Mexico	1992	9	Dog bites	Postexposure vaccination (combination)	Severe sequelae[a]	Alvarez et al [49]
India	2000	6	Dog bites	Postexposure vaccination (combination)	Severe sequelae[b]	Madhusudana et al [50]
United States	2004	15	Bat bite	none	Mild to moderate sequelae	Willoughby et al [39]

[a] Patient died less than 4 years after developing rabies with marked neurologic sequelae (L. Alvarez, personal communication, 1997).
[b] Patient died about 2 years after developing rabies with marked neurologic sequelae (S. Mahusudana, personal communication, 2005).
From Jackson AC. Human disease. In: Jackson AC, Wunner WH, editors. Rabies. San Diego: Academic Press; 2007; with permission.

reaction (RT–PCR) of saliva, tears, and skin biopsies [28–30], and this test is less useful for evaluation of CSF [30]. RT-PCR is fairly sensitive for rabies diagnosis; however, any diagnostic test, unless performed on multiple brain samples, does not completely exclude a diagnosis of rabies, and repeat testing may be necessary. Rabies virus variants from rabies cases can be characterized with RT-PCR and sequencing, which usually helps identify the origin of the variant (eg, rabies virus associated with silver-haired and eastern pipistrelle bats).

Transmission by transplantation

There have been eight well-documented cases of human rabies transmitted by corneal transplantation [4]. In 2004 transplantation was associated with the subsequent development of rabies in recipients of organs and a vascular conduit in the United States (four cases) [31,32] and in recipients of organs in Germany (three cases) [33]. The 20-year-old donor in Texas presented with fever, mental status changes, and throat pain [31,32], whereas the 26-year-old donor in Germany had traveled in India and newspapers reported that she had visited three hospitals with headache and aggressive and bizarre behavior before a cardiac arrest [34]. Rabies virus is present within small nerves of organs and arteries, and organ recipients are immunosuppressed to prevent organ rejection, compromising host defenses and creating a favorable environment for viral replication.

Laboratory screening for rabies has been recommended to prevent future transmissions by organ transplantation [35]; however, this premature recommendation was made without full consideration of many complex issues associated with organ transplantation. Clinical screening of organ donors should include a history of animal bites, evaluation for the presence of clinical features of rabies [4], and a travel history to rabies-endemic areas within a period of months. The physician responsible for the screening process must be knowledgeable about the diverse clinical features of rabies [4]. Laboratory screening of the donor would ideally be performed before organ transplantation. Rabies can only be reliably excluded on the basis of a laboratory evaluation performed on brain tissue. In the case of an organ donor, the best option would be to obtain the specimens immediately after harvesting the organs. The brain or brain tissues would need to be removed under the supervision of a pathologist and subsequently transported to a location where reliable rabies diagnostic laboratory testing could be performed. Experienced diagnostic laboratories probably have false-positive results on at least 1 per 500 cases on the initial evaluation, and false-positive results would be expected to be much higher for hospital laboratories performing infrequent testing. The organs and tissues would not be transplanted if a false-positive result is obtained. Because the demand for organs far outweighs the supply, many potential organ recipients die each year while they are on a waiting list for procurement of suitable organs from donors. Hence, false-positive results could have very serious consequences that need to be weighed against a very low risk of rabies virus transmission. Less hurried laboratory evaluation of the donor after organ transplantation is another possibility, but the management options are more limited and include removal of the transplanted organ (eg, kidney) in some cases, and initiation of postexposure rabies prophylaxis in an immunosuppressed patient, which may be unsatisfactory. Pre-exposure rabies immunization (with three doses of rabies vaccine) of potential organ recipients while they are on the waiting list for transplantation is an alternative approach to help prevent transmission with organ transplantation. The costs of rabies immunization are high, however, and there is uncertainty about the protection provided by this approach. Hence, appropriate clinical screening of potential organ donors is of paramount importance in preventing reoccurrences of these tragic situations.

Prevention of rabies

Rabies is a preventable disease after a recognized rabies exposure. A decision to initiate rabies postexposure prophylaxis depends on the details of the exposure, the species and clinical status of the involved animal and whether the animal is available for observation (in case of dog, cat, or ferret) or laboratory testing, and the local epidemiologic situation. Wounds should be thoroughly cleansed. Both active immunization with a modern cell culture rabies vaccine and passive immunization with human rabies immune globulin should be given to a previously unimmunized individual [15,36,37]. Licensed vaccines in the United States and Canada include purified chick embryo cell vaccine and human diploid cell vaccine [36]. After a rabies exposure five doses are given intramuscularly in the deltoid muscle on days 0, 3, 7, 14, and 28. On day 0, human rabies immune globulin (HRIG) should be given at a dosage of 20 international units/kg body weight into and around the wounds, and any remaining volume should be given intramuscularly into the buttocks.

Therapy of human rabies

Unlike the situation for rabies prophylaxis, there is no known effective therapy for rabies. In 2003, a viewpoint article was published that summarized a consensus of physicians with expertise in rabies and basic scientists knowledgeable in rabies pathogenesis [38]. It was felt that aggressive therapy of rabies should be considered, particularly in young and healthy patients, in patients who have early clinical disease, and in patients who have received some rabies vaccine before the onset of the disease. Diagnostic tests may not be positive or the results not yet available at the time of initiation of therapy, but an improved outcome would be expected after initiation of an effective therapy at an earlier time in the course of the disease. Rabies vaccine, human rabies immune globulin, ribavirin, interferon-α, and ketamine are all presently available therapeutic agents that could be considered for administration. It was felt that combination therapy might be more useful than a single agent, as shown for therapy of human immunodeficiency virus infection, chronic hepatitis C virus infection, and cancer.

In 2004 a 15-year-girl in Wisconsin developed rabies after a bat bite, with the early appearance of neutralizing anti-rabies virus antibodies [39]. This patient was treated in a critical care unit in Milwaukee with a therapeutic coma, intravenous ketamine, intravenous ribavirin, and amantadine by the enteric route. Benzodiazepines and supplemental barbiturates were given to deliberately maintain a burst-suppression pattern on her electroencephalogram. It is unclear whether the specific therapy she received played an important role in her survival [40,41]. The rationale for therapeutic coma is questionable. Subsequently, at least four similar approaches to therapy have been unsuccessful, including a detailed report of a case from Thailand [42]. Subsequent experimental studies performed in a mouse model and in primary neuronal cultures have cast doubt on the therapeutic efficacy of ketamine therapy [43], and more experimental work is needed on therapy with this drug. The bat virus variant, which was not isolated, that caused the infection of the Wisconsin girl may have been an attenuated rabies virus, and this may have been much more important in determining the outcome than the medical therapy she received [44]. The only previous human rabies case with an excellent outcome was a boy from Ohio who was also infected by a bat [45]. It is of concern that therapy used in Milwaukee, including therapeutic coma and ketamine, might be repeated countless times throughout the world, and the vast majority of this experience would go unpublished. This therapy may actually reduce the chance of a successful outcome in rabies, as well as consuming resources that could be much better used elsewhere, especially in developing countries with a large burden of human rabies [42]. For this reason, caution was recommended in using this therapy outside the context of a clinical trial with the usual safeguards [42]. Clearly, more basic research needs to be done to find effective therapeutic agents for rabies, including studies in cell culture and in animal models. The Wisconsin case establishes that survival from rabies is possible even without prior administration of rabies vaccine. The challenge is to find effective therapeutic agents for the management of rabies, but achieving this goal may require a better understanding of rabies pathogenesis.

References

[1] Conomy JP, Leibovitz A, McCombs W, et al. Airborne rabies encephalitis: demonstration of rabies virus in the human central nervous system. Neurology 1977;27:67–9.

[2] Constantine DG. Rabies transmission by air in bat caves. Atlanta (GA): National Communicable Disease Center, U.S. Government Printing Office, Washington, DC; 1967.

[3] Gibbons RV. Cryptogenic rabies, bats, and the question of aerosol transmission. Ann Emerg Med 2002;39(5):528–36.
[4] Jackson AC. Human disease. In: Jackson AC, Wunner WH, editors. Rabies. 2nd edition. Amsterdam: Elsevier Academic Press; 2007. p. 309–40.
[5] Smith JS, Fishbein DB, Rupprecht CE, et al. Unexplained rabies in three immigrants in the United States: a virologic investigation. N Engl J Med 1991;324:205–11.
[6] Lentz TL, Burrage TG, Smith AL, et al. Is the acetylcholine receptor a rabies virus receptor? Science 1982;215:182–4.
[7] Baer GM, Shaddock JH, Quirion R, et al. Rabies susceptibility and acetylcholine receptor [Letter]. Lancet 1990;335:664–5.
[8] Tsiang H. Evidence for an intraaxonal transport of fixed and street rabies virus. J Neuropathol Exp Neurol 1979;38:286–96.
[9] Gillet JP, Derer P, Tsiang H. Axonal transport of rabies virus in the central nervous system of the rat. J Neuropathol Exp Neurol 1986;45:619–34.
[10] Smart NL, Charlton KM. The distribution of challenge virus standard rabies virus versus skunk street rabies virus in the brains of experimentally infected rabid skunks. Acta Neuropathol 1992;84:501–8.
[11] Jackson AC. Pathogenesis. In: Jackson AC, Wunner WH, editors. Rabies. 2nd edition. Amsterdam: Elsevier Academic Press; 2007. p. 341–81.
[12] Fu ZF, Jackson AC. Neuronal dysfunction and death in rabies virus infection. J Neurovirol 2005;11(1):101–6.
[13] Bryceson ADM, Greenwood BM, Warrell DA, et al. Demonstration during life of rabies antigen in humans. J Infect Dis 1975;131(1):71–4.
[14] Jackson AC, Ye H, Phelan CC, et al. Extraneural organ involvement in human rabies. Lab Invest 1999;79(8):945–51.
[15] World Health Organization. WHO expert consultation on rabies: first report. Geneva: WHO; 2005.
[16] Coleman PG. Estimating the public health impact of rabies. Emerg Infect Dis 2004;10(1): 140–2.
[17] Blanton JD, Hanlon CA, Rupprecht CE. Rabies surveillance in the United States during 2006. J Am Vet Med Assoc 2007;231(4):540–56.
[18] Silverstein MA, Salgado CD, Bassin S, et al. First human death associated with raccoon rabies—Virginia, 2003. MMWR Morb Mortal Wkly Rep 2003;52(45):1102–3.
[19] Warrell DA. The clinical picture of rabies in man. Trans R Soc Trop Med Hyg 1976;70: 188–95.
[20] Warrell DA, Davidson NM, Pope HM, et al. Pathophysiologic studies in human rabies. Am J Med 1976;60(2):180–90.
[21] Hattwick MAW. Human rabies. Public Health Rev 1974;3:229–74.
[22] Mrak RE, Young L. Rabies encephalitis in a patient with no history of exposure. Hum Pathol 1993;24:109–10.
[23] Sing TM, Soo MY. Imaging findings in rabies. Australas Radiol 1996;40(3):338–41.
[24] Hantson P, Guerit JM, de Tourtchaninoff M, et al. Rabies encephalitis mimicking the electrophysiological pattern of brain death. A case report. Eur Neurol 1993;33:212–7.
[25] Awasthi M, Parmar H, Patankar T, et al. Imaging findings in rabies encephalitis. AJNR Am J Neuroradiol 2001;22(4):677–80.
[26] Pleasure SJ, Fischbein NJ. Correlation of clinical and neuroimaging findings in a case of rabies encephalitis. Arch Neurol 2000;57(12):1765–9.
[27] Anderson LJ, Nicholson KG, Tauxe RV, et al. Human rabies in the United States, 1960 to 1979: epidemiology, diagnosis, and prevention. Ann Intern Med 1984;100:728–35.
[28] Noah DL, Drenzek CL, Smith JS, et al. Epidemiology of human rabies in the United States, 1980 to 1996. Ann Intern Med 1998;128(11):922–30.
[29] Kamolvarin N, Tirawatnpong T, Rattanasiwamoke R, et al. Diagnosis of rabies by polymerase chain reaction with nested primers. J Infect Dis 1993;167:207–10.

[30] Crepin P, Audry L, Rotivel Y, et al. Intravitam diagnosis of human rabies by PCR using saliva and cerebrospinal fluid. J Clin Microbiol 1998;36(4):1117–21.
[31] Srinivasan A, Burton EC, Kuehnert MJ, et al. Transmission of rabies virus from an organ donor to four transplant recipients. N Engl J Med 2005;352(11):1103–11.
[32] Burton EC, Burns DK, Opatowsky MJ, et al. Rabies encephalomyelitis: clinical, neuroradiological, and pathological findings in 4 transplant recipients. Arch Neurol 2005;62(6):873–82.
[33] Johnson N, Brookes SM, Fooks AR, et al. Review of human rabies cases in the UK and in Germany. Vet Rec 2005;157(22):715.
[34] Brockmann S. Rabies, human, organ transplantation—Germany (04): rabies in transplant organ recipients: more information is needed about infectious diseases and organ transplantation. ProMED-mail 2005;20050221.0561. www.promedmail.org [accessed April 18, 2008].
[35] Dietzschold B, Koprowski H. Rabies transmission from organ transplants in the USA. Lancet 2004;364(9435):648–9.
[36] Centers for Disease Control and Prevention. Human rabies prevention—United States, 1999: recommendations of the Advisory Committee on Immunization Practices (ACIP). MMWR Recomm Rep 1999;48(No. RR-1):1–21.
[37] Jackson AC. Rabies. Current Treatment Options in Infectious Diseases 2003;5(1):35–40.
[38] Jackson AC, Warrell MJ, Rupprecht CE, et al. Management of rabies in humans. Clin Infect Dis 2003;36(1):60–3.
[39] Willoughby RE Jr, Tieves KS, Hoffman GM, et al. Survival after treatment of rabies with induction of coma. N Engl J Med 2005;352(24):2508–14.
[40] Jackson AC. Rabies: new insights into pathogenesis and treatment. Curr Opin Neurol 2006; 19(3):267–70.
[41] Jackson AC. Recovery from rabies [editorial]. N Engl J Med 2005;352(24):2549–50.
[42] Hemachudha T, Sunsaneewitayakul B, Desudchit T, et al. Failure of therapeutic coma and ketamine for therapy of human rabies. J Neurovirol 2006;12:407–9.
[43] Weli SC, Scott CA, Ward CA, et al. Rabies virus infection of primary neuronal cultures and adult mice: failure to demonstrate evidence of excitotoxicity. J Virol 2006;80(20):10270–3.
[44] Lafon M. Bat rabies—the Achilles heel of a viral killer? Lancet 2005;366(9489):876–7.
[45] Hattwick MAW, Weis TT, Stechschulte CJ, et al. Recovery from rabies: a case report. Ann Intern Med 1972;76:931–42.
[46] Baer GM, Shaddock JH, Houff SA, et al. Human rabies transmitted by corneal transplant. Arch Neurol 1982;39:103–7.
[47] Porras C, Barboza JJ, Fuenzalida E, et al. Recovery from rabies in man. Ann Intern Med 1976;85:44–8.
[48] Tillotson JR, Axelrod D, Lyman DO. Rabies in a laboratory worker—New York. MMWR Morb Mortal Wkly Rep 1977;26:183–4.
[49] Alvarez L, Fajardo R, Lopez E, et al. Partial recovery from rabies in a nine-year-old boy. Pediatr Infect Dis J 1994;13:1154–5.
[50] Madhusudana SN, Nagaraj D, Uday M, et al. Partial recovery from rabies in a six-year-old girl [letter]. Int J Infect Dis 2002;6(1):85–6.

ELSEVIER
SAUNDERS

Neurol Clin 26 (2008) 727–757

NEUROLOGIC
CLINICS

North American Encephalitic Arboviruses

Larry E. Davis, MD[a,b,*], J. David Beckham, MD[c,d], Kenneth L. Tyler, MD[d,e]

[a]*New Mexico Veterans Affairs Health Care System, 1500 San Pedro Drive SE, Albuquerque, NM 87108, USA*

[b]*Department of Neurology, MSC105620, 1 University of New Mexico, Albuquerque, NM 87131-0001, USA*

[c]*Infectious Diseases B168, University of Colorado Health Sciences Center, 4200 East 9th Avenue, Denver, CO 80262, USA*

[d]*Denver Veterans Affairs Medical Center, Denver, CO, USA*

[e]*Neurology B-182, University of Colorado Health Sciences Center, 4200 East 9th Avenue, Denver, CO 80262, USA*

"Arbovirus" is an informal term that refers to viruses that are transmitted by biting arthropods including mosquitoes, ticks, and flies. Arboviruses are more formally classified in different families and genera, based on their morphology (size, shape, capsid symmetry, presence or absence of an envelope); physical properties (genome structure and antigens); and biologic properties (mode of replication and transmission, host range, pathogenicity). All arboviruses causing human central nervous system (CNS) disease fall within four families: (1) Togaviridae, (2) Flaviviridae, (3) Bunyaviridae, and (4) Reoviridae (Table 1). Over 500 different arboviruses are distributed around the world, mainly in the tropics. A much smaller number cause human disease, however, and only a subset of these involve the CNS and cause meningitis, encephalitis, or myelitis (neuroinvasive disease). In North America, there are fewer than 10 arboviruses that cause significant neurologic disease (Table 2).

KLT is supported by R01 grants (NS051403, NS050138) from the National Institutes of Health and a MERIT Grant from the Department of Veterans Affairs and by the Reuler-Lewin Family Professorship of Neurology. JDB is supported by a Department of Veterans Affairs Research Career Development Award and has been supported by a National Institutes of Health postdoctoral fellowship training grant (T32AI07537).

* Corresponding author. New Mexico Veterans Affairs Health Care System, 1500 San Pedro Drive SE, Albuquerque, NM 87108.

E-mail address: ledavis@unm.edu (L.E. Davis).

0733-8619/08/$ - see front matter. Published by Elsevier Inc.
doi:10.1016/j.ncl.2008.03.012 *neurologic.theclinics.com*

Table 1
Families and genera that contain North American arboviruses that cause encephalitis

Family	Genera
Togaviridae	Alphavirus
	Eastern equine encephalitis
	Western equine encephalitis
	Venezuelan equine encephalitis
Flaviviridae	Flavivirus
	West Nile
	St. Louis encephalitis
	Powassan
	Dengue
Bunyaviridae	Bunyavirus
	La Crosse
	Jamestown Canyon
	Other California group viruses
Reoviridae	Coltivirus
	Colorado tick fever

All arboviruses are small spherical RNA viruses ranging in size from 40 to 50 nm for Togaviridae and Flaviviridae to 80 to 120 nm for Bunyaviridae [1]. Alphaviruses and flaviviruses are enveloped spherical viruses whose genome consists of positive-sense (+), single-stranded (ss) RNA molecule of 10 to 12 kb. Bunyaviridae are enveloped spherical viruses whose genome consists of three linear minus-sense (−) ssRNA segments of approximately 11 kb, and Reoviridae are nonenveloped icosahedral-shaped viruses that can contain up to 12 double-stranded linear RNA segments totaling 19 to 32 kb [2]. All arboviruses, except those belonging to the family Reoviridae, possess a lipid envelope derived from host cells in which several envelope or membrane proteins are inserted and play an essential role in viral attachment and cell entry. In the case of the Reoviridae, these proteins are part of the virion's outer protein shell. Antibodies to these proteins form the basis for host immunity and serologic diagnosis of infections [3].

In most cases, human infection by arboviruses results in an asymptomatic infection associated with seroconversion, and such cases typically vastly outnumber cases of neuroinvasive disease by 100 to 1000:1. Symptomatic disease generally results in one of three syndromes: (1) a self-limited flulike or denguelike illness, (2) hemorrhagic fever, or (3) CNS disease. This article focuses exclusively on arbovirus-induced CNS disease, with an emphasis on those agents important and endemic in North America.

It has been difficult to predict when and where arbovirus outbreaks or epidemics will appear, but several environmental factors increase the probability of human infections. With the exception of infections caused by Colorado tick fever (CTF) virus and Powassan virus (POWV), which are tick-borne, most arboviral infections in the United States are caused by mosquito vectors. Outside the United States, sandflies (*Phlebotomus* sp)

Table 2
Common arboviruses of North America

Virus	Distribution	Insect vector	Intermediate vertebrate host
West Nile	United States, Canada, Mexico	Mosquito (*Culex* sp)	Passiform birds (jays, blackbirds, crows, finches, sparrows)
St. Louis encephalitis	United States, Canada, Mexico	Mosquito (*Culex* sp)	Passerine birds (sparrows, house finches)
Eastern equine encephalitis	Atlantic and Gulf coast states, upper New York, Michigan, eastern Canada	Mosquito (*Culex* sp)	Fresh water swamp birds
Western equine encephalitis	Western United States, Canada	Mosquito (*Culex* sp)	Passerine birds, jackrabbit
Venezuelan encephalitis	Mexico, Florida, Texas	Mosquito (*Culex* and *Aedes* sp)	Rodents, aquatic birds
Powassan encephalitis	Northern United States, Canada	Ixodes ticks	Squirrels, mice, ground hogs, voles
La Crosse or Jamestown encephalitis	North central and northeast United States	Mosquito (*Aedes triseriatus*)	Chipmunks, squirrels
Colorado tick fever	Rocky Mountain states, Canada	Tick (*Dermacentor andersoni*)	Chipmunks, squirrels, small mammals
Dengue	Mexico and Florida	Mosquito (*Aedes* sp)	Humans and nonhuman primates

are an important vector for arboviruses, such as Toscana, a bunyavirus that is an important cause of summertime meningitis in Mediterranean Europe. For an epidemic to occur, local mosquito vectors must be reasonably abundant, capable of becoming infected, and capable of transmitting virus to mammalian hosts [4]. Wet warm spring seasons tend to enable mosquitoes to expand early, and increase the potential for infection of newly hatched birds, which are typically more susceptible to infection than their adult counterparts. A period of heavy precipitation followed by dry weather facilitates development of small stagnant water pools ideal for mosquito breeding and enhances the potential for arbovirus epidemics [4].

Most arboviruses are maintained in nature in bird or small mammal populations; therefore, local bird or rodent populations must be susceptible to viral infection. Although high levels of immunity generated by prior arbovirus exposure would be expected to reduce the availability of bird or mammal reservoirs, from a practical viewpoint this situation has not yet developed to the point that it significantly affects patterns of arbovirus infection in the United States. Finally, the amplifying cycle of infection must be within the transmission zone of populations of susceptible humans who are exposed to the infected mosquitoes.

Although human seroprevalence rates against arboviruses increase in regions affected by epidemics, there are currently no areas in the United States where these rates approach levels that have a significant impact on the spread or development of arbovirus epidemics. The ability of public health departments and organizations rapidly to identify encephalitis outbreaks and to track their progression in real-time has improved dramatically in the past several years. The Centers for Disease Control and Prevention (CDC) and state public health departments all participate in ArboNet and contribute data on arbovirus infections derived from trapping of mosquito pools; susceptible animals that can serve as sentinels for human infection (eg, birds, horses); and human disease. These data for the United States are available through the US Geological Survey Web site (http://diseasemaps.usgs.gov). This site provides real-time data at county-level resolution for human, mosquito, bird, veterinary, and sentinel infections caused by West Nile virus (WNV), St. Louis encephalitis virus (SLEV), La Crosse virus, western equine encephalomyelitis virus (WEEV), eastern equine encephalomyelitis virus (EEEV), and POWV.

Pathogenesis

All arboviruses have geographic and ecologic limitations imposed by the host ranges of their vectors and natural reservoirs, and by the temporal features of vector life cycles and breeding patterns. Additionally, arboviruses exhibit variations in replicative capacity between different species of vectors. The enzootic life cycle of arboviruses usually does not involve humans, which are considered dead-end hosts because the level of viremia in infected humans is seldom high enough to allow continued transmission of infection to mosquitoes or ticks. Exceptions to this general rule include dengue and Venezuelan equine encephalitis virus (VEEV), both of which may generate high titer viremia in humans. Human-to-human transmission of other arbovirus infections can occur when blood or organs from an infected individual are transfused or transplanted into a naive host [5,6]. Perinatal and transplacental human-to-human transmission of arbovirus infection can also occur [7,8].

WNV infection of *Culex pipiens* and *quinquefasciatus* mosquitoes provides an example of the replication of arboviruses in arthropods. Following ingestion of infected blood by a susceptible female mosquito, viral replication first occurs in the midgut around day 2 and rapidly expands until almost all midgut epithelial cells, adjacent muscle, and fat cells are infected [9]. The virus then appears in salivary glands and accumulates in salivary gland ducts by day 14. Virus also appears in abdominal, thoracic, and cephalic ganglia and persists in all infected organs. Virus inoculation into a host rapidly occurs after the proboscis breaks the skin and biting can inject up to 10^4 plaque-forming units (pfu) of virus into dermis, subcutaneous tissue, and directly into blood vessels [10].

Mechanisms of arbovirus survival through winter months (overwintering) in arthropod populations are incompletely understood [11]. Not all mosquitoes die with onset of cold weather, and some infected adult mosquitoes instead become dormant, becoming active again with warmer weather [12]. Infected female mosquitoes can also transmit virus to their eggs (transovarian transmission), which survive through winter and then hatch into infected pupa, which mature into infected adults [13]. Like infected mosquitoes, infected ticks can also overwinter, and transmit virus by transovarian transmission with infected larvae developing into infected nymphs and infected adults (transtadial transmission) [14].

Although arthropods are required for efficient transmission of arboviruses, the major natural reservoir for these viruses is species of birds or small mammals. These animals share the common trait of being able to develop and sustain sufficient viremia to transmit infection back to arthropods when bitten. For most arboviruses, replication in the natural host occurs without causing illness or death. WNV infection of passeriform birds (crows, jays, and so forth) results in viremias as high as 10^{10} pfu/mL, however, with associated multiorgan system pathology and high morbidity and mortality. A viremia of this magnitude leads to subsequent efficient transmission in more than 80% of biting mosquitoes [15]. The duration of viremia in birds is variable, but frequently lasts weeks and even months allowing for repeated cycles of mosquito infection [16]. Infected birds can also shed considerable amounts of virus from the cloaca and nasopharynx allowing for viral spread between birds independent of mosquitoes [17]. In passeriform birds, WNV infects the heart, kidney, spleen, gut, adrenal glands, liver, and brain, and these tissues are then infectious to raptors that may prey on these birds [18]. In birds, infection is associated with development of protective immune responses resulting in immunity to repeat challenges of virus. Consequently, new populations of naive birds are required to maintain viral endemicity.

In humans, once WNV is inoculated into the dermis, viral replication begins in Langerhans dendritic cells, which then migrate to regional lymph nodes. The primary viremia disseminates virus to the entire reticuloendothelial system where subsequent replication augments the viremia. A low-level viremia can be detected as early as 1 to 2 days postinoculation, persists for about 1 week, and terminates in association with the appearance of serum IgM neutralizing antibodies (Fig. 1) [19]. When viremia is rapidly and efficiently cleared by the host's immune system, the patient experiences either an asymptomatic infection or a flulike syndrome without encephalitis. When immune clearance of the viremia is delayed, as may occur in the elderly or immunosuppressed, the risk of virus entering the CNS increases [6]. The precise method by which WNV enters the CNS is unknown, but other flaviviruses have been shown to enter the CNS following infection of cerebral microvascular endothelial cells or by infection of the olfactory bulb with subsequent transneuronal spread to the brain [19].

Once an arbovirus reaches the CNS, it causes widespread, patchy, infection of neurons in the cerebral cortex, basal ganglia, brainstem, cerebellum, and occasionally spinal cord [20,21]. The specific topographic patterns of infection vary between viruses, although there is substantial overlap. The mechanism by which arboviruses kill infected neurons is incompletely understood, although several flaviviruses including WNV and Japanese

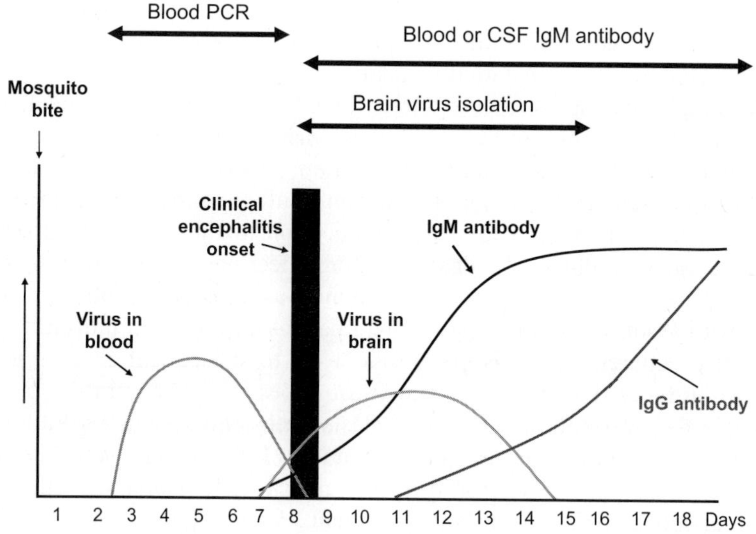

Fig. 1. Typical time course for development of West Nile virus in blood and brain and occurrence of IgG and IgM antibodies in humans after West Nile virus infection. CSF, cerebrospinal fluid; PCR, polymerase chain reaction.

encephalitis virus can induce apoptosis in infected neurons [22]. Arboviruses do not generally infect oligodendrocytes, so acute demyelination is not a typical feature of arbovirus encephalitis, although rare cases of postinfectious acute disseminated encephalomyelitis have been reported after dengue and SLEV and following immunization with the vaccine strain of Japanese B encephalitis virus [23,24].

The amount of brain inflammation varies by arbovirus but seldom is overwhelming. More severe parenchymal inflammatory responses generally parallel the degree of pleocytosis induced in the cerebrospinal fluid (CSF). The leptomeninges appear normal grossly but histologically show a lymphocytic infiltration. The brain may appear grossly normal, edematous, or may be congested with focal petechiae in cortical gray matter, deep gray matter, and white matter. Multiple focal areas of brain softening may be present, which histologically demonstrate perivascular cuffing, infiltration of mononuclear and polymorphonuclear cells, and neuronophagia of infected cells. The inflammatory cells are mainly $CD4^+$ and $CD8^+$ cells with occasional B cells. In fatal cases, a heavy concentration of viral lesions is often seen in the brainstem [20]. When the spinal cord is involved, viral antigen, neuronophagia, neuronal disappearance, perivascular cuffing, and microglia proliferation are seen in the anterior horns of the lumbar and cervical areas [25].

Studies of experimental WNV infection in mice indicate humoral, cell-mediated, and innate immune responses all contribute to control and clearance of virus from infected organs including brain. Mice deficient in their ability to develop IgG, IgM, CD4, or CD8 immune responses all show increased severity of WNV infection, as do mice deficient in interferon, interferon receptor, and complement [26–31].

Syndromes caused by arboviruses

Asymptomatic infections

Most individuals infected with arboviruses develop asymptomatic infection, although the exact ratio of neuroinvasive disease cases to asymptomatic infection varies for specific viruses. For most arboviruses, asymptomatic infections occur mainly in children and young adults. The resulting infection produces prolonged immunity, except in the case of dengue. It is important to remember that asymptomatically infected individuals can still transmit infection to others through blood transfusion or less commonly organ transplantation. Blood donors are typically asymptomatic at the time of donation. Nonetheless, in 2003, 877 units of WNV-infected blood were identified among the 2.5 million units screened by the American blood centers [32]. Currently, American blood banks screen pooled lots of donated blood units by polymerase chain reaction (PCR) assay for WNV nucleic acid. Positive pools are then subjected to individual unit-by-unit assay with positive units discarded. Screening has dramatically reduced the

decreased recruitment of motor unit action potentials). Compound muscle action potentials are reduced in amplitude and sensorineural action potentials are preserved consistent with involvement of anterior horns or roots. Demyelination is not a typical feature of arbovirus infections and as a result conduction velocities are not slowed and conduction block does not occur [38]. Although cases of WNV-associated AFP were originally attributed to a Guillain-Barré like syndrome, it is now generally believed that more than 80% of WNV-associated AFP cases represent as an anterior poliomyelitis. Well-documented WNV-associated Guillain-Barré like syndrome cases are rare.

Diagnosis of arbovirus encephalitis

Diagnosis of almost all arbovirus infections is most commonly based on serum or CSF serology. In acute infections, virus-specific IgM antibodies can be detected by IgM capture ELISA [39]. For many arboviruses, over 50% of patients with neuroinvasive disease have IgM antibody present at the time of hospitalization and over 90% have the antibody by 1 week of symptoms (see Fig. 1). Closely related arboviruses share epitopes, and antibody responses developed following infection with one virus may cross-react with antigens from other related viruses (heterologous antibody response) [39]. Diagnosis of the specific virus involved may require performance of ELISAs against several candidate viruses. The offending virus typically is the one with the highest titer, although variations in the methods for coating ELISA plates and other technical aspects of assay performance makes this conclusion a relative rather than absolute one. More traditional assays for arboviral antibodies (hemagglutination inhibition, complement fixation, indirect immunofluorescence, and neutralization assays) are only performed by the CDC or qualified reference laboratories because they require use of live virus under appropriate biosafety containment conditions. Diagnosis of a specific arbovirus infection using these assays is generally established by detecting IgM antibodies or a fourfold rise in IgG antibody titer between acute and convalescent serum. Unlike herpesviruses, which are ubiquitous agents with generally high basal seroprevalence levels in the general population, seroprevalence levels for individual arboviruses in the United States are generally low. Following infection arbovirus antibodies persist for life, however, so the mere presence of IgG antibody in the absence of a rising titer or IgM antibody may simply be indicative of past exposure rather than acute infection. Although IgM antibodies generally reflect acute infection, cases of IgM antibodies against WNV persisting in serum or CSF for greater than a year have been documented [40]. In some cases serial determination of both IgG and IgM antibody titers may be needed to differentiate between acute and prior infection. It is important to recognize that individuals who have been immunized against Japanese encephalitis virus or yellow fever virus may develop heterologous cross-reacting antibodies to WNV and SLEV that may confound diagnosis.

As a general rule the presence of IgM antibodies in CSF is indicative of intrathecal synthesis and the associated presence of the inciting antigen in the CNS, and as such are diagnostic not only of infection but of neurologic disease. An elevated CSF/serum antibody ratio, when corrected for breakdown of the blood-brain barrier through use of concomitant CSF/serum albumin ratios, can also be indicative of intrathecal antibody synthesis and CNS infection.

Isolation of arboviruses in culture is difficult and not routinely performed. Human viremia is usually of low titer and is terminated by IgM antibody that appears around the onset of the clinical symptoms as shown in Fig. 1. For most patients, virus is rarely isolated from blood or CSF and not usually detected by PCR assay from blood, with the exception of infections caused by VEEV and dengue virus. In individuals who are donating blood with asymptomatic early WNV infections, PCR assays successfully detect the viremia in most individuals [32].

The sensitivity of CSF PCR for diagnosis of arboviral infections is significantly below that reported for both enteroviruses and herpesviruses, in which this technique is the diagnostic procedure of choice. Comparative data are available predominantly for WNV infection, for which CSF PCR sensitivity is generally around 70% when compared with demonstration of CSF IgM as a diagnostic standard. CSF PCR and serologic tests should be considered complementary rather than competing diagnostic methods. CSF PCR may be particularly useful in immunocompromised patients who may have attenuated or delayed antibody responses and in the early stages after onset of illness, when antibody responses may not have fully developed. The specificity of CSF PCR approaches 100%, and the amplification of arboviral nucleic acid from CSF in an appropriate clinical setting by an established laboratory is generally diagnostic of infection. Isolation of virus from brain tissue or the demonstration of arbovirus nucleic acid by PCR or antigen by immunohistochemistry is also diagnostic [20].

Management and prognosis

Currently, there is no antiviral treatment of proved efficacy available for any arbovirus infection. The few drugs that have shown efficacy in experimental animals and infected cell culture, particularly when administered early after viral infection, include interferon [41], ribavirin [42], monoclonal antibodies [43,44], immune globulin [45], and antisense oligomers that bind to arboviral RNA and inhibit translation of viral proteins and viral RNA replication [46,47].

In the absence of specific antiviral therapy the mainstay of treatment for arboviral encephalitides remains optimal supportive care. Patients with encephalitis should be hospitalized because disease can progress for several days to a week or more following presentation. To avoid contributing to cerebral edema and increased intracranial pressure, patients require careful monitoring of fluid balance, and should not receive hypotonic solutions. Accurate intracranial pressure monitoring in severely ill patients may require placement of an intracranial pressure bolt. Seizures can complicate many types of encephalitis, although their frequency in arbovirus infections is generally low enough that anticonvulsants are not given prophylactically, but only therapeutically. Patients should receive appropriate prophylaxis for prevention of pressure sores and deep venous thrombosis. Alimentary support should be provided for patients who are unable to sustain adequate oral caloric and nutritional intake. Intubation may be required for airway protection or for ensuring adequate oxygenation in patients with depressed consciousness or respiratory muscle weakness. Because of the limited risk of human-to-human transmission of infection, patient isolation is not required for individuals with arbovirus encephalitis, although universal precautions should be observed because both blood and CSF are potentially infectious. Stool, urine, and respiratory secretions are generally not infectious. Many patients with encephalitis require treatment in rehabilitation facilities following acute hospitalization.

Prevention

There are currently no human vaccines available for prevention of infection by arboviruses prevalent in the United States. Effective vaccines for Japanese encephalitis virus and yellow fever virus exist. It is unclear whether vaccination for these two flaviviruses offers cross-protection against the North American flaviviruses [48,49]. Prevention of arbovirus infections in the United States focuses on preventing the infection. Health departments in many states and Canada use methods to identify mosquitoes and birds that are infected with an arbovirus. The earliest warning system comes from mosquito traps with subsequent PCR identification of arboviruses. Later warning systems include sentinel bird flocks and monitoring and assaying dead birds for arboviruses. Finally, veterinarians participate in late warning systems by identifying arbovirus horse infections, such as WNV, EEEV, and WEEV. Other public health measures include the elimination of standing water pools that harbor mosquito larvae or administration of larvicides or adulticides. Even urban areas may support infected mosquitoes. In Phoenix, Arizona, abandoned swimming pools with stagnant water were found to serve as a breeding ground for mosquitoes [50]. Finally, aerial insecticide spraying of rural and urban communities can dramatically reduce the number of infected adult mosquitoes. Costs to conduct surveillance can be high. Various insecticide applications can cost as much as $1.4 million [51].

Personal protection for individuals comes from fixing house window screens to prevent mosquitoes from entering; draining neighborhood collections of standing water; clearing brush for tick removal; wearing long sleeved shirts and long pants outdoors; minimizing being outdoors from dusk to dawn; use of mosquito nets during sleep outdoors; spraying permethrin on clothing when hiking in tick- or mosquito-infested areas; and applying insect repellants, such as DEET and Picaridin, on exposed skin or wearing permethrin-treated clothing [51,52].

West Nile virus (family: Flaviviridae)

Epidemiology

WNV has been recognized in Africa, the Middle East, and Asia since 1937. The virus abruptly appeared in New York City in 1999 and has rapidly spread across the United States [11]. Over 24,000 cases have been reported by the CDC since 1999 [51]. The rapid spread across the United States and Canada has been attributed to the capacity of WNV to infect multiple strains of mosquitoes and many species of birds.

Virus appears in blood of infected patients within 1 to 2 days following a mosquito bite and persists for about 1 week [19]. Over 89% of patients develop an asymptomatic infection. Over 1820 asymptomatic viremic blood donors have been reported to the CDC since 2002 with virus titers ranging from 25 to 200 RNA viral copies/mL [51]. Fifty-eight percent of cases reported to the CDC are classified as West Nile fever and 42% as West Nile neuroinvasive disease [51]. The incidence of WNV AFP is about 5% to 10% of the neuroinvasive disease cases.

Clinical features

Patients with West Nile fever experience an abrupt onset of fever, headache, and fatigue. A maculopapular generalized rash occurs with variable frequency. Patients may also develop anorexia, nausea, myalgia, and lymphadenopathy. West Nile neuroinvasive disease includes patients with meningitis, encephalitis, and AFP [11,53–56]. These patients often have a flulike prodrome that is quickly followed by development of neurologic signs and symptoms. It has been suggested that approximately 35% to 40% of patients with WNV neuroinvasive disease have meningitis, 55% to 60% have encephalitis, and 5% to 10% AFP [56], although these numbers are quite variable in different case series. Patients with encephalitis typically have fever (85%–100%); headache (47%–90%); and an altered mental status (46%–74%). Following infection, many patients can develop movement disorders, which include tremors of various types, myoclonus, and parkinsonism. Cerebellar signs and symptoms occur in a variable number. Weakness is also common and may be generalized and nonspecific or a lower motor neuron pattern of flaccid paralysis associated with absent or reduced reflexes and preserved sensation. Cranial nerve palsies have been reported in 10% or more of cases and most commonly involve the ophthalmic (2nd) and facial (7th) nerves. Visual problems in patients with West Nile neuroinvasive disease have become increasingly recognized. Patients often complain of blurry vision, trouble seeing, and photophobia. Clinically, reduced vision, active nongranulomatous uveitis, vitreitis, multifocal chorioretinitis, and optic neuritis have been described [34,57]. Immunosuppressed patients, such as patients with transplants or AIDS, often develop a prolonged serious encephalitis with a higher mortality [54,58].

Laboratory features

Patients have a normal complete blood count or mild leukocytosis. Hyponatremia is seen in one third of patients and abnormal liver functions tests occur in one fourth [59]. About 40% of patients have an initial CSF that shows a pleocytosis with a neutrophil predominance that then becomes predominately lymphocytes [36]. Cranial CT scans are usually normal but MRI scans may show subtle abnormalities on diffusion-weighted images [60]. A few patients show T2-weighted hyperintense abnormalities in the basal ganglia, thalami, brainstem, and cerebellum, often appearing at about 1 week after onset of illness [61]. MRI imaging of spinal cord in patients with AFP may be normal or show abnormalities involving the anterior horn area.

In fatal cases, WNV antigen was detected by immunohistochemistry 60% of the time in the brainstem and 13% in the cortex [20]. When AFP was present, viral antigen was found in 40% of spinal cords. At postmortem viral antigen may also be detected with variable frequency in other organs including kidney, lung, pancreas, and gastrointestinal tract [62].

Diagnosis

The diagnosis of WNV infection is usually made by detection of WNV IgM antibody in serum or CSF [39]. WNV ELISA detects cross-reacting antibodies from other flaviviruses infections including SLEV, Japanese encephalitis virus, dengue, yellow fever virus, and POWV requiring titer comparisons with appropriate other arboviruses to confirm the diagnosis [39]. Prior vaccination with yellow fever or Japanese encephalitis viruses may produce cross-reacting antibodies. WNV IgM antibody first appears in serum 6 to 9 days after infection, which is typically within a day or two of the onset of neuroinvasive illness [39]. Initial false-negative IgM antibody tests can occur if serum is taken very early in the illness [39]. An initial negative serum WNV IgM antibody test in a patient with suspected WNV disease should be repeated 3 to 7 days later. Most immunosuppressed transplant patients with West Nile neuroinvasive disease do develop IgM antibody in serum and CSF but may exhibit a prolonged time to seroconversion [58]. IgM antibody persists in blood or CSF for more than 2 months in 50% of patients and for a year or more in 10% [40].

Virus can be isolated from brain of fatal patients and rarely from CSF, but this is rarely attempted except at the CDC or in reference laboratories because of the requirement for appropriate biosafety containment procedures. PCR amplification of WNV nucleic acid from CSF is positive in 57% to 70% of patients but is less sensitive than detection of WNV-specific CSF IgM antibody [63]. Less than 15% of patients with neuroinvasive disease have blood positive for virus by PCR assay [63].

Therapy and prognosis

There is no treatment of proved efficacy for WNV infection. Uncontrolled studies of individual or small groups of cases have reported both possible beneficial effects and no utility for treatment with interferon-α. A randomized controlled multicenter trial of an Israeli IVIG preparation (Omr-IgG-am) with high titer anti-WNV antibodies was recently completed, but the results are not yet available. The study was initiated on the basis of reports of possible efficacy in isolated cases. Several effective and safe equine vaccines using either formalin inactivated or chimeric recombinant viruses are licensed for veterinary use. Phase I/II trials of a human chimeric vaccine ChimeriVax-West Nile (Acambis) containing WNV envelope and PrM proteins on a yellow fever 17D vaccine strain virus background are currently in process. In initial studies the vaccine was found to be safe and immunogenic in humans [64].

Mortality from WNV is confined almost exclusively to patients with encephalitis or AFP and approaches 20% in severely ill patients. The nature, frequency, and severity of neurologic sequelae are poorly understood and comprehensive studies particularly of cognitive impairment are still very limited [56]. About 40% of patients with movement disorders including tremors, myoclonus, and parkinsonism have persisting symptoms at 6 months postillness, and approximately 20% have residual symptoms at 18 months postillness [56]. Recovery from AFP is variable. It has been estimated that one third of patients return to baseline, one third have no significant recovery of strength, and one third have partial recovery, with the brunt of improvement occurring within 4 months of onset of illness [56]. Recent studies suggest that many patients with encephalitis may suffer from residual cognitive problems. In one survey 50% of encephalitis patients reported such symptoms at 3 months postillness. Deficits can also involve motor speed and dexterity [56], as might be expected in patients with striatal-thalamic pathology.

St. Louis encephalitis virus (family: Flaviviridae)

Epidemiology

Until WNV arrived, SLEV was the most important arbovirus in North America. Cases of encephalitis caused by SLEV were first recognized in the 1930s. SLEV is now widespread in North America ranging from Canada through the United States (except the New England coast) to Central and South America. Strains of virus circulating in the United States differ from those in Central and South America. In the eastern United States, strains are transmitted by *C pipiens*/*C quinquefasciatus* and *Culex nigrapalpus* mosquitoes and have a greater epidemic potential [65]. In the western United States, the virus is primarily transmitted by *Culex tarsalis* mosquitoes and is endemic causing sporadic cases. The highest incidence rates have been in Ohio and Mississippi River basins and the Gulf Coast states. Over 4605 cases of SLE have been reported to the CDC since 1964, with most years ranging from 20 to 50 cases [51]. In 2005, the most recent year for which complete data are available, only seven cases were reported in the United States. Outbreaks occurred in 1975 with an estimated 2800 cases in 31 states [1], in 1990 with 222 cases reported from Florida [66], in 1995 in Texas [67], and in 2001 with 70 cases in Louisiana [68]. Like other arboviruses in the United States, cases have typically occurred between July and September.

Clinical features

In children the rate of asymptomatic to symptomatic infections is about 800:1, in adults it is 300:1 to 85:1, and in older adults may be only 16:1 [65,69]. The incubation period is from 4 to 21 days. Patients with the flulike syndrome often have fever, myalgias, and headaches. Rashes are uncommon [68]. In young adults, meningitis occurs in 40% and encephalitis in 60%, whereas in those over the age of 60, 90% develop encephalitis [70]. Common signs of encephalitis include depressed consciousness from lethargy and confusion to coma and tremors of eyelids, lips, and limbs [71]. Some patients develop myoclonic jerks, opsoclonus, nystagmus, lower facial weakness, and cerebellar ataxia. Seizures were reported in 4 (36%) of 11 cases in one recent series with one patient developing status epilepticus [67]. AFP has been reported in 6% of the encephalitis cases [72] and transverse myelitis may also occur [69]. The acute illness lasts 1 to 2 weeks but recovery from such symptoms as forgetfulness, tremors, unsteadiness, weakness, and headaches may take months to years.

Laboratory studies

In one series of 11 cases CSF lymphocytic pleocytosis occurred in all (mean, 107 cells/mm^3; range, 5–446); protein elevation in 64% (mean,

67 mg/dL; range, 39–143); and all had a normal CSF glucose concentration [67]. EEG is usually abnormal with the most common finding being diffuse slowing, although rarely seizures and periodic lateralizing epileptiform discharges have been reported [67]. MRI is more sensitive than CT in detecting abnormalities but is often normal. In one series of six patients with MRIs performed, none had abnormalities on T1 sequences and only two showed T2 hyperintensities, both involving the substantia nigra [67]. Electromyograms may demonstrate evidence of denervation in limb muscles suggesting viral involvement of anterior horn neurons in the spinal cord. Patients usually have a peripheral leukocytosis and occasionally sterile pyuria [71]. Hyponatremia from syndrome of inappropriate antidiuretic hormone occurs in about one third of cases. Mild elevations in liver function tests and muscle enzymes can occur.

Diagnosis

Diagnosis is typically based on the demonstration of IgM in serum or CSF by capture ELISA [73]. An increase between acute and convalescent neutralizing serum antibody titer to SLEV also confirms the diagnosis. CSF cultures are usually negative, although virus can be isolated from brain tissue by biopsy or at autopsy.

Therapy and prognosis

No specific therapy of proved efficacy is available. In one nonrandomized unblinded study, interferon-α2b, 3 million units intravenously x 1 followed by the same dose subcutaneously 12 hours later and then daily for 14 days, was reported to improve outcome compared with nontreated controls [74]. Mortality in recent studies has varied from 4% to 27% [67,68].

Eastern equine encephalitis virus (family: Togaviridae)

Epidemiology

Of the four lineages of EEEV, group I causes the most disease in humans, whereas groups IIA, IIB, and III cause primarily equine disease in Central and South America [75]. In the United States, most cases are located along the eastern seaboard and sporadically located along the Gulf coast, typically within 5 miles of swamp or marsh lands that help to maintain the enzootic cycle. EEEV causes sporadic infections in human populations during the summer months and occasional epidemic outbreaks. The last United States outbreak occurred in September, 2005, in Massachusetts and New Hampshire and involved 11 patients with four deaths (36%) [75]. The primary enzootic mosquito vector associated with EEEV transmission is the mosquito *Culiseta melanura* [76]. Birds serve as the primary reservoir host and amplifying hosts, and humans are incidentally infected by a variety of mosquito bridging vectors including *Culex* and *Aedes* species [77].

Clinical features

The clinical features of EEEV are nonspecific. Patients initially develop a prodrome of fever, chills, malaise, and myalgias lasting for 1 to 2 weeks [78]. The prodrome is followed by recovery or onset of encephalitis characterized by severe headache, confusion, nausea, and vomiting. Seizures, focal neurologic deficits, such as cranial nerve palsies or focal weakness, and meningismus are common findings in patients that develop encephalitis [78]. Severely affected cases often progress to coma and death.

Laboratory studies

Basic serum laboratory studies are remarkable for a leukocytosis and hyponatremia in as many as 60% of patients [78]. Cerebral spinal fluid analysis is abnormal in patients with EEEV with a median pleocytosis of 370 leukocytes/mm^3 (range, 0–2400) and a median neutrophil predominance of 70% [78]. The protein is elevated with a median value of 97 mg/dL and the glucose ratio may be normal to decreased [78,79].

As for many cases of arbovirus-induced encephalitis, MRI is typically more sensitive than CT for abnormalities associated with infection. Although imaging may be normal, EEEV often shows increased signal intensity on T2-weighted images of the basal ganglia and thalami [78]. Noncontiguous lesions may also be seen in the brainstem, cortex, and periventricular white matter.

Diagnosis

In most cases, diagnosis is made with serology. The most common method is the IgM capture ELISA, which detects antibody in a single blood or CSF sample, providing earlier diagnosis. Detection of IgM against EEEV antigen in the CSF is diagnostic of CSF infection and encephalitis when associated with the appropriate clinical syndrome.

Therapy and prognosis

No proved antiviral therapy exists for EEEV. Treatment is focused on supportive care and managing complications, such as seizures and increased intracranial pressure. There is no commercial vaccine for EEEV. In evaluating laboratory and imaging studies for prognostic value, one study found that CSF leukocytosis greater than 500 cells/mm^3 and hyponatremia less than 130 meq/L were predictive of a poor outcome [78]. Typically, the mortality rate represents about one third of infections [78].

Western equine encephalitis virus (family: Togaviridae)

Epidemiology

Although WEEV was the first arbovirus isolated in America (1930), cases of human disease are now exceedingly rare. WEEV activity occurs mainly in western and midwestern United States and western Canada. There have been 639 human cases reported by the CDC since 1964 but only five confirmed cases in the United States in the last 20 years and none since a single case in Minnesota in 1999 [51]. Major epidemics occurred in Minnesota in 1941 (>3000 human cases and hundreds of thousands of equine cases) [1,80] and in Saskatchewan in 1965 [81]. Reasons for the decline in both horse and human infections are unclear.

Clinical features

The human incubation period ranges from 2 to 10 days. Very young and old individuals are at highest risk for severe encephalitis. Most children and young adults develop a flulike syndrome with fever and malaise lasting about a week rather than neuroinvasive disease. Meningismus is common. In those developing encephalitis, hemiparesis, stupor, cranial nerve palsies, generalized rigidity, transient tremors, parkinsonism, and occasional opisthotonus develop [1,65]. Fewer than 10% develop coma, which is usually transient. The clinical course usually runs 2 weeks and convalescence occurs over several weeks.

Laboratory studies

A leukocytosis and hyponatremia caused by syndrome of inappropriate antidiuretic hormone may develop. CSF findings are typical for arbovirus encephalitis. There are few reports of neuroimaging studies. EEG typically shows diffuse slowing, although rare cases with focal temporal slowing that can mimic that seen in herpes simplex encephalitis have been reported [46].

Diagnosis

Diagnosis is usually made by detection of IgM antibody to WEEV in serum or CSF [1], but virus can be occasionally isolated from brain and CSF [82].

Therapy and prognosis

No specific therapy is available. There is a single case report over half a century ago describing the successful use of equine immune serum in an infected laboratory technician [83]. The overall case mortality is 3% to 8% but is higher in the elderly and infants [81,84]. Survivors typically recover fully, except for infants and elderly who may be left with neurologic sequelae including mental retardation, dementia, and parkinsonism [85].

Venezuelan equine encephalitis virus (family: Togaviridae)

Epidemiology

VEEV is roughly divided into enzootic and epizootic strains. Enzootic strains occur mainly in Venezuela, Central America, and Equador, but an Everglade strain exists in Southern Florida that rarely causes human disease [1,86]. Enzootic viral strains cycle between mosquitoes and small rodents, such as spiny or cotton rats, causing low viremias and mild disease in horses, and occasionally illness in humans. A recent focus in southern Mexico demonstrated over 60% of older adults had VEEV antibodies but the virus appeared to cause only rare cases of encephalitis [87].

Epizootic strains have a wider range of mosquito hosts, causing high viremias and serious illness in horses, and frequent flulike illness in humans. These strains are capable of becoming epidemics that spread outside their normal viral territory. From 1969 to 1972, an outbreak in Guatemala spread as far north as southern Texas causing several hundred human cases and over 1500 horse deaths in Texas [1,88]. Currently, the CDC reports only rare human cases occurring within the United States.

Viremia in patients with VEEV can be as high as 10^8 x the intracranial LD_{50} for suckling mice, which is of sufficient magnitude to infect biting mosquitoes. Human to mosquito to human transmission, however, does not seem to play a significant role in disease spread or pathogenesis [89].

Clinical features

Over 90% of infected individuals develop an inapparent infection [89]. The incubation period in humans is usually 1 to 5 days, and patients with clinical symptoms usually develop the typical arbovirus flulike syndrome often with pharyngitis and cervical adenopathy [65,86]. Neurologic features from encephalitis occur in 4% to 25% of symptomatic patients, mainly in the young and old. Clinical features include confusion, somnolence and seizures, cerebellar ataxia, and cranial nerve palsies [65,89]. Seizures, which are often focal, can be common.

Laboratory features

CSF shows a lymphocytic pleocytosis with a normal or elevated protein level and a normal glucose. Leukopenia and elevated liver transaminase levels are commonly seen.

Diagnosis

VEEV can be isolated from blood, CSF, and pharyngeal secretions, and viral shedding is often prolonged for a week or more. Detection of VEEV nucleic acid in CSF or blood by PCR is diagnostic [89]. Diagnosis can also be made by detection of VEEV IgM antibody in CSF or serum by IgM capture ELISA assay [90].

Therapy and prognosis

There is no specific therapy for VEEV. The US Army developed a live attenuated VEEV vaccine in 1961(TC-83) and a formalin inactivated derivative in 1974 (C-84). Vaccination of humans with the live attenuated vaccine followed by a booster dose of inactivated virus results in immunogenicity and is well tolerated [91]. Because of ongoing concerns about the potential use of VEEV as a biowarfare weapon, impetus to develop a safe and effective VEEV vaccine has intensified and several candidate vaccines are in active development [92]. Mortality in patients with VEEV is typically reported as 10% to 25%; by contrast, in symptomatic patients without encephalitis mortality is less than 1%. The prognosis for full recovery in survivors is excellent.

Powassan encephalitis virus (family: Flaviviridae)

Epidemiology

This virus is part of the tick-borne encephalitis group that has been renamed the mammalian group of tick-borne flaviviruses. Most of these viruses are endemic in Europe and Asia and often cause viral hemorrhagic fever or encephalitis.

POWV was first isolated from the brain of a 5-year-old child dying of encephalitis in Ontario, Canada, in 1958. Since that time isolated additional cases of encephalitis have occurred both in Canada and in the northern United States [93]. Less than 10 cases of POWV encephalitis have been reported to the CDC since 2001 [51]. Between September 1999 and July 2001 a small outbreak of four cases was reported from Maine and Vermont [94], the first reported United States cases since 1994.

The POWV life cycle is mainly between small rodents (squirrels, groundhogs, mice, voles) and hard *Ixodes* ticks, although POWV can also infect mosquitoes. Infected ticks live for months to over a year, and virus can persist in the tick population by transtatial transmission and in the offspring by transovarial transmission [14]. The principle cause of human disease results from the adult ticks taking a blood meal from humans walking outdoors in forests during the spring to fall months. Infections typically occur between June and September when ticks are actively feeding. Despite the importance of ticks in pathogenesis, patients typically do not recall a bite or finding a tick.

Clinical features

The incubation period ranges from 8 to 34 days. Patients with encephalitis typically have fever, headache, retro-orbital pain, and photophobia associated with seizures, focal weakness, and depressed level of consciousness. In the most recent series of four cases the patients ranged in age from 25 to 70 with three being age 53 years or older. Additional features reported in individual patients in this small series included ophthalmoplegia, gaze paresis, dysarthria, ataxia, muscle twitching, and bilateral proximal leg weakness.

Laboratory studies

In the four cases from the most recent outbreak, CSF cell counts ranged from 47 to 920 cells/mm^3, with three of four patients having greater than 74% lymphocytes. Protein was elevated in the three cases in which data were available (range, 67–96 mg/dL). Two patients had areas of increased T2 signal consistent with microvascular ischemia or demyelination, but it was not clear these were related to the acute infection. EEG was diffusely slow in the three patients in which it was performed [94].

Diagnosis

Virus has been isolated from brain at autopsy but rarely from other sites [93]. Diagnosis is made by detection of CSF or serum IgM antibodies or by a fourfold rise in IgG antibody between acute and convalescent sera. These tests are not available commercially, but can be requested through state public health laboratories and are performed at the CDC.

Therapy and prognosis

There is no specific therapy for POWV. The mortality in patients with encephalitis is 10% to 15%, and sequelae in survivors include hemiparesis, aphasia, and flaccid limb paralysis [95,96]. At autopsy the brain shows multifocal areas of infection in the gray matter of the cerebral cortex, brainstem, and spinal cord [93].

California encephalitis group (La Crosse virus) (family: Bunyaviridae)

Within the family of Bunyaviridae and genus *Bunyavirus*, California encephalitis virus, La Crosse virus, Jamestown Canyon virus, and Tahyna virus comprise the major causes of encephalitis in the California encephalitis group. Of these viruses, La Crosse virus, California encephalitis virus, and Jamestown Canyon virus are causes of disease in the United States and Tahyna virus is predominantly a cause of encephalitis in Russia.

Epidemiology

La Crosse virus is the most common cause of disease in the California encephalitis group. It was originally described in 1965 following a postmortem examination of a child that died because of encephalitis in La Crosse, Wisconsin [97]. La Crosse virus is transmitted in an enzootic pattern between squirrels and chipmunks by the mosquito *Aedes triseriatus* in areas of the Mississippi and Ohio River basins [98]. The *A triseriatus* vector is a forest-dwelling mosquito that inhabits tree-holes and overwinters by transovarian transmission into *Aedes* eggs [98]. Like the other viruses in this group, infection is related to contact with wooded forests during warm summer months. In cases of La Crosse virus infection, the ratio of asymptomatic to symptomatic infections is approximately 1000:1 [99]. California encephalitis virus was originally isolated in 1941 [100], and although the group of diseases are named after this virus, it is relatively rare with disease occurrence mainly located in the western United States and Canada. Although La Crosse and California encephalitis viruses cause most of disease in children, Jamestown Canyon virus affects predominantly the elderly in regions of the northern United States with seroprevalence in some areas reaching 10% [101].

Clinical features

In children infected with La Crosse virus, the mean age is 7.5 years and symptoms typically consist of fever, headache, vomiting in 70%, seizures in 46%, and altered mental status in 42% [99]. Additionally, such focal neurologic signs as hemiparesis, aphasia, dysarthria, and chorea occur. About 10% of patients develop increased intracranial pressure and rarely cerebral herniation can occur [99]. Jamestown Canyon virus has similar clinical features.

Laboratory diagnosis

Peripheral leukocytosis is common and patients can develop severe hyponatremia caused by the syndrome of inappropriate antidiuretic hormone syndrome. CSF often reveals a lymphocytic pleocytosis of 600 cells/mm^3, with a normal glucose, and an increased protein in 30% of patients [99]. As with other arboviruses, diagnosis is made with serologic techniques because viral isolation is not sensitive. Indirect immunofluorescence is the method of choice to detect IgG and IgM. IgM detection in the CSF or a fourfold rise in paired sera for IgG is considered diagnostic for infection.

Prognosis and therapy

In general, children infected with La Crosse virus recover with a case fatality rate less than 1% [99]. Although about 6% to 15% of patients that recover have recurrent seizures, most return to normal function with approximately 10% of patients suffering from sequelae following discharge from the hospital. Predictors of a poor outcome include hyponatremia, persistently elevated body temperature, and a Glasgow Coma Score of less than 13 [100,102]. No antiviral therapy currently exists for the California encephalitis group of viruses and no vaccine is available. Physicians are relegated to treating the complications, such as seizures, elevated intracranial pressure, and other complications.

Colorado tick fever (family: Reoviridae)

All CTF viruses have common structural features consisting of segmented, double-stranded RNA surrounded by double-capsid proteins that form a nonenveloped viral particle. Although viruses in this family cause a variety of diseases, CTF virus causes an acute febrile syndrome that is occasionally recurrent and can be complicated by encephalitis.

Epidemiology

CTF is transmitted by the wood tick (*Dermacentor andersoni*) mainly in the western mountain regions of the United States and Canada above 4000 ft in elevation [103]. The enzootic pattern of infection exists between small mammals, such as ground squirrels, marmots, or chipmunks, and *D andersoni*. Once infected, the tick is infected for life (transtadially) throughout the three main stages of development: (1) laval, (2) nymphal, and (3) adult. There is no transovarial (vertical) transmission but sustained viremias in the amplifying, vertebrate hosts allow for successful viral overwintering [103]. The younger stages of the tick (larval and nymphal) typically feed on small mammals; however, the adult tick often feeds on larger mammals including humans. Humans are typically exposed to the habitat of *D andersoni* during summer months while hiking, fishing, or camping and are infected as dead-end hosts following a blood meal from an infected tick. Occasional cases of CTF have occurred in patients exposed to the tick from a traveling family member by way of clothing or equipment [104]. Lastly, occasional cases of CTF in California, outside the *D andersoni* range, have been reported and attributed to infection of a different tick vector, *Dermacentor variabilis* [105].

Clinical features

A history of tick bite is obtained in approximately 90% of patients, and the incubation period is 0 to 14 days, mean 3 days [103]. Patients develop abrupt onset of fever, chills, generalized myalgias, severe headache, and hyperesthetic skin leading to severe malaise that often confines patients to bedrest. Gastrointestinal symptoms, such as nausea, vomiting, or diarrhea, may be present but are not prominent. About 15% of patients develop a maculopapular or petechial rash. Otherwise, physical findings are elusive and include occasional pharyngitis, mild lymphadenopathy, or mild splenomegaly. In 50% of patients, the fever resolves after 2 days and then recurs in the biphasic or saddleback pattern [103]. Following convalescence, patients over the age of 30 may have persistent fatigue for as long as 3 weeks. Complications, such as encephalitis or meningitis, predominantly occur in children in about 5% to 10% of the cases; yet, CTF is rarely fatal [106]. In these patients, neurologic signs and symptoms include nuchal rigidity, photophobia, and mild altered mental status.

Laboratory studies

CTF virus infects hematopoietic cells resulting in a persistent infection and subsequent neutropenia, thrombocytopenia, and mild anemia [107]. Occasionally, other laboratory abnormalities are found, such as elevated liver transaminases and elevated creatinine phosphokinase levels. Analysis of CSF may exhibit a mild lymphocytic pleocytosis with a normal glucose and normal or slightly elevated protein. Although virus can be isolated from peripheral blood or by blood smear using indirect immunofluorescence of red blood cells [108], the diagnosis of CTF is usually made using IgM capture ELISA, neutralization, or complement fixation [109].

Prognosis and therapy

There is no established treatment for CTF. Most patients recover following infection, and only three deaths have been reported caused by intravascular coagulopathy following CTF infection [106]. Therapy is largely supportive consisting of bed rest, fluid maintenance, and antipyretics with particular avoidance of aspirin and nonsteroidal anti-inflammatory drugs because of their exacerbation of platelet dysfunction and theoretic increased risk of coagulopathy.

Dengue virus (family: Flaviviridae)

Epidemiology

Dengue virus is widely endemic in tropics around the world. Parts of Mexico have dengue and indigenous dengue infections have occurred in Texas, Hawaii, and Puerto Rico [110,111]. In addition, over 100 American foreign travelers return to the United States each year with dengue [112].

Clinical features

Most patients infected with one of the four dengue viruses develop a fever-arthralgia-rash syndrome or viral hemorrhagic fever. Some patients, however, present with a decreased level of consciousness mainly caused by encephalopathy but occasionally encephalitis [113,114]. Patients with encephalitis may develop coma; meningismus; seizures; and focal neurologic signs, such as hemiparesis [113,115].

Laboratory studies

The blood typically shows leukopenia and thrombocytopenia. CSF has a mild lymphocytic pleocytosis in 30% of patients with diminished mental status.

Diagnosis

Dengue virus has been isolated or detected by PCR in CSF, and IgM antibody to dengue has also been found in CSF [113,114]. For the diagnosis of dengue, one should obtain acute and convalescent serum for antibody titers plus an acute serum for virus isolation [112].

Prognosis and therapy

The overall prognosis is good unless hemorrhagic fever develops.

Summary

Arboviruses continue to be a major cause of encephalitis in North America. With the appearance of WNV in 1999, West Nile neuroinvasive disease has become the dominant cause of encephalitis. All North American arbovirus are transmitted to humans by infected mosquitoes or ticks. Most infections are asymptomatic or produce a flulike illness. Elderly, immunosuppressed individuals, and infants for some arboviruses have the most severe encephalitis. Rapid serum or CSF IgM antibody capture ELISA assays are available to diagnosis the acute infection for all North American arboviruses. Unfortunately, no antiviral drugs are available for treatment of the encephalitis, so therapy is symptomatic.

References

[1] Calisher CH. Medically important arboviruses of the United States and Canada. Clin Microbiol Rev 1994;7:89–116.

[2] Condit RC. Principles of virology. In: Fields virology. 5th edition. Philadelphia: Wolters Kluwer-Lippincott William & Wilkins; 2007. p. 25–57.

[3] Roehrig JT, Staudinger LA, Hunt AR, et al. Antibody prophylaxis and therapy for flavivirus encephalitis infections. Ann N Y Acad Sci 2001;951:286–97.

[4] Day JF. Predicting St. Louis encephalitis virus epidemics: lessons from recent, and not so recent, outbreaks. Annu Rev Entomol 2001;46:111–38.

[5] Pealer L, Marfin A, Petersen L, et al. Transmission of West Nile virus through blood transfusion in the United States in 2002. N Engl J Med 2003;349:1236–45.
[6] Iwamoto M, Jernigan D, Guasch A, et al. Transmission of West Nile virus from an organ donor to four transplant recipients. N Engl J Med 2003;348:2196–203.
[7] CDC. Possible West Nile virus transmission to an infant through breast-feeding—Michigan 2002. MMWR Morb Mortal Wkly Rep 2002;51:877–8.
[8] CDC. Intrauterine West Nile virus infection—New York 2002. MMWR Morb Mortal Wkly Rep 2002;51:1135–6.
[9] Girard YA, Klingeler KA, Higgs S. West Nile virus dissemination and tissue tropisms in orally infected Culex pipiens quinquefasciatus. Vector Borne Zoonotic Dis 2004;4:109–22.
[10] Van Landingham DL, Schneider BS, Klinger K, et al. Real-time reverse transcriptase-polymerase chain reaction quantification of West Nile virus transmitted by Culex pipiens quinquefasciatus. Am J Trop Med Hyg 2004;71:120–3.
[11] Davis LE, DeBiasi R, Goade DE, et al. West Nile virus neuroinvasive disease. Ann Neurol 2006;60:286–300.
[12] Nasci R, Savage H, White D, et al. West Nile virus in overwintering Culex mosquitoes, New York City, 2000. Emerg Infect Dis 2001;7:742–4.
[13] Miller B, Nasci R, Godsey M, et al. First field evidence for natural vertical transmission of West Nile virus in Culex univittatus complex mosquitoes from Rift Valley Province, Kenya. Am J Trop Med Hyg 2000;62:240–6.
[14] Costero A, Grayson MA. Experimental transmission of Powassan virus (Flaviviridae) by Ixodes scapularis ticks (Acari:Ixodidae). Am J Trop Med Hyg 1996;55:536–46.
[15] Turell M, O'Guinn M, Oliver J. Potential for New York Mosquitoes to transmit West Nile virus. Am J Trop Med Hyg 2000;62:413–4.
[16] Komar N, Langevin S, Hinten S, et al. Experimental infection of North American birds with New York 1999 strain of West Nile virus. Emerg Infect Dis 2003;9:311–22.
[17] Komar N, Lanciotti R, Bowen R, et al. Detection of West Nile virus in oral and cloacal swabs collected from bird carcasses. Emerg Infect Dis 2002;8:741–2.
[18] Steele KE, Linn MJ, Schoepp RJ, et al. Pathology of fatal West Nile virus infections in native and exotic birds during the 1999 outbreak in New York City, New York. Vet Pathol 2000;37:208–24.
[19] Chambers TJ, Diamond MS. Pathogenesis of flavivirus encephalitis. Adv Virus Res 2003; 60:273–342.
[20] Guarner J, Shieh W, Hunter S, et al. Clinicopathologic study and laboratory diagnosis of 23 cases with West Nile virus encephalomyelitis. Hum Pathol 2004;35:983–90.
[21] Solomon T. Flavivirus encephalitis. N Engl J Med 2004;351:370–8.
[22] Samuel MA, Morrey JD, Diamond MS. Caspase 3-dependent cell death of neurons contributes to the pathogenesis of West Nile virus encephalitis. J Virol 2007;81:2614–23.
[23] Ohtaki E, Matsuishi T, Hirano Y, et al. Acute disseminated encephalomyelitis after treatment with Japanese B encephalitis vaccine. J Neurol Neurosurg Psychiatry 1995;59:316–7.
[24] Sejvar JJ, Bode AV, Curiel M, et al. Post-infectious encephalomyelitis associated with St. Louis encephalitis virus infection. Neurology 2004;63:1719–21.
[25] Fratkin J, Leis A, Stokic D, et al. Spinal cord neuropathology in human West Nile virus infection. Arch Pathol Lab Med 2004;128:533–7.
[26] Wang Y, Lobigs M, Lee E, et al. $CD8^+$ T cells mediate recovery and immunopathology in West Nile virus encephalitis. J Virol 2003;77:13323–34.
[27] Shrestha B, Diamond MS. Role of $CD8^+$ T cells in control of West Nile virus infection. J Virol 2004;78:8312–21.
[28] Shrestha B, Wang T, Samuel MA, et al. Gamma interferon plays a crucial early antiviral role in protection against West Nile virus infection. J Virol 2006;80:5338–48.
[29] Shrestha B, Samuel MA, Diamond MS. $CD8^+$ T cells require perforin to clear West Nile virus from infected neurons. J Virol 2006;80:119–29.

[30] Sitati EM, Diamond MS. $CD4^{+}$ T-cell responses are required for clearance of West Nile virus from the central nervous system. J Virol 2006;80:12060–9.
[31] Purtha WE, Myers N, Mitaksov V, et al. Antigen-specific cytotoxic T lymphocytes protect against lethal West Nile virus encephalitis. Eur J Immunol 2007;37:1845–54.
[32] Kleinman S, Glynn SA, Busch M, et al. The 2003 West Nile virus United States epidemic: the America's Blood Centers experience. Transfusion 2005;45:469–79.
[33] Sejvar J, Haddad M, Tierney B, et al. Neurologic manifestations and outcome of West Nile virus infection. JAMA 2003;290:511–5.
[34] Khairallah M, Ben Yahia S, Ladjimi A, et al. Chorioretinal involvement in patients with West Nile virus infection. Ophthalmology 2004;111:2065–70.
[35] Sejvar JJ, Bode AV, Marfin AA, et al. West Nile virus-associated flaccid paralysis. Emerg Infect Dis 2005;11:1021–7.
[36] Tyler KL, Pape J, Goody RJ, et al. CSF findings in 250 patients with serologically confirmed West Nile virus meningitis and encephalitis. Neurology 2006;66:361–5.
[37] Gandelman-Marton R, Kimiagar I, Itzhaki A, et al. Electroencephalography findings in adult patients with West Nile virus-associated meningitis and meningoencephalitis. Clin Infect Dis 2003;37:1573–8.
[38] Al-Shekhlee A, Katirji B. Electrodiagnostic features of acute paralytic poliomyelitis associated with West Nile virus infection. Muscle Nerve 2004;29:376–80.
[39] Shi P-Y, Wong SJ. Serologic diagnosis of West Nile virus infection. Expert Rev Mol Diagn 2003;3:733–41.
[40] Roehrig JT, Nash D, Maldin B, et al. Persistence of virus-reactive serum immunoglobulin M antibody in confirmed West Nile virus encephalitis cases. Emerg Infect Dis 2003;9:376–9.
[41] Brooks TJ, Phillpotts RJ. Interferon-alpha protects mice against lethal infection with St Louis encephalitis virus delivered by the aerosol and subcutaneous routes. Antiviral Res 1999;41:57–64.
[42] Jordan I, Briese T, Fischer N, et al. Ribavirin inhibits West Nile virus replication and cytopathic effect in neural cells. J Infect Dis 2000;182:1214–7.
[43] Kimura-Kuroda J, Yasui K. Protection of mice against Japanese encephalitis virus by passive administration with monoclonal antibodies. J Immunol 1988;141:3606–10.
[44] Morrey JD, Siddharthan V, Olsen AL, et al. Defining limits of treatment with humanized neutralizing monoclonal antibody for West Nile virus neurological infection in a hamster model. Antimicrob Agents Chemother 2007;51:2396–402.
[45] Ben-Nathan D, Lustig S, Tam G, et al. Prophylactic and therapeutic efficacy of human intravenous immunoglobulin in treating West Nile virus infection in mice. J Infect Dis 2003; 188:5–12.
[46] Bia FJ, Thornton GF, Main AJ, et al. Western equine encephalitis mimicking herpes simplex encephalitis. JAMA 1980;244:367–9.
[47] Deas TS, Binduga-Gajewska I, Tilgner M, et al. Inhibition of flavivirus infections by antisense oligomers specifically suppressing viral translation and RNA replication. J Virol 2005;79:4599–609.
[48] Tesh RB, Travassos da Rosa RP, Guzman H, et al. Immunization with heterologous flaviviruses protective against fatal West Nile encephalitis. Emerg Infect Dis 2002;8:245–51.
[49] Kanesa-Thasan N, Putnak JR, Mangiafico JA, et al. Absence of protective neutralizing antibodies to West Nile virus in subjects following vaccination with Japanese encephalitis or dengue vaccines. Am J Trop Med Hyg 2002;66:115–6.
[50] Levy C. West Nile virus in Arizona. Available at: www.cdc.gov/ncidod/dvbid/westnile/conf/2005pdf/AZWNVUpdateCraigLevyVersion.pdf. Accessed August 20, 2007.
[51] CDC. Arboviral encephalitis. Available at: http://www.cdc.gov/ncidod/dvbid/arbor/. Accessed August 21, 2007.
[52] Fradin MS, Day JF. Comparative efficacy of insect repellents against mosquito bites. N Engl J Med 2002;347:13–8.

[53] DeBiasi RL, Tyler KL. West Nile virus meningoencephalitis. Nat Clin Pract Neurol 2006;2: 264–75.
[54] Bode AV, Sejvar JJ, Pape WJ, et al. West Nile virus disease: a descriptive study of 228 patients hospitalized in a 4-county region of Colorado in 2003. Clin Infect Dis 2006;42: 1234–40.
[55] Kramer LD, Li J, Shi PY. West Nile virus. Lancet Neurol 2007;6:171–81.
[56] Sejvar JJ. The long-term outcomes of human West Nile virus infection. Clin Infect Dis 2007; 44:1617–24.
[57] Bakri SJ, Kaiser PK. Ocular manifestations of West Nile virus. Curr Opin Ophthalmol 2004;15:537–40.
[58] Kleinschmidt-DeMasters BK, Marder BA, Levi ME, et al. Naturally acquired West Nile virus encephalomyelitis in transplant recipients: clinical, laboratory, diagnostic, and neuropathological features. Arch Neurol 2004;61:1210–20.
[59] Weiss D, Carr D, Kellachan J, et al. Clinical findings of West Nile virus infection in hospitalized patients, New York and New Jersey, 2000. Emerg Infect Dis 2001;7:654–8.
[60] Petropoulou KA, Gordon SM, Prayson RA, et al. West Nile virus meningoencephalitis: MR imaging findings. AJNR Am J Neuroradiol 2005;26:1986–95.
[61] Ali M, Safriel Y, Sohi J, et al. West Nile virus infection: MR imaging findings in the nervous system. AJNR Am J Neuroradiol 2005;26:289–97.
[62] Armah HB, Wang G, Omalu BI, et al. Systemic distribution of West Nile virus infection: Postmortem immunohistochemical study of six cases. Brain Pathol 2007;17:345–62.
[63] Lanciotti RS, Kerst AJ, Nasci RS, et al. Rapid detection of West Nile virus from human clinical specimens, field collected mosquitoes, and avian samples by a TaqMan reverse transcriptase-PCR assay. J Clin Microbiol 2000;38:4066–71.
[64] Monath TP, Liu J, Kanesa-Thasan N, et al. A live, attenuated recombinant West Nile virus vaccine. Proc Natl Acad Sci U S A 2006;103:6694–9.
[65] Solomon T, Whitley RJ. Arthropod-borne viral encephalitides. In: Scheld WM, Whitley RJ, Marra CM, editors. Infections of the central nervous system. 3rd edition. Philadelphia: Lippincott Williams & Wilkins; 2004. p. 205–30.
[66] Meehan PJ, Wells DL, Paul W, et al. Epidemiological features and public health response to a St. Louis encephalitis epidemic in Florida, 1990–1991. Epidemiol Infect 2000;125:181–8.
[67] Wasay M, Diaz-Arrastia R, Suss RA, et al. St. Louis encephalitis: a review of 11 cases in a 1995 Dallas, Tex, epidemic. Arch Neurol 2000;57:114–8.
[68] Jones SC, Morris J, Hill G, et al. St. Louis encephalitis outbreak in Louisiana in 2001. J La State Med Soc 2002;154:303–6.
[69] Monath TP. Epidemiology. In: Monath TP, editor. St. Louis encephalitis. Washington, DC: American Public Health Association; 1980. p. 239–312.
[70] Brinker KR, Monath TP. The acute disease. In: Monath TP, editor. St. Louis encephalitis. Washington, DC: American Public Health Association; 1980. p. 503–34.
[71] Southern PM, Smith JW, Luby JP, et al. Clinical and laboratory features of epidemic St. Louis encephalitis. Ann Intern Med 1969;71:681–9.
[72] Quick DT, Thompson JM, Bond JO. The 1962 epidemic of St. Louis encephalitis in Florida. IV. Clinical features of cases occurring in the Tampa Bay area. Am J Epidemiol 1965;81: 415–27.
[73] Monath TP, Nystrom RR, Bailey RE, et al. Immunoglobulin M antibody capture enzyme-linked immunosorbent assay for diagnosis of St. Louis encephalitis. J Clin Microbiol 1984; 20:784–90.
[74] Rahal JJ, Anderson J, Rosenberg C, et al. Effect of interferon-alpha2b therapy on St. Louis viral meningoencephalitis: clinical and laboratory results of a pilot study. J Infect Dis 2004; 190:1084–7.
[75] CDC. Eastern equine encephalitis—New Hampshire and Massachusetts, August–September 2005. MMWR Morb Mortal Wkly Rep 2006;55:697–700.

[76] Loftin KC, Diallo AA, Herbert MW, et al. Five-year surveillance of West Nile and eastern equine encephalitis viruses in Southeastern Virginia. J Environ Health 2006;68:33–40.
[77] Mitchell CJ, Niebylski ML, Smith GC, et al. Isolation of eastern equine encephalitis virus from *Aedes albopictus* in Florida. Science 1992;257:526–7.
[78] Deresiewicz RL, Thaler SJ, Hsu L, et al. Clinical and neuroradiographic manifestations of eastern equine encephalitis. N Engl J Med 1997;336:1867–74.
[79] Przelomski MM, O'Rourken E, Grady GF, et al. Eastern equine encephalitis in Massachusetts: a report of 16 cases, 1970–1984. Neurology 1988;38:736–9.
[80] Eklund CM. Human encephalitis of the western equine type in Minnesota in 1941. Clinical and epidemiological study of serologically proven cases. Am J Hyg 1946;43:171–93.
[81] Rozdilsky B, Robertson HE, Chorney J. Western encephalitis: report of eight fatal cases. Saskatchewan epidemic, 1965. Can Med Assoc J 1968;98:79–86.
[82] Sciple GW, Ray G, La Motte LC, et al. Western encephalitis with recovery of virus from cerebrospinal fluid. Neurology 1967;17:169–71.
[83] Gold H, Hampil B. Equine encephalomyelitis in a laboratory technician with recovery. Ann Intern Med 1942;16:556–69.
[84] Longshore WA, Stevens IM, Hollister AC, et al. Epidemiologic observations on acute infectious encephalitis in California, with special reference to the 1952 outbreak. Am J Hyg 1956;63:69–86.
[85] Earnest MP, Goolishian HA, Calverley JR, et al. Neurologic, intellectual and psychologic sequelae following western encephalitis: a follow-up study of 35 cases. Neurology 1971;21: 969–74.
[86] Weaver SC, Ferro C, Barrera R, et al. Venezuelan equine encephalitis. Annu Rev Entomol 2004;49:141–74.
[87] Estrada-Franco JG, Navarro-Lopez R, Freier JE, et al. Venezuelan equine encephalitis virus, Southern Mexico. Emerg Infect Dis 2004;10:2113–21.
[88] Walton TE, Grayson MA. Venezuelan equine encephalomyelitis. In: Monath TP, editor. The arboviruses: epidemiology and ecology, vol. IVBoca Raton Florida: CRC Press; 1988. p. 203–31.
[89] Bowen GS, Calisher CH. Virological and serological studies of Venezuelan equine encephalomyelitis in humans. J Clin Microbiol 1976;4:22–7.
[90] Watts DM, Callahan J, Rossi C, et al. Venezuelan equine encephalitis febrile cases among humans in the Peruvian Amazon River region. Am J Trop Med Hyg 1998;58:35–40.
[91] Pittman PR, Makuch RS, Mangiafico JA, et al. Long-term duration of detectable neutralizing antibodies after administration of live-attenuated VEE vaccine and following booster vaccination with inactivated VEE vaccine. Vaccine 1996;14:337–43.
[92] Rao V, Hinz ME, Roberts BA, et al. Toxicity assessment of Venezuelan equine encephalitis virus vaccine candidate strain V3526. Vaccine 2006;24:1710–5.
[93] McLean DM, Donohue WL. Powassan virus: isolation of virus from a fatal case of encephalitis. Can Med Assoc J 1959;80:708–11.
[94] CDC. Outbreak of Powassan encephalitis—Maine and Vermont, 1999-2001. MMWR Morb Mortal Wkly Rep 2001;50:761–4.
[95] Smith R, Woodall JP, Whitney E, et al. Powassan virus infection: a report of three human cases of encephalitis. Am J Dis Child 1974;127:691–3.
[96] Rossier E, Harrison RJ, Lemieux B. A case of Powassan virus encephalitis. Can Med Assoc J 1974;110:1173–80.
[97] Thompson WH, Kalfayan B, Anslow RO. Isolation of California encephalitis group virus from a fatal human illness. Am J Epidemiol 1965;81:245–53.
[98] Watts DM, Pantuwatana S, DeFoliart GR, et al. Transovarial transmission of LaCrosse virus (California encephalitis group) in the mosquito, *Aedes triseriatus*. Science 1973;182: 1140–1.
[99] McJunkin JE, de los Reyes EC, Irazuzta JE, et al. La Crosse encephalitis in children. N Engl J Med 2001;344:801–7.

[100] Cramblett HG, Stegmiller H, Spencer C. California encephalitis virus infections in children: clinical and laboratory studies. JAMA 1966;198:108–12.

[101] Mayo D, Karabatsos N, Scarano FJ, et al. Jamestown Canyon virus: seroprevalence in Connecticut. Emerg Infect Dis 2001;7:911–2.

[102] Hilty MD, Haynes RE, Azimi PH, et al. California encephalitis in children. Am J Dis Child 1972;124:530–3.

[103] Goodpasture HC, Poland JD, Francy DB, et al. Colorado tick fever: clinical, epidemiologic, and laboratory aspects of 228 cases in Colorado in 1973–1974. Ann Intern Med 1978;88:303–10.

[104] Midoneck SR, Richard J, Murray HW. Colorado tick fever in a resident of New York City. Arch Fam Med 1994;3:731–2.

[105] Attoui H, Mohd Jaafar F, Biagini P, et al. Genus Coltivirus (family Reoviridae): genomic and morphologic characterization of Old World and New World viruses. Arch Virol 2002; 147:533–61.

[106] Spruance SL, Bailey A. Colorado tick fever: a review of 115 laboratory confirmed cases. Arch Intern Med 1973;131:288–93.

[107] Philipp CS, Callaway C, Chu MC, et al. Replication of Colorado tick fever virus within human hematopoietic progenitor cells. J Virol 1993;67:2389–95.

[108] Emmons RW, Lennette EH. Immunofluorescent staining in the laboratory diagnosis of Colorado tick fever. J Lab Clin Med 1966;68:923–9.

[109] Mohd Jaafar F, Attoui H, Gallian P, et al. Recombinant VP7-based enzyme-linked immunosorbent assay for detection of immunoglobulin G antibodies to Colorado tick fever virus. J Clin Microbiol 2003;41:2102–5.

[110] CDC. Dengue hemorrhagic fever—U.S.-Mexico border, 2005. MMWR Morb Mortal Wkly Rep 2007;56:785–9.

[111] Effler PV, Pang L, Kitsutani P, et al. Dengue fever, Hawaii, 2001–2002. Emerg Infect Dis 2005;11:742–9.

[112] CDC. Travel-associated dengue—United States, 2005. MMWR Morb Mortal Wkly Rep 2006;55:700–2.

[113] Solomon T, Dung NM, Vaughn DW, et al. Neurological manifestations of dengue infection. Lancet 2000;355:1053–9.

[114] Lum LC, Lam SK, Choy YS, et al. Dengue encephalitis: a true entity? Am J Trop Med Hyg 1996;54:256–9.

[115] Pancharoen C, Thisyakorn U. Neurological manifestations of dengue patients. Southeast Asian J Trop Med Public Health 2001;32:341–5.

ELSEVIER
SAUNDERS

NEUROLOGIC CLINICS

Neurol Clin 26 (2008) 759–780

Acute Disseminated Encephalomyelitis: Clinical and Pathogenesis Features

Farshid Noorbakhsh, MD, PhD[a],
Richard T. Johnson, MD[b], Derek Emery, MD[c],
Christopher Power, MD[a,*]

[a]Department of Medicine (Neurology), University of Alberta, Edmonton, Alberta, T6G 2S2, Canada
[b]Department of Neurology, Johns Hopkins University School of Medicine and Bloomberg School of Public Health, Baltimore, MD 21287, USA
[c]Department of Radiology and Diagnostic Imaging, University of Alberta, Edmonton, Alberta, T6G 2S2, Canada

Acute disseminated encephalomyelitis (ADEM) is an immune-mediated disorder of the central nervous system (CNS). Disease typically starts with an abrupt onset of neurologic symptoms and signs within days to weeks after a viral infection or immunization. ADEM is also known as "postinfectious," "parainfectious," "postexanthematous," or "postvaccinal" encephalomyelitis [1]. Although many viral agents associated with ADEM also cause acute viral encephalitis, ADEM usually occurs much later after the onset of infection and differs clinically by virtue of greater white matter involvement with respective neurologic symptoms [2]. Contrary to acute viral encephalitis, attempts to isolate the virus from postmortem ADEM brains have often failed, implying mechanisms other than direct invasion of CNS by the infectious agent. Neuropathologic examination of ADEM consistently discloses widespread perivenular inflammation and myelin disruption, giving rise to the pathologically derived terms "perivascular myelinoclasis," "perivenous encephalitis," and "acute demyelinating encephalomyelitis" [3].

The first clues to the possible pathogenic mechanisms underlying ADEM came from studies on demyelinating encephalomyelitis cases that occasionally complicated smallpox vaccination [4]. Rivers and Schwentker [5]

F.N. is an Alberta Heritage Foundation for Medical Research and Canadian Institutes of Health Research Fellow. C.P. is an Alberta Heritage Foundation for Medical Research Scholar and holds a Canada Research Chair (T1) in Neurologic Infection and Immunity.

* Corresponding author. HMRC 611, Department of Medicine (Neurology), University of Alberta, Edmonton, AB, Canada, T6G 2S2.

E-mail address: chris.power@ualberta.ca (C. Power).

doi:10.1016/j.ncl.2008.03.009

reported histologic similarities between postvaccinal encephalomyelitis brains and fatal neuroparalytic accidents following rabies vaccination. Perivenular inflammation and demyelination were obvious in both cases. To examine the contribution of brain-derived proteins present in rabies vaccine to the induction of lesions, Rivers and Schwentker [5] repeatedly injected homogenates of normal rabbit brains into monkeys. Several monkeys receiving the virus-free brain homogenates developed an inflammatory, demyelinating brain disease, closely resembling clinical and pathologic features of postvaccinal encephalomyelitis [6]. Indeed, this was the first example of "experimental autoimmune encephalomyelitis" (EAE), which came to be studied extensively in the ensuing decades as a prototype autoimmune disease model, particularly multiple sclerosis (MS) [7–9].

EAE was later shown to be inducible by single injections of homologous brain tissue (ie, brain tissue removed from the same animal) when it was emulsified in complete Freund's adjuvant. The myelin protein antigens able to induce disease were subsequently identified and shown to have common sequences across species. Interestingly, disease was later shown to be passively transferable to healthy animals by injecting $CD4^+$ T lymphocytes from immunized animals [10]. EAE signs vary depending on species; strain (genetic background); age; gender; and the immunization protocol. Disease typically starts with weakness and paralysis, 7 to 21 days after inoculation of brain homogenate or myelin components. On neuropathologic analysis, mononuclear cell infiltrates consisting of lymphocytes and monocytoid cells are seen in meninges and in perivenular areas in the white matter. These findings are accompanied by activation of resident microglia and are followed by demyelination and axonal injury. Although mainly considered a disease of the CNS, inflammatory demyelinating lesions have also been described in the peripheral nervous system (ie, dorsal root ganglia in rabbits induced with EAE) [11,12]. Repair of demyelinated foci with recurrent inflammation and demyelination gives rise to a relapsing-remitting disease course in some species. A pathogenically similar but clinically distinct disease termed "experimental autoimmune neuritis" can be induced in the peripheral nervous system. Reported by Waksman and Adams [13] in 1955, rabbits receiving peripheral nerve emulsions in adjuvant develop an acute demyelinating neuropathy after a period of about 2 weeks. Similar to EAE, experimental autoimmune neuritis is characterized by mononuclear cells infiltrating out of endoneural venules followed by demyelination.

The presence of brain-derived components in rabies vaccines at the time, the striking histologic similarities between vaccine-induced encephalomyelitis and EAE, and similar latency times and clinical disease course after immunizations led to the belief that postvaccinal encephalomyelitis was a disease of an autoimmune nature. This assumption was reinforced by the observation that lymphocytes from post–rabies vaccine encephalomyelitis patients could be stimulated in vitro with myelin antigens, a feature similar to the lymphocytes isolated from EAE animals. Indeed, the incidence of postvaccinal encephalomyelitis dramatically decreased with the introduction of rabies vaccines devoid of myelin components (ie, virus grown in embryonated eggs or cell cultures, rather than infected rabbit or sheep brain) [14]. Growing rabies virus in newborn mice brains, however, which are largely unmyelinated, failed completely to eliminate neurologic complications of vaccine [15]. Nevertheless, many of the patients receiving these vaccines developed polyradicular neuropathies with cranial nerve involvement, which have been attributed to the presence of peripheral nerve myelin components in the cranial nerves isolated with newborn mice brains [15].

Clinical presentation

The incidence of ADEM from different causes has been reported to be between 0.4 and 0.8 per 100,000 of population [16,17] with a median age of onset of 4.5 to 7.5 in pediatric studies [18,19] and 33.5 in a study of adult patients [20]. Disease has a seasonal peak in winter and spring, consistent with its putative infectious etiologies. ADEM typically appears with the abrupt onset of neurologic symptoms 2 to 30 days after the occurrence of a preceding infection or vaccination [19]. A clear preceding infection or vaccination is not found, however, in around one third of children and half of adults presenting with disease [16,19,21]. The contribution of different infections or vaccination to disease varies based on regional immunization programs, with the natural infections still reported to constitute most cases (93%) in countries with widespread immunization against childhood viral infections [16,22]. Systemic symptoms including fever (43%–52%), headache (45%–58%), malaise, and myalgias may occur shortly before the appearance of neurologic signs and symptoms [22,23]. ADEM usually presents as a monophasic demyelinating disease, and neurologic manifestations depend on which region of the CNS is affected, with the most common signs including obtundation and depressed consciousness; unilateral or bilateral long tract signs (85%); acute hemiparesis (76%); and ataxia (59%) [19]. Depressed mental status exists in cases of ADEM. Meningismus (26%–31%), caused by inflammation in subarachnoid space, can also be found on neurologic examination [22,23]. Although motor deficits occur in both adult and pediatric cases, sensory deficits are more frequent in adults but seizures predominate in pediatric cases; one study has reported prolonged focal motors seizures in 70% of children with a high tendency to develop status epilepticus. Involvement of peripheral nervous system is rare in childhood ADEM but more common in adult patients, usually in the form of acute polyradiculoneuropathy [19,20]. Although ADEM displays a monophasic disease course, rare cases of disease relapse have been described in some studies. Long-term clinical and imaging follow-up has shown the resolution of lesions with no long-lasting neurologic impairments in most of these multiphasic cases [22].

Acute hemorrhagic leukoencephalitis (AHLE) is considered a hyperacute form of ADEM and has been reported to occur in 2% of pediatric cases.

Rare adult cases have also been reported [24]. Prodromal symptoms including fever, malaise, and myalgia are more common than conventional ADEM and are followed by rapidly progressive hemorrhagic demyelination of white matter. Although there are case reports of recovery after high-dose steroid therapy or neurosurgical interventions [25–28], most patients die with brain edema within the first week after the onset of neurologic symptoms [19,28].

Acute transverse myelitis (ATM) is a common clinical presentation of postinfectious or postvaccinal ADEM that is characterized by focal inflammation of the spinal cord with subsequent neural injury [29,30] leading to sensory, motor, and autonomic dysfunction. About one third of patients have pain in the distribution of the involved segments of the spinal cord, before the development of sensory-motor or autonomic symptoms. At the peak of the disease about 50% of patients with ATM are paraplegic; 80% to 94% have been reported to have paresthesia, dysesthesia, or numbness; with almost all the patients suffering from bladder dysfunction [31–34]. Although ATM can occur in the context of multifocal CNS disease or as a part of multisystemic autoimmune disorders (eg, systemic lupus erythematosus or sarcoidosis) [35,36], it can also present as an isolated idiopathic entity [29]. The initial events leading to detrimental autoimmune responses targeting spinal cord are still a matter of debate, but disease has been pathogenically linked to postinfectious demyelinating disorders [29,37]. Indeed, in 30% to 60% of idiopathic cases there is a preceding respiratory, gastrointestinal, or systemic illness [31,32,38]. Like cerebral ADEM, ATM has also been reported following measles, rubella, influenza, and hepatitis B vaccinations (see below) [39–42]. In about 5% of cases, transverse myelitis represents the first attack of MS. Unlike ATM, however, sensory-motor impairments following myelopathic MS are usually asymmetric, making the two entities distinguishable [43]. Moreover, monosegmental involvement of the spinal cord is more commonly seen in myelopathic MS, compared with other ATM etiologies [44]. Cerebrospinal fluid (CSF) pleocytosis and abnormal IgG index can be observed in both MS and ATM. Initial severity of weakness and evidence of denervation on electromyography has been considered poor prognostic indicators for ATM [45]. Approximately one third of patients with ATM recover completely, one third show partial recovery with moderate disabilities, and the rest of the cases lead to permanent severe disabilities [46].

Pathology

ADEM usually affects white matter; however, lesions in cortical gray matter and basal ganglia have also been reported [23,47]. Pathologic features of ADEM in CNS white matter closely resemble that of EAE, with infiltration of monocytoid cells and perivenous areas of demyelination. Axons in the areas of demyelination are relatively preserved and neuronal soma are less affected. Hyperemia and periventricular edema followed by fibrosis in the later stages of disease are also seen [48]. AHLE is characterized by petechial hemorrhages around blood vessels with infiltrates containing high numbers of polymorphonuclear cells, perivascular demyelination, and fibrosis [49]. AHLE is believed to be the outcome of the same pathogenic process as ADEM but with a more severe clinical course and poorer prognosis [50].

ATM pathology can vary based on underlying etiology and precise pathogenic process. When occurring as part of a systemic autoimmune disorder, specific pathologic findings (eg, vasculitis in systemic lupus erythematosus or granulomatosis in sarcoidosis) can be found within the spinal cord [51,52]. In idiopathic cases, which are presumably more closely related to ADEM, perivascular and intraparenchymal inflammatory infiltrates followed by demyelination and neuronal injury have been described [29].

Etiology

Viral etiologies

ADEM is more commonly preceded by a viral infection, with measles, varicella, rubella, mumps, and influenza being the more frequently reported infections (Box 1). Despite the availability of vaccines in many countries, measles virus still remains one of the most common global infectious causes of childhood mortality and neurologic morbidity. Measles infection can cause a transient but severe suppression of cell-mediated immunity, which is the chief reason for postmeasles fatal complications, including disseminated pneumonitis and bacterial infections [53]. Paradoxically, measles infection can also lead to ADEM, an autoimmune phenomenon. The incidence of ADEM is 1 to 2 per 1000 measles infections, being more common in children above 5 years of age [54]. Postmeasles ADEM varies in its time of onset, with most cases developing when the fever and rash from the primary infection are diminishing. Return of the fever together with headache and signs of meningeal irritation herald the start of inflammation in the brain. Focal or generalized convulsions that can be followed by coma have been reported in about half of the patients. Gradual or abrupt depression of consciousness may occur. Focal neurologic deficits indicative of the involvement of cerebral hemispheres, cerebellum, or spinal cord may develop. CSF shows a mild mononuclear pleocytosis with elevated protein content in most patients [55]. Mortality rate is high, reported to be between 10% and 40% in different studies, with a substantial number of survivors suffering permanent neurologic sequelae. The length of stupor or coma has been generally considered a poor prognostic factor [54,56]. On pathologic examination of postmortem brains, perivenular infiltration of mononuclear cells together with demyelination is observed, features that are reminiscent of EAE neuropathology. ADEM comprises about 95% of postmeasles neurologic complications with the rest being myelitis, polyneuritis, and toxic encephalopathy.

Box 1. Causes of postinfectious and postvaccinal encephalomyelitis

Viral infections

- Measles
- Varicella-zoster
- Rubella
- Mumps
- Influenza A and B
- Hepatitis A
- Hepatitis C
- Epstein-Barr virus
- HIV
- Nonspecific upper respiratory tract infection
- Human herpsevirus-6[a]
- Herpes simplex virus[a]
- Dengue virus[a]
- Coxsackie B[a]
- Coronavirus[a]

Nonviral infections

- Group A β-hemolytic streptococci
- *Legionella pneumophila*
- *Salmonella typhi*
- Leptospirosis
- *Plasmodium falciparum*
- *Mycoplasma pneumoniae*
- *Rickettsia rickettsii*
- *Borrelia burgdorferi*

Postvaccinal ADEM

- Rabies vaccine made in brain or spinal cord preparations
- Measles
- Japanese encephalitis virus
- Oral polivirus
- Tetanus toxoid
- Influenza
- Hepatitis B recombinant vaccine
- Tick-borne encephalitis[a]

[a] Denotes single case-reports.

Compared with measles, neurologic complications of acute varicella-zoster virus infection are much less common (1:10,000 of infections), with acute cerebellar ataxia and acute toxic encephalopathy being the most common forms. The former has a very good prognosis [57], whereas the latter, also known as "Reye's syndrome" can be fatal. Postvaricella encephalomyelitis is rare, and it typically starts 1 to 2 weeks after the onset of rash. Clinical manifestations are similar to measles ADEM, and include the return of fever with headache, meningismus, convulsions, and depression of consciousness [58,59]. CSF exhibits mild lymphocytic pleocytosis and elevated protein content. On pathologic examination perivenular demyelination has been reported, with cells from patients with neurologic complications showing proliferative response to myelin antigens. These indicate a pathophysiologic process similar to postmeasles ADEM and EAE [58,59].

Neurologic complications of rubella are even less common that varicella, with an incidence of around 1:20,000 infections, but with a high mortality rate of approximately 20% [6]. Occurrence of fever with headache, convulsions, and decreased consciousness occur 1 week after the appearance of rash. Unlike measles, focal neurologic signs are not common and those who recover from disease are largely free of long-lasting neurologic sequelae. On pathologic analysis of postmortem brains, perivenular demyelination is rare, more likely caused by short period of disease rather than an alternative pathogenic process. Congestion and multiple petechial hemorrhages are the principal pathologic findings [60,61].

Before the availability of vaccines, mumps was the most common nonexanthematous disease followed by neurologic complications. Mumps frequently causes acute viral meningitis and mild encephalitis, both with generally good prognoses. In rare fatal cases of mumps encephalitis, perivenular demyelination has been reported in half of the brains. Whether this merely reflects severe inflammation, or is caused by an autoimmune process and hence representing true ADEM is not clearly known. Unlike postmeasles, varicella, and rubella ADEMs, postmumps encephalomyelitis is not readily distinguishable from more common acute viral encephalitis [62,63].

In the postvaccine era, nonspecific flulike upper respiratory tract infections have been more frequently linked etiologically to ADEMs. Benign postinfectious encephalitides have been reported in influenza virus A and B epidemics, but reports linking influenza infections to inflammatory perivenular demyelination are rare [64,65]. Other neurologic complications, however, including toxic encephalopathies, myelitis, and polyradiculoneuropathies have been more reliably associated with influenza infections [66]. Acute necrotizing encephalopathy is another neurologic complication of influenza A virus infection, which is frequently reported from East Asia but is rare in western countries [67]. Nonspecific upper respiratory tract infections have been associated more commonly with the most severe form of disease (AHLE). Interestingly, this association has existed even before the advent of vaccines for measles, varicella, and rubella, indicating the participation of a different group of viruses in AHLE induction, as compared with classical ADEM [28,50]. ADEM has also been reported in association with hepatitis A [68–70], hepatitis C [71], and Epstein-Barr virus infections [72]. There are few, but pathogenically interesting, reports of ADEM as a manifestation of primary HIV infection [73,74].

Bacterial etiologies

Streptococcal infections of childhood are more likely to cause neurologic complications including Sydenham's chorea and autoimmune neuropsychiatric symptoms; however, ADEM has also been reported after infection with group A β-hemolytic streptococci [75]. The disease phenotype in reported cases was shown to be a novel extrapyramidal movement disorder and other ADEM clinical features [75].

ADEM has also been reported after *Legionella pneumophila* infections [76], typhoid fever, and leptospirosis [77]. ADEM has been reported as a neurologic syndrome after malaria infection, and distinguishable from cerebral malaria. The latter syndrome occurs during parasitemia, whereas a negative blood smear together with convulsions, confusion, tremor, and ataxia suggests ADEM [78].

Nonmicrobial etiologies

Encephalomyelitis following noninfectious events, including autologous stem cell transplantation [79], autoimmune hemolytic anemia [80], and systemic lupus erythematosus [81], has also been described. Although the inherent immune alterations in any of these disease processes, perhaps in the presence of subclinical infections, might set the stage for development of an autoimmune demyelinating response, clear pathologic association with autoimmune brain disease has not been shown for any of these entities.

Encephalomyelitis following antiviral therapies

Postvaccinal acute disseminated encephalomyelitis

Vaccines produced in CNS tissue pose a higher risk of postvaccinal encephalomyelitis. With the introduction of nonneural human diploid cell vaccines for rabies, ADEM induced by rabies neural vaccine (Semple form) is now only of historical interest [82–84]. Vaccines to Japanese encephalitis virus prepared from mouse brain–derived virus, however, is still the principal form of vaccine used for this mosquito-borne encephalitis that occurs throughout East Asia and Australia. Neurologic disorders, sometimes with clear-cut diagnosis of encephalomyelitis, have been described after Japanese encephalitis virus vaccinations [85].

The occurrence of postvaccinal encephalomyelitis following vaccination with live attenuated measles vaccine is not well documented. Some children develop fever, rash, and conjunctivitis in the second week after immunization. This is rarely followed by neurologic manifestations of encephalomyelitis. Indeed, the reported incidence is one to two per million doses, which is not above the background incidence. Nevertheless, there is a much higher incidence of encephalomyelitis following measles infection (20–30 per million infections), which can be prevented by vaccination [86]. Live attenuated varicella vaccine might cause a mild case of chickenpox, which can be followed by cerebellar ataxia. This resolves completely, however, as is the case for natural varicella infection [86]. Rubella vaccine has not been implicated in encephalomyelitis, although transient paresthesias and mild signs of neuropathy have been reported 1 to 3 weeks after vaccination. Encephalomyelitis associated with tetanus toxoid [87], oral polio [88], influenza [42,89], and hepatitis B recombinant vaccines [90] have also been described [86] (see also [91] for adverse events associated with childhood vaccines). The pathogenic process leading to development of postvaccinal encephalomyelitis is generally believed to be the same as the virus-associated ADEM (ie, molecular mimicry or altered immunoregulation). In the case of oral polio–associated paralytic events, however, it might be simply recapitulating poliomyelitis disease course (see Box 1).

Central nervous system involvement in immune reconstitution inflammatory syndrome

Paradoxical deterioration of clinical status following highly active antiretroviral therapy (HAART) in HIV-AIDS patients is an increasingly recognized clinical entity [92]. This phenomenon has been ascribed to partial restoration of immune function following HAART with ensuing infiltration of different organs by reactivated immunocompetent cells; hence the nomination "immune reconstitution inflammatory syndrome" (IRIS). IRIS has also been recognized in the context of other infectious diseases including tuberculosis, where commencement of therapy leads to a transient deterioration of clinical status. With the increasingly widespread availability of HAART and its impact on restitution of immune status, however, reports of HAART-associated IRIS now constitute the bulk of reported cases.

Diagnosis of HAART-induced IRIS usually relies on the occurrence of a clinical event following the initiation of therapy with concurrent improvement in host immune status, as reflected by diminished viral loads. Presence of an external pathogen against which the exuberant immune-inflammatory response is directed is usually required, although IRIS can also be directed toward innate antigens. Infection with *Mycobacterium avium* complex, *Cryptococcus neoformans*, JC virus, and cytomegalovirus are the more closely associated microbial etiologies [92–94]. Exacerbation of rheumatoid arthritis and systemic lupus erythematosus are two examples of IRIS-mediated reaction against self-antigens. As expected from the underlying immunopathogenesis, clinical IRIS might represent as an anatomically compartmentalized phenomenon (ie, where the microbial or self-antigen is already present). Lower $CD4^+$ T cell counts and higher plasma viral load at the time of HAART initiation together with younger age at the onset of HAART have been reported as risk factors for IRIS development [95–97].

Reports of reactivated cryptococcal meningitis after the initiation of HAART were among the first examples that raised the possibility of CNS involvement in IRIS. Other opportunistic infections of CNS were later reported, however, to be more likely to put the CNS parenchyma at risk of exaggerated IRIS immune responses [98,99].

Progressive multifocal leukoencephalopathy is a demyelinating disease of the CNS, caused by the JC virus, a human polyomavirus infecting and replicating in human glial cells. Before the advent of HAART, there were no effective therapies for progressive multifocal leukoencephalopathy. Some progressive multifocal leukoencephalopathy patients' clinical status deteriorates after the start of HAART, however, a phenomenon attributed to the restoration of immune function (ie, IRIS) [94,100]. Indeed, NeuroIRIS occurs in the context of other CNS disorders associated with HIV infection including CNS tuberculosis, cryptococcal meningitis, and HIV-associated dementia. Imaging of NeuroIRIS shows contrast enhancement in all cases of clinical deterioration. Neuropathologic analyses in these cases have shown perivascular and intraparenchymal infiltration of T lymphocytes [101] and severe demyelinating lesions [98]. In fatal cases, the inflammatory infiltrates are chiefly composed of $CD8^+$ cytotoxic T lymphocytes, whereas in less severe cases $CD4^+$ cells with macrophage activation have been more evident [98]. NeuroIRIS has also been reported in conjunction with cytomegalovirus infection of brain [93]. Although chiefly described in association with opportunistic infections, few cases of NeuroIRIS have been reported as acute deterioration of HIV-associated dementia following the initiation of HAART [102,103]. Perivascular and intraparenchymal infiltration of inflammatory cells, chiefly $CD8^+$ cytotoxic T lymphocytes, have been described in neuropathologic examinations in these cases of NeuroIRIS [102,103].

Although supportive care is usually all that is needed in nonnervous system IRIS, immunomodulatory treatment including short periods of therapy with corticosteroids has been suggested to help resolve NeuroIRIS [99], which usually has a good clinical outcome unless the underlying clinical disorder (ie, progressive multifocal leukoencephalopathy) continues to progress.

Acute disseminated encephalomyelitis pathogenesis

Multiple observations have convincingly demonstrated the presence of pathogenic correlates between experimentally induced autoimmune disorders in the nervous system and post–rabies vaccine encephalomyelitis. The missing link with regard to other virus-induced ADEMs, however, is that patients have not been inoculated with myelin components. Instead, they are usually recovering from exanthematous or nonspecific upper respiratory tract viral infections. Although immune-activation is a major contributor to the pathogenesis, initial events induced by the infectious agent might be more subtle and alternative nonimmune mechanisms are not unlikely to contribute. To investigate the chain of events leading from an infection to CNS inflammation more precisely, several animal models of acute and chronic virus-induced demyelination have been extensively studied. Among murine coronaviruses, the JHM strain of murine hepatitis virus represents an interesting example. Murine hepatitis virus induces acute inflammatory lesions in the brains and spinal cords of mice with concomitant demyelination [104]. Although direct infection and lysis of oligodendrocytes takes place in murine hepatitis virus infection, the pathogenic process seems to involve immune-mediated mechanisms. Immunodeficient SCID mice do not develop demyelination after murine hepatitis virus infection [105] and immunosuppression with irradiation also inhibits demyelination [106]. Theiler's murine encephalomyelitis virus infection in mice is another example of infection-related demyelination in rodents. Like murine hepatitis virus, direct Theiler's murine encephalomyelitis virus infection and apoptosis of oligodendrocytes have been described in the spinal cord in cases of chronic infection. Nevertheless, depletion of macrophages but not T lymphocytes diminishes demyelination [107]. Involvement of different humoral or cellular arms of the immune system and their interplay with direct viral effects on the CNS has also been shown for murine rhabdoviruses [108] and togaviruses [109] and mammalian lentiviruses [110–112].

In the face of extensive observations and the evidence derived from vaccine and animal studies, there remain substantial gaps in the understanding of the exact disease mechanisms underlying ADEM. Here are categorized mechanisms proposed for ADEM that are most relevant to human pathogens.

Nonimmune mechanisms

Most pathogens that are associated with human ADEM have not been shown to cause myelin damage by direct infection and injury of myelin-producing cells (ie, oligodendrocytes in CNS). Myelin damage by viral products has been proposed, however, for some of the viruses. An interesting example is the susceptibility of myelin basic protein to vaccinia virus core protein kinase [113]. Incorporation of viral proteins into the myelin membranes can also alter membrane biology and function.

Immune-mediated mechanisms

Molecular mimicry is one of the proposed mechanisms by which pathogens might lead to autoimmune responses. If self- and non–self-pathogen-derived antigens share the same epitopes, presentation of the epitope to the immune system with concomitant activation of a primary innate immune-mediated inflammatory reaction might lead to activation of self-reactive lymphocytes, with subsequent infiltration of the target organ. Although subject to thymic negative selection, some self-reactive lymphocytes including lymphocytes reactive to different components of myelin still persist in adult immune system [114]. Sequence similarity searches to find common linear epitopes between different ADEM-inducing pathogens and myelin basic protein have yielded some results, including a sequence similarity between myelin basic protein and hepatitis B nonstructural polymerase [115]. Epitope sharing is more likely to be in tertiary structures of antigens, however, which necessitates three-dimensional structural homology searches using currently available software and algorithms.

Even in the absence of epitopes common with self-antigens, pathogens can cause autoimmunity by perturbing the intrinsic balance of the immune system, the so-called "immunoregulatory mechanisms." This could take place in the peripheral immune system, leading to a breakdown in the so-called "self-tolerance" to self-antigens. Interestingly, most of the viral infections associated with ADEM cause a transient period of mild to moderate immunosuppression. It remains to be elucidated whether recovery from immunosuppression caused by measles infection might lead to perturbed immunoregulation or perhaps reactivation of self-reactive lymphocytes.

Although involvement of adaptive immune system with generation or activation of self-reactive antigen-specific cells is a major aspect of autoimmune processes, the involvement of innate immune system in disease process is also pivotal to pathogenesis. Infection of cells within the nervous system (eg, monocytoid cells, astrocytes) with subsequent release of factors compromising oligodendrocyte physiology or myelin integrity has been shown for nonhuman lentiviruses and remains a less-explored possibility in the case of rare HIV-associated ADEM [112].

Differential diagnosis

Diagnosis of ADEM is based on the clinical history, neurologic and neuroimaging findings, and CSF analysis with the principal diseases to be considered in the differential diagnosis being MS and acute viral encephalitis. Although ADEM and MS share common pathophysiologic aspects, they are usually distinguishable based on clinical features and disease course. ADEM is typically a monophasic disease of children with a slight male preponderance, whereas MS is usually a chronic relapsing-remitting disease that has its first onset in young adults, with a female predilection. A history of a preceding infection followed by the return of fever, and systemic symptoms, altered level of consciousness, multifocal neurologic dysfunction, seizures, and movement disorders are cardinal features of ADEM, but as a syndrome are not seen together in MS. From a clinical viewpoint, however, ADEM can be difficult to distinguish from the first attack of MS and a clouded sensorium is often the most prominent distinguishing feature (Table 1).

CSF immunologic analysis usually provides clues to the diagnosis. Elevated CSF/serum IgG index and the presence of oligoclonal bands in CSF, which are indicative of endogenous immunoglobulin production in CNS, are seen in MS. Transient appearance of oligoclonal bands have been rarely reported in ADEM, unlike the persistent presence of these bands in MS cases, although pleocytosis is common in both MS and ADEM, albeit greater in ADEM.

Neuroimaging plays a key role in the diagnosis of ADEM from other similar entities. Both CT scanning and MRI have been used to diagnose ADEM; however, MRI is much more informative. CT scan of brain can be normal at the onset and start to reveal abnormalities as late as 5 to 14 days after the start of the disease, showing multifocal subcortical lesions in the white matter [116]. Hemorrhage and edema can be detected in CT scans of AHLE cases [116]. MRI changes occur much earlier, usually

Table 1
Differential clinical and diagnostic features of acute disseminated encephalomyelitis and multiple sclerosis

	Acute disseminated encephalomyelitis	Multiple sclerosis
Predominant age of onset	Childhood (5–8 y)	Young adults
Prodromal febrile illness or vaccination	50%–75% cases	Rare
F:M ratio	Slight male preponderance (F:M 0.8)	Strong female preponderance (F:M 2:1)
Clinical manifestations	Fever, headache, malaise, altered mental status, meningismus Multifocal neurologic dysfunction	Monosymptomatic presentation, sensory symptoms, unilateral optic neuritis, pyramidal signs, ataxia, chronic fatigue
Cerebrospinal fluid analysis	Lymphocytic pleocytosis Elevated protein levels transient if any oligoclonal bands	Intrathecal Ig synthesis (persistent oligoclonal bands, increased cerebrospinal fluid/serum IgG index)
Neuroimaging	Multiple ill-defined lesions on MRI, lacking any orientation Basal ganglia gray matter involvement Less variability in contrast enhancement Rare appearance of new lesions	Single or few sharply demarcated lesions Long axis perpendicular to lesions Different levels of enhancement between lesions Common appearance of new lesions
Recurrence	Chiefly monophasic, recovery over 1–6 months, self-limiting disease even in the absence of therapy	Recurrent, relapsing-remitting disease course

when the neurologic signs and symptoms appear. MRI abnormalities are more likely to be seen on T2-weighted and fluid-attenuated inversion recovery images. ADEM MRI typically shows multiple, large lesions in subcortical and central white matter in cerebral hemispheres, cerebellum, brainstem, and spinal cord [23]. Lesions might involve gray matter of basal ganglia and the junction of gray-white matter in cerebral hemispheres [19,117]. The margins of ADEM lesions are usually blurry, unlike the well-defined sharply demarcated lesions of MS. ADEM lesions are more amorphous than MS plaques and they lack the vertical orientation to the midline axis of brain, which is frequently seen in MS plaques. Gadolinium enhancement in T1-weighted images has been reported in 30% to 100% of patients in different studies, likely reflecting the stage and severity of inflammation. Unlike MS, different lesions in ADEM appear at the same time, hence less variability in contrast enhancement between lesions [118]. Complete resolution of lesions after treatment has been reported in 37% to 75% of patients and partial resolution in 25% to 53% of cases [23,119]. More advanced imaging techniques have also been used in cases with the suspicion of ADEM. Positive emission tomography scanning in one study has shown decreased global and bilateral cerebral metabolism and reduced cerebral blood flow [120]. Reduced blood flow and metabolism has been linked to ADEM severity and clinical course in this study [120].Tc-99m single photon emission CT shows areas of hypoperfusion, whereas MRI has been showing more limited lesions in the same cases [121]. SPECT with acetazolamide has shown longer-lasting lesions compared with MRI [122]. It is likely that longer-lasting cerebral circulatory impairment evident in SPECT might underlie the neurocognitive deficits observed after the resolution of neurologic abnormalities and MRI lesions [122].

Specific neurologic syndromes including optic neuritis, demyelinative transverse myelitis, and Devic disease may occur as manifestations of either MS or ADEM; hence, both disorders should be considered and assessed for as the underlying disease entity.

ADEM should be distinguished from acute viral encephalitis, acute noninfectious encephalitis, and toxic encephalopathies [123]. Acute viral encephalitis occurs as a result of direct infection of brain parenchyma by an infectious agent and can occur at any age. It generally happens as part of a systemic infectious disease and signs and symptoms of other organs' involvement might be present. The clinical profile is variable, depending on the infectious agent. A common clinical feature of viral encephalitis is of an abrupt onset of a febrile disease, accompanied by headache, altered consciousness, and cognitive and behavioral disturbances [123]. Herpes simplex virus is the commonest agent for acute sporadic viral encephalitis. Other common viral agents are varicella-zoster virus, measles, mumps, and enteroviruses, hence the same causative agents as ADEM [123,124]. Having common clinical features and etiologic agents with ADEM, the diagnosis is not always straightforward. Younger age of onset, history of prodromal infection or vaccination, focal neurologic signs including visual loss, spinal cord symptoms, and involvement of peripheral nerves point to ADEM. If not accompanied by meningitis, signs of meningeal irritation are less common in acute viral encephalitis. CSF analysis of ADEM usually shows features similar to that of acute viral encephalitis (ie, lymphocytic pleocytosis, elevated protein levels, normal glucose, with negative cultures). Unlike viral encephalitis, however, virus culture or PCR detection of pathogens shows negative results. Red blood cells appear in CSF most commonly in the cases of AHLE and herpes encephalitis [123–125]. MRI in acute viral encephalitis can identify characteristic changes caused by specific pathogens in brain (eg, frontotemporal changes in herpes simplex encephalitis or striatal-thalamic lesions in Japanese or West Nile encephalitis). T2-weighted images of acute viral encephalitis usually show one or more diffuse areas of increased intensity affecting the cortical gray matter and subjacent white matter [123], although the gray matter of basal ganglia or brainstem might be involved but to a lesser extent [123].

Treatment

Vaccination against viral etiologies has significantly decreased the incidence of viral cases of ADEM, and the risk imposed by vaccines themselves is significantly less than natural infections. There is a lack of placebo-controlled double-blind studies to evaluate different treatment options for ADEM. Nevertheless, high-dose glucocorticoid therapy has been the most widely used treatment for ADEM. Adrenocorticotropic hormone and prednisone have been successfully used in the past [126,127]. Currently, intravenous methylprednisolone (10–30 mg/kg/d) or dexamethasone (1 mg/kg) for 3 to 5 days are the most commonly used steroids in pediatric cases [18,23]. Intravenous glucocorticoid therapy needs to be tapered with administration of oral steroids for 4 to 6 weeks, following intravenous therapy. Shorter tapering might increase the risk of relapses [128]. In the first few days of disease, intravenous methylprednisolone might be combined with intravenous acyclovir if there is still a suspicion of acute viral encephalitis. Intravenous immunoglobulin has been used alone or in combination with glucocorticoids with a dose of 1 to 2 g/kg single-dose or divided over 3 to 5 days [129,130]. Although reports of success exist for intravenous immunoglobulin treatment of autoimmune disorders, intravenous immunoglobulin therapy for ADEM remains at the level of case-reports. One study recommends high-dose intravenous immunoglobulin, given separately or in combination with high-dose methylprednisolone, in cases of severe debilitating pediatric-onset acute encephalomyelitis [131]. The mechanism of intravenous immunoglobulin action is not well-known; it might have immunomodulatory effects through binding to pathogenic antibodies or myelin basic protein–mimicking antigens, thereby inhibiting the generation or activation of myelin-reactive T cells.

Plasma exchange has been reported to improve the clinical status in some patients who fail to respond to high-dose intravenous glucocorticoids. This is perhaps through removal of antibodies that contribute to demyelination or by partly modifying the cytokine milieu in peripheral immune system, where self-reactive lymphocytes are being generated or reactivated. Plasmapheresis is usually regarded as a last resort, however, and there are no studies assessing its efficacy if used early in the course of disease [132].

Prognosis

The long-term prognosis of ADEM varies with etiology, with postmeasles cases having a high mortality rate and a high rate of neurologic sequelae in survivors. The prognosis of nonmeasles cases is generally favorable. Many studies have reported a full recovery in 50% to 75% of patients, in a period of 1 to 6 months after the appearance of symptoms and signs [22,23]. The most common neurologic sequelae following ADEM are focal motor deficits. This could range from mild ataxia to hemiparesis. It is generally believed that duration and severity of inflammation in brain with the extent of neuronal and axonal damage are determinants of clinical outcome. Sudden onset, severe neurologic symptoms, and unresponsiveness to glucocorticoids have been considered poor prognostic factors.

Gradual recovery from ADEM over a period of a few weeks has been reported with 50% to 70% of patients completely recovering without neurologic sequelae. Some studies have described a correlation between untreated ADEM outcome and the type of precedent infection. Cases diagnosed with postvaricella and postrubella ADEM showed 54% and 43% full recovery, respectively, whereas approximately 70% of ADEM cases following nonspecific infections fully recovered. Muliphasic ADEM has been generally associated with longer recovery periods [119].

Although resolution of neurologic symptoms and signs together with normal imaging profile have been considered indicative of full recovery, minor neuropsychologic abnormalities have been reported in pediatric cases years after the disease. One study evaluated intellectual, educational, and social functioning of children who had recovered from ADEM. Impairments in cognitive and social domains with a higher incidence of severe behavioral and emotional problems were found in children who had experienced ADEM below 5 years of age [133]. Of interest, younger age of onset in pediatric cases of MS has also been closely associated with subsequent neurocognitive deficits. This evidence underscores the sensitivity of the developing brain to transient inflammatory demyelinating events.

References

[1] Johnson RT. The pathogenesis of acute viral encephalitis and postinfectious encephalomyelitis. J Infect Dis 1987;155(3):359–64.

[2] Hartung HP, Grossman RI. ADEM: distinct disease or part of the MS spectrum? Neurology 2001;56(10):1257–60.
[3] Johnson RT, Griffin DE. Postinfectious encephalomyelitis. In: Infections of the nervous system. London: Butterworth; 1987. p. 209–26.
[4] Gurvich EB, Vilesova IS. Vaccinia virus in postvaccinal encephalitis. Acta Virol 1983;27(2): 154–9.
[5] Rivers T, Schwentker F. Encephalomyelitis accompanied by myelin destruction experimentally produced in monkeys. J Exp Med 1935;61:689–702.
[6] Johnson RT. Postinfectious demyelinating diseases. In: Johnson RT, editor. Viral infections of the nervous system. 2nd ed. Philadelphia: Raven-Lippincott; p. 181–210.
[7] Sospedra M, Martin R. Immunology of multiple sclerosis. Annu Rev Immunol 2005;23: 683–747.
[8] Steinman L, Zamvil SS. How to successfully apply animal studies in experimental allergic encephalomyelitis to research on multiple sclerosis. Ann Neurol 2006;60(1):12–21.
[9] Steinman L, Zamvil SS. Virtues and pitfalls of EAE for the development of therapies for multiple sclerosis. Trends Immunol 2005;26(11):565–71.
[10] Whitacre CC, Paterson PY. Transfer of experimental allergic encephalomyelitis in Lewis rats using supernates of incubated sensitized lymph node cells. J Exp Med 1977;145(5):1405–10.
[11] Pender MP, Sears TA. Vulnerability of the dorsal root ganglion in experimental allergic encephalomyelitis. Clin Exp Neurol 1985;21:211–23.
[12] Pender MP, Sears TA. The pathophysiology of acute experimental allergic encephalomyelitis in the rabbit. Brain 1984;107(Pt 3):699–726.
[13] Waksman BH, Adams RD. Allergic neuritis: an experimental disease of rabbits induced by the injection of peripheral nervous tissue and adjuvants. J Exp Med 1955;102(2):213–36.
[14] Dreesen DW. A global review of rabies vaccines for human use. Vaccine 1997;15(Suppl): S2–6.
[15] Cabrera J, Griffin DE, Johnson RT. Unusual features of the Guillain-Barré syndrome after rabies vaccine prepared in suckling mouse brain. J Neurol Sci 1987;81(2–3):239–45.
[16] Leake JA, Albani S, Kao AS, et al. Acute disseminated encephalomyelitis in childhood: epidemiologic, clinical and laboratory features. Pediatr Infect Dis J 2004;23(8):756–64.
[17] Menge T, Hemmer B, Nessler S, et al. Acute disseminated encephalomyelitis: an update. Arch Neurol 2005;62(11):1673–80.
[18] Gupte G, Stonehouse M, Wassmer E, et al. Acute disseminated encephalomyelitis: a review of 18 cases in childhood. J Paediatr Child Health 2003;39(5):336–42.
[19] Tenembaum S, Chamoles N, Fejerman N. Acute disseminated encephalomyelitis: a long-term follow-up study of 84 pediatric patients. Neurology 2002;59(8):1224–31.
[20] Schwarz S, Mohr A, Knauth M, et al. Acute disseminated encephalomyelitis: a follow-up study of 40 adult patients. Neurology 2001;56(10):1313–8.
[21] Murthy SN, Faden HS, Cohen ME, et al. Acute disseminated encephalomyelitis in children. Pediatrics 2002;110(2 Pt 1):e21–7.
[22] Dale RC, de Sousa C, Chong WK, et al. Acute disseminated encephalomyelitis, multiphasic disseminated encephalomyelitis and multiple sclerosis in children. Brain 2000;123(Pt 12): 2407–22.
[23] Hynson JL, Kornberg AJ, Coleman LT, et al. Clinical and neuroradiologic features of acute disseminated encephalomyelitis in children. Neurology 2001;56(10):1308–12.
[24] Barontini F, Di Lollo S, Toscani L. Acute hemorrhagic leukoencephalopathy: clinical and CT diagnosis and histopathological confirmation in an elderly patient. Ital J Neurol Sci 1984;5(2):215–8.
[25] Rosman NP, Gottlieb SM, Bernstein CA. Acute hemorrhagic leukoencephalitis: recovery and reversal of magnetic resonance imaging findings in a child. J Child Neurol 1997;12(7): 448–54.
[26] Meilof JF, Hijdra A, Vermeulen M. Successful recovery after high-dose intravenous methylprednisolone in acute hemorrhagic leukoencephalitis. J Neurol 2001;248(10):898–9.

[27] Payne ET, Rutka JT, Ho TK, et al. Treatment leading to dramatic recovery in acute hemorrhagic leukoencephalitis. J Child Neurol 2007;22(1):109–13.
[28] Seales D, Greer M. Acute hemorrhagic leukoencephalitis: a successful recovery. Arch Neurol 1991;48(10):1086–8.
[29] Kerr DA, Ayetey H. Immunopathogenesis of acute transverse myelitis. Curr Opin Neurol 2002;15(3):339–47.
[30] Krishnan C, Kaplin AI, Deshpande DM, et al. Transverse myelitis: pathogenesis, diagnosis and treatment. Front Biosci 2004;9:1483–99.
[31] Berman M, Feldman S, Alter M, et al. Acute transverse myelitis: incidence and etiologic considerations. Neurology 1981;31(8):966–71.
[32] Jeffery DR, Mandler RN, Davis LE. Transverse myelitis: retrospective analysis of 33 cases, with differentiation of cases associated with multiple sclerosis and parainfectious events. Arch Neurol 1993;50(5):532–5.
[33] Sakakibara R, Hattori T, Yasuda K, et al. Micturition disturbance in acute transverse myelitis. Spinal Cord 1996;34(8):481–5.
[34] Krishnan C, Kaplin AI, Pardo CA, et al. Demyelinating disorders: update on transverse myelitis. Curr Neurol Neurosci Rep 2006;6(3):236–43.
[35] Krishnan AV, Halmagyi GM. Acute transverse myelitis in SLE. Neurology 2004;62(11): 2087.
[36] Baum J, Solomon M, Alba A. Sarcoidosis as a cause of transverse myelitis: case report. Paraplegia 1981;19(3):167–9.
[37] Pittock SJ, Lucchinetti CF. Inflammatory transverse myelitis: evolving concepts. Curr Opin Neurol 2006;19(4):362–8.
[38] Christensen PB, Wermuth L, Hinge HH, et al. Clinical course and long-term prognosis of acute transverse myelopathy. Acta Neurol Scand 1990;81(5):431–5.
[39] Pidcock FS, Krishnan C, Crawford TO, et al. Acute transverse myelitis in childhood: center-based analysis of 47 cases. Neurology 2007;68(18):1474–80.
[40] Lim S, Park SM, Choi HS, et al. Transverse myelitis after measles and rubella vaccination. J Paediatr Child Health 2004;40(9–10):583–4.
[41] Fonseca LF, Noce TR, Teixeira ML, et al. Early-onset acute transverse myelitis following hepatitis B vaccination and respiratory infection: case report. Arq Neuropsiquiatr 2003; 61(2A):265–8.
[42] Nakamura N, Nokura K, Zettsu T, et al. Neurologic complications associated with influenza vaccination: two adult cases. Intern Med 2003;42(2):191–4.
[43] Scott TF, Bhagavatula K, Snyder PJ, et al. Transverse myelitis: comparison with spinal cord presentations of multiple sclerosis. Neurology 1998;50(2):429–33.
[44] Harzheim M, Schlegel U, Urbach H, et al. Discriminatory features of acute transverse myelitis: a retrospective analysis of 45 patients. J Neurol Sci 2004;217(2):217–23.
[45] Kalita J, Misra UK, Mandal SK. Prognostic predictors of acute transverse myelitis. Acta Neurol Scand 1998;98(1):60–3.
[46] Transverse Myelitis Consortium Working Group. Proposed diagnostic criteria and nosology of acute transverse myelitis. Neurology 2002;59(4):499–505.
[47] Mikaeloff Y, Adamsbaum C, Husson B, et al. MRI prognostic factors for relapse after acute CNS inflammatory demyelination in childhood. Brain 2004;127(Pt 9):1942–7.
[48] Roos K, Miravalle A. Postinfectious encephalomyelitis. In: Infections of the central nervous system. 3rd edition. p. 323–9.
[49] Kuperan S, Ostrow P, Landi MK, et al. Acute hemorrhagic leukoencephalitis vs ADEM: FLAIR MRI and neuropathology findings. Neurology 2003;60(4):721–2.
[50] Geerts Y, Dehaene I, Lammens M. Acute hemorrhagic leukoencephalitis. Acta Neurol Belg 1991;91(4):201–11.
[51] Ayala L, Barber DB, Lomba MR, et al. Intramedullary sarcoidosis presenting as incomplete paraplegia: case report and literature review. J Spinal Cord Med 2000;23(2):96–9.

[52] Nakano I, Mannen T, Mizutani T, et al. Peripheral white matter lesions of the spinal cord with changes in small arachnoid arteries in systemic lupus erythematosus. Clin Neuropathol 1989;8(2):102–8.
[53] Griffin DE. Immune responses during measles virus infection. Curr Top Microbiol Immunol 1995;191:117–34.
[54] Gibbons JL, Miller HG, Stanton JB. Para-infectious encephalomyelitis and related syndromes: a critical review of the neurological complications of certain specific fevers. Q J Med 1956;25(100):427–505.
[55] Johnson RT, Griffin DE, Hirsch RL, et al. Measles encephalomyelitis: clinical and immunologic studies. N Engl J Med 1984;310(3):137–41.
[56] Tyler HR. Neurological complications of rubeola (measles). Medicine (Baltimore) 1957; 36(2):147–67.
[57] Connolly AM, Dodson WE, Prensky AL, et al. Course and outcome of acute cerebellar ataxia. Ann Neurol 1994;35(6):673–9.
[58] Takashima S, Becker LE. Neuropathology of fatal varicella. Arch Pathol Lab Med 1979; 103(5):209–13.
[59] Johnson R, Milbourn PE. Central nervous system manifestations of chickenpox. Can Med Assoc J 1970;102(8):831–4.
[60] Connolly JH, Hutchinson WM, Allen IV, et al. Carotid artery thrombosis, encephalitis, myelitis and optic neuritis associated with rubella virus infections. Brain 1975;98(4):583–94.
[61] Frey TK. Neurological aspects of rubella virus infection. Intervirology 1997;40(2–3): 167–75.
[62] Sonmez FM, Odemis E, Ahmetoglu A, et al. Brainstem encephalitis and acute disseminated encephalomyelitis following mumps. Pediatr Neurol 2004;30(2):132–4.
[63] Unal A, Emre U, Atasoy HT, et al. Encephalomyelitis following mumps. Spinal Cord 2005; 43(7):441–4.
[64] Hawkins SA, Lyttle JA, Connolly JH. Two cases of influenza B encephalitis. J Neurol Neurosurg Psychiatr 1987;50(9):1236–7.
[65] Sulkava R, Rissanen A, Pyhala R. Post-influenzal encephalitis during the influenza A outbreak in 1979/1980. J Neurol Neurosurg Psychiatr 1981;44(2):161–3.
[66] Hayase Y, Tobita K. Influenza virus and neurological diseases. Psychiatry Clin Neurosci 1997;51(4):181–4.
[67] Olgar S, Ertugrul T, Nisli K, et al. Influenza A-associated acute necrotizing encephalopathy. Neuropediatrics 2006;37(3):166–8.
[68] Tan H, Kilicaslan B, Onbas O, et al. Acute disseminated encephalomyelitis following hepatitis A virus infection. Pediatr Neurol 2004;30(3):207–9.
[69] Quaranta L, Batocchi AP, Sabatelli M, et al. Monophasic demyelinating disease of the central nervous system associated with hepatitis A infection. J Neurol 2006;253(7):944–5.
[70] Alehan FK, Kahveci S, Uslu Y, et al. Acute disseminated encephalomyelitis associated with hepatitis A virus infection. Ann Trop Paediatr 2004;24(2):141–4.
[71] Sacconi S, Salviati L, Merelli E. Acute disseminated encephalomyelitis associated with hepatitis C virus infection. Arch Neurol 2001;58(10):1679–81.
[72] Shoji H, Kusuhara T, Honda Y, et al. Relapsing acute disseminated encephalomyelitis associated with chronic Epstein-Barr virus infection: MRI findings. Neuroradiology 1992; 34(4):340–2.
[73] Allen SH, Malik O, Lipman MC, et al. Acute demyelinating encephalomyelitis (ADEM) in a patient with HIV infection. J Infect 2002;45(1):62–4.
[74] Narciso P, Galgani S, Del Grosso B, et al. Acute disseminated encephalomyelitis as manifestation of primary HIV infection. Neurology 2001;57(8):1493–6.
[75] Dale RC, Church AJ, Cardoso F, et al. Poststreptococcal acute disseminated encephalomyelitis with basal ganglia involvement and auto-reactive antibasal ganglia antibodies. Ann Neurol 2001;50(5):588–95.

[76] Spieker S, Petersen D, Rolfs A, et al. Acute disseminated encephalomyelitis following Pontiac fever. Eur Neurol 1998;40(3):169–72.
[77] Alonso-Valle H, Munoz R, Hernandez JL, et al. Acute disseminated encephalomyelitis following *Leptospira* infection. Eur Neurol 2001;46(2):104–5.
[78] Koibuchi T, Nakamura T, Miura T, et al. Acute disseminated encephalomyelitis following *Plasmodium vivax* malaria. J Infect Chemother 2003;9(3):254–6.
[79] Re A, Giachetti R. Acute disseminated encephalomyelitis (ADEM) after autologous peripheral blood stem cell transplant for non-Hodgkin's lymphoma. Bone Marrow Transplant 1999;24(12):1351–4.
[80] Jaing TH, Lin KL, Chiu CH, et al. Acute disseminated encephalomyelitis in autoimmune hemolytic anemia. Pediatr Neurol 2001;24(4):303–5.
[81] Dubreuil F, Cabre P, Smadja D, et al. Acute disseminated encephalomyelitis preceding cutaneous lupus). Rev Med Interne 1998;19(2):128–30 [in French].
[82] Hemachudha T, Griffin DE, Johnson RT, et al. Immunologic studies of patients with chronic encephalitis induced by post-exposure Semple rabies vaccine. Neurology 1988; 38(1):42–4.
[83] Swaddiwuthipong W, Weniger BG, Wattanasri S, et al. A high rate of neurological complications following Semple anti-rabies vaccine. Trans R Soc Trop Med Hyg 1988;82(3): 472–5.
[84] Javier RS, Kunishita T, Koike F, et al. Semple rabies vaccine: presence of myelin basic protein and proteolipid protein and its activity in experimental allergic encephalomyelitis. J Neurol Sci 1989;93(2–3):221–30.
[85] Plesner AM, Arlien-Soborg P, Herning M. Neurological complications to vaccination against Japanese encephalitis. Eur J Neurol 1998;5(5):479–85.
[86] Fenichel GM. Neurological complications of immunization. Ann Neurol 1982;12(2): 119–28.
[87] Bolukbasi O, Ozmenoglu M. Acute disseminated encephalomyelitis associated with tetanus vaccination. Eur Neurol 1999;41(4):231–2.
[88] Shibazaki K, Murakami T, Kushida R, et al. Acute disseminated encephalomyelitis associated with oral polio vaccine. Intern Med 2006;45(20):1143–6.
[89] Saito H, Endo M, Takase S, et al. Acute disseminated encephalomyelitis after influenza vaccination. Arch Neurol 1980;37(9):564–6.
[90] Tourbah A, Gout O, Liblau R, et al. Encephalitis after hepatitis B vaccination: recurrent disseminated encephalitis or MS? Neurology 1999;53(2):396–401.
[91] Stratton KR, Howe CJ, Johnston RB Jr. Adverse events associated with childhood vaccines other than pertussis and rubella. Summary of a report from the Institute of Medicine. JAMA 1994;271(20):1602–5.
[92] Lipman M, Breen R. Immune reconstitution inflammatory syndrome in HIV. Curr Opin Infect Dis 2006;19(1):20–5.
[93] Murdoch DM, Venter WD, Van Rie A, et al. Immune reconstitution inflammatory syndrome (IRIS): review of common infectious manifestations and treatment options. AIDS Res Ther 2007;4:9–18.
[94] Martinez JV, Mazziotti JV, Efron ED, et al. Immune reconstitution inflammatory syndrome associated with PML in AIDS: a treatable disorder. Neurology 2006;67(9):1692–4.
[95] Shelburne SA, Visnegarwala F, Darcourt J, et al. Incidence and risk factors for immune reconstitution inflammatory syndrome during highly active antiretroviral therapy. AIDS 2005;19(4):399–406.
[96] Ratnam I, Chiu C, Kandala NB, et al. Incidence and risk factors for immune reconstitution inflammatory syndrome in an ethnically diverse HIV type 1-infected cohort. Clin Infect Dis 2006;42(3):418–27.
[97] Jevtovic DJ, Salemovic D, Ranin J, et al. The prevalence and risk of immune restoration disease in HIV-infected patients treated with highly active antiretroviral therapy. HIV Med 2005;6(2):140–3.

[98] Gray F, Bazille C, Adle-Biassette H, et al. Central nervous system immune reconstitution disease in acquired immunodeficiency syndrome patients receiving highly active antiretroviral treatment. J Neurovirol 2005;11(Suppl 3):16–22.
[99] Riedel DJ, Pardo CA, McArthur J, et al. Therapy insight: CNS manifestations of HIV-associated immune reconstitution inflammatory syndrome. Nat Clin Pract Neurol 2006; 2(10):557–65.
[100] Du Pasquier RA, Koralnik IJ. Inflammatory reaction in progressive multifocal leukoencephalopathy: harmful or beneficial? J Neurovirol 2003;9(Suppl 1):25–31.
[101] Vendrely A, Bienvenu B, Gasnault J, et al. Fulminant inflammatory leukoencephalopathy associated with HAART-induced immune restoration in AIDS-related progressive multifocal leukoencephalopathy. Acta Neuropathol (Berl) 2005;109(4):449–55.
[102] Miller RF, Isaacson PG, Hall-Craggs M, et al. Cerebral $CD8^+$ lymphocytosis in HIV-1 infected patients with immune restoration induced by HAART. Acta Neuropathol (Berl) 2004;108(1):17–23.
[103] Langford TD, Letendre SL, Marcotte TD, et al. Severe, demyelinating leukoencephalopathy in AIDS patients on antiretroviral therapy. AIDS 2002;16(7):1019–29.
[104] Weiner LP. Pathogenesis of demyelination induced by a mouse hepatitis. Arch Neurol 1973;28(5):298–303.
[105] Houtman JJ, Hinze HC, Fleming JO. Demyelination induced by murine coronavirus JHM infection of congenitally immunodeficient mice. Adv Exp Med Biol 1995;380:159–63.
[106] Wang FI, Stohlman SA, Fleming JO. Demyelination induced by murine hepatitis virus JHM strain (MHV-4) is immunologically mediated. J Neuroimmunol 1990;30(1):31–41.
[107] Rossi CP, Delcroix M, Huitinga I, et al. Role of macrophages during Theiler's virus infection. J Virol 1997;71(4):3336–40.
[108] Dal Canto MC, Rabinowitz SG. Murine central nervous system infection by a viral temperature-sensitive mutant: a subacute disease leading to demyelination. Am J Pathol 1981; 102(3):412–26.
[109] Dal Canto MC, Rabinowitz SG. Central nervous system demyelination in Venezuelan equine encephalomyelitis infection. J Neurol Sci 1981;49(3):397–418.
[110] Norman S, Smith MC. Caprine arthritis-encephalitis: review of the neurologic form in 30 cases. J Am Vet Med Assoc 1983;182(12):1342–5.
[111] Chebloune Y, Karr BM, Raghavan R, et al. Neuroinvasion by ovine lentivirus in infected sheep mediated by inflammatory cells associated with experimental allergic encephalomyelitis. J Neurovirol 1998;4(1):38–48.
[112] Georgsson G. Neuropathologic aspects of lentiviral infections. Ann N Y Acad Sci 1994; 724:50–67.
[113] Steck AJ, Siegrist P, Herschkowitz N, et al. Phosphorylation of myelin basic protein by vaccinia virus cores. Nature 1976;263(5576):436–8.
[114] Steinman L, Martin R, Bernard C, et al. Multiple sclerosis: deeper understanding of its pathogenesis reveals new targets for therapy. Annu Rev Neurosci 2002;25:491–505.
[115] Fujinami RS, Oldstone MB. Amino acid homology between the encephalitogenic site of myelin basic protein and virus: mechanism for autoimmunity. Science 1985;230(4729): 1043–5.
[116] Mihai C, Jubelt B. Post-infectious encephalomyelitis. Curr Neurol Neurosci Rep 2005;5(6): 440–5.
[117] Baum PA, Barkovich AJ, Koch TK, et al. Deep gray matter involvement in children with acute disseminated encephalomyelitis. AJNR Am J Neuroradiol 1994;15(7):1275–83.
[118] Rust RS. Multiple sclerosis, acute disseminated encephalomyelitis, and related conditions. Semin Pediatr Neurol 2000;7(2):66–90.
[119] Tenembaum S, Chitnis T, Ness J, et al. Acute disseminated encephalomyelitis. Neurology 2007;68(16 Suppl 2):S23–36.
[120] Tabata K, Shishido F, Uemura K, et al. Positron emission tomography in acute disseminated encephalomyelitis: a case report). Kaku Igaku 1990;27(3):261–5 [in Japanese].

[121] San Pedro EC, Mountz JM, Liu HG, et al. Postinfectious cerebellitis: clinical significance of Tc-99m HMPAO brain SPECT compared with MRI. Clin Nucl Med 1998;23(4):212–6.
[122] Okamoto M, Ashida KI, Imaizumi M. Hypoperfusion following encephalitis: SPECT with acetazolamide. Eur J Neurol 2001;8(5):471–4.
[123] Kennedy PG. Viral encephalitis: causes, differential diagnosis, and management. J Neurol Neurosurg Psychiatr 2004;75(Suppl 1):i10–5.
[124] Chaudhuri A, Kennedy PG. Diagnosis and treatment of viral encephalitis. Postgrad Med J 2002;78(924):575–83.
[125] Kennedy PG. Viral encephalitis. J Neurol 2005;252(3):268–72.
[126] Miller HG. Acute disseminated encephalomyelitis treated with A.C.T.H. Br Med J 1953; 1(4803):177–82.
[127] Miller HG, Gibbons JL. Acute disseminated encephalomyelitis and acute disseminated sclerosis; results of treatment with A.C.T.H. Br Med J 1953;2(4850):1345–8.
[128] Anlar B, Basaran C, Kose G, et al. Acute disseminated encephalomyelitis in children: outcome and prognosis. Neuropediatrics 2003;34(4):194–9.
[129] Straussberg R, Schonfeld T, Weitz R, et al. Improvement of atypical acute disseminated encephalomyelitis with steroids and intravenous immunoglobulins. Pediatr Neurol 2001; 24(2):139–43.
[130] Nishikawa M, Ichiyama T, Hayashi T, et al. Intravenous immunoglobulin therapy in acute disseminated encephalomyelitis. Pediatr Neurol 1999;21(2):583–6.
[131] Shahar E, Andraus J, Savitzki D, et al. Outcome of severe encephalomyelitis in children: effect of high-dose methylprednisolone and immunoglobulins. J Child Neurol 2002; 17(11):810–4.
[132] Lin CH, Jeng JS, Yip PK. Plasmapheresis in acute disseminated encephalomyelitis. J Clin Apheresis 2004;19(3):154–9.
[133] Jacobs RK, Anderson VA, Neale JL, et al. Neuropsychological outcome after acute disseminated encephalomyelitis: impact of age at illness onset. Pediatr Neurol 2004;31(3):191–7.

ELSEVIER
SAUNDERS

Neurol Clin 26 (2008) 781–797

NEUROLOGIC
CLINICS

Treatment of HTLV-I-Associated Myelopathy/Tropical Spastic Paraparesis: Toward Rational Targeted Therapy

Unsong Oh, MD, Steven Jacobson, PhD*

Viral Immunology Section, Neuroimmunology Branch, National Institute of Neurological Diseases and Stroke, 10 Center Drive, Bldg 10 Rm 5C103, Bethesda, MD 20892, USA

Human T-cell lymphotropic virus type I (HTLV-I) is the etiologic agent in a progressive neurologic disease termed HTLV-I–associated myelopathy/tropical spastic paraparesis (HAM/TSP) [1,2]. HAM/TSP is a chronic inflammatory disease of the central nervous system (CNS) characterized by unremitting myelopathic symptoms such as spastic paraparesis, bowel/bladder dysfunction, and lower limb sensory changes. Our understanding of the pathogenesis of HAM/TSP has advanced since its first characterization in the mid 1980s. This review discusses the current and future approaches to the treatment of HAM/TSP within the context of our current understanding of the underlying pathogenic mechanisms.

HTLV-I is the first human oncogenic retrovirus to be identified and isolated [3,4]. In the peripheral blood, HTLV-I preferentially infects CD4+ T cells and to a lesser extent CD8+ T cells in vivo [5,6]. A wider range of vertebrate cells are susceptible to HTLV-I infection in vitro [7], and the localization of HTLV-I RNA in astrocyes suggests a broader tropism that may include resident cells of the CNS [8].

HTLV-I is transmitted via three major routes: (1) transmission from mother to child through breast milk, (2) transmission by sexual contact, and (3) transmission by way of infected blood. In endemic areas, maternal-child transmission via breast milk is a major mode of transmission of HTLV-I, followed by sexual contact. Transmission also occurs through unscreened blood transfusions and the use of contaminated intravenous needles.

This work was supported by the National Institutes of Health intramural fund.

* Corresponding author.

E-mail address: jacobsons@ninds.nih.gov (S. Jacobson).

0733-8619/08/$ - see front matter. Published by Elsevier Inc.
doi:10.1016/j.ncl.2008.03.008

Approximately 20 to 40 million people are infected worldwide. Endemic areas of HTLV-I infection include the Caribbean, southern Japan, Central and South America, the Middle East, Melanesia, and equatorial Africa [9,10]. Most infected individuals remain lifelong asymptomatic carriers. Approximately 0.25% to 3.8% develop HAM/TSP. Another 2% to 3% of infected individuals develop an aggressive mature T-cell malignancy termed adult T-cell leukemia (ATL). In Japan, the lifetime risk of HAM/TSP among the nearly 1 million infected individuals is 0.25% [11]. The lifetime risk of HAM/TSP in the estimated 22,000 infected individuals in England is 3% [12]. Seroprevalence of HTLV-I in US blood donors is 1 per 10,000. A recent study estimates that approximately 266,000 individuals are infected with HTLV-I or -II, and that there are likely more than 3600 people in the United States with unrecognized HAM/TSP [13]. HTLV-I infections have also been associated with other inflammatory disorders including myositis [14,15], peripheral neuropathy [16,17], uveitis [18,19], arthritis [20], Sjogren syndrome [21], and alveolitis [22].

Understanding the pathogenesis of the virus within the host has important implications for therapies directed against the virus itself. The maintenance of HTLV-I may occur through several, non–mutually exclusive, mechanisms. Proliferation of infected cells is likely a major mechanism by which the expansion of the number of HTLV provirus copies occurs. Alternatively, new infection may occur through virion production or through the formation of a virus-induced synapse [23]. Reverse transcriptase activity is required for the latter two mechanisms of viral propagation. The contribution of reverse transcriptase activity to maintaining HTLV-I proviral load in vivo is controversial. Conflicting data from clinical studies with regard to the efficacy of nucleoside analogs to reduce proviral load in HAM/TSP leaves this issue unsettled [24,25].

Among the HTLV-I encoded viral genes, the *Tax* gene encodes a protein that activates viral transcription and also transactivates a large number of human genes [26]. The Tax protein also contains immunodominant epitopes for MHC class I restricted cytotoxic cellular immune responses [27]. Thus, the transactivating Tax protein is not only the target but also modifies the host immune response. The Tax protein and its downstream effects have been implicated in the persistent immune activation that contributes to the pathogenesis of HAM/TSP, and therefore represents an important target for therapy in HAM/TSP [28].

Risk factors for development of HAM/TSP

An elevated HTLV-I proviral load remains the single best marker of symptomatic neurologic disease in individuals infected with HTLV-I. In cross-sectional studies, the HTLV-I proviral load, determined by real-time polymerase chain reaction (PCR), was consistently higher in individuals with HAM/TSP than in asymptomatic carriers [29,30]. In practice, the typically low proviral loads (mean of ~2% of peripheral blood mononuclear cells in the NIH cohort) of asymptomatic carriers tend to be reassuring, but the occasional asymptomatic carrier who demonstrates intermediate range proviral load presents a prognostic challenge. Often, such carriers are family members of individuals with HAM/TSP [29].

A number of host susceptibility genes have been identified that modify the risk of HAM/TSP. Many of these genes play a role in the host immunologic response and support the hypothesis that host immune response to the virus is a critical determinant in the development of HAM/TSP. The presence of HLA-A*02 allele has been reported to be protective, whereas the HLA-DRB1*0101 appears to confer susceptibility [31]. Polymorphisms in the genes for the inflammatory cytokines tumor necrosis factor (TNF)-α and interleukin (IL)-15, and in the chemokine stromal cell derived factor 1 has been reported to confer susceptibility [32].

In practice, the most reassuring data to the individual with incidental finding of positive HTLV-I serology are that the over 90% of HTLV-I infected individuals are asymptomatic carriers and that the lifetime risk of developing HAM/TSP is likely less than 1% [11].

Diagnosis

The diagnosis of HAM/TSP is seldom in doubt in a seropositive individual who presents with a chronic progressive spastic paraparesis in the setting of elevated HTLV proviral load. The onset is typically insidious. When subacute (relatively rapid) progression occurs, it is typically seen in the first 2 to 3 years of symptomatic disease. Most experience slowly progressive disease, but some remain stable for years or rarely improve to milder disability. The initial symptoms are typically back pain, gait disturbance, and bladder/bowel or sexual dysfunction. Sensory symptoms including paresthesias and neuropathic pain of the lower extremities, are often present. Clinical signs include weakness of the legs, spasticity, hyperreflixia, gait atxia, and loss of vibratory sensation. Less common signs and symptoms include cerebellar signs, optic neuritis and atrophy, nystagnus, and depressed ankle reflexes. In general, the burden of disability tends to be high with a majority of symptomatic individuals requiring assistance with ambulation [33].

Diagnosis is often delayed due to slow development of the full constellation of symptoms or misdiagnosis. The World Health Organization (WHO) diagnostic guidelines for HAM/TSP outline the full spectrum of disease without codifying levels of ascertainment [34]. A recent recommendation proposes to refine the WHO diagnostic guidelines by formulating levels of ascertainment as "definite," "probable," and "possible" HAM/TSP, where a patient with "definite" HAM/TSP manifests nonremitting progressive spastic paraparesis, positive serology, and/or detection of proviral HTLV-I DNA and the exclusion of other disorders [35].

The differential diagnosis for HAM/TSP includes primary progressive multiple sclerosis (MS), primary lateral sclerosis, hereditary spastic paraplegias, subacute combined degeneration secondary to vitamin B12 deficiency myelopathy associated with copper deficiency, HIV vacuolar myelopathy, syphilis, and Lyme disease. Some have reported higher prevalence of positive HTLV-I serology in patients with Sjogren syndrome, and the findings of some elements of Sjogren syndrome may not be mutually exclusive with the diagnosis of HAM/TSP [21].

Differentiating HAM/TSP from primary progressive MS is occasionally a diagnostic challenge, since the two are clinically indistinguishable and the mere presence of positive HTLV-I serology does not always lead to neurologic disease. The difficulty is compounded by the fact that it is not unusual to find white matter abnormalities on brain magnetic resonance images (MRI) of patients with HAM/TSP (Fig. 1) [36–38]. Cerebrospinal fluid (CSF) pleocytosis, when present, typically falls within a similar range and oligoclonal bands are present in both. A recent study suggests that a high ratio of CSF to peripheral blood proviral load may distinguish HAM/TSP from individuals with MS [39]. In general, HTLV-I proviral loads measured in the CSF of individuals with HAM/TSP are typically greater than twice their load in the peripheral blood [40], whereas the ratio of CSF to peripheral blood HTLV-I proviral loads are typically lower in asymptomatic carriers [39,41], reflecting either recruitment or expansion of HTLV-I–infected cells in the CNS.

Pathology

Examination of the spinal cord demonstrates degeneration of the lateral columns, with variable damage to the anterior and posterior columns [42], with the thoracic cord being typically most severely affected [43]. Inflammatory changes are prominent histopathologic features of the HAM/TSP spinal cord (Fig. 2), characterized by perivascular and parenchymal lymphocytic infiltrates in the white and gray matter along with activated microglia/macrophages and severe degeneration and gliosis of the white matter [44]. The lymphocytic infiltrate in the spinal cord is a mixture of CD4+ and CD8+ T cells in early disease leading to a predominance of CD8+ T cells in later disease [44–46]. Similar perivascular inflammatory infiltrates are observed in the brains of patients with HAM/TSP [47]. Infiltrating cells include HTLV-I–specific CD8+ T cells [48], demonstrated in the spinal cord of HAM/TSP patients by the use of $Tax_{11\text{-}19}$ peptide-loaded HLA pentamers (Eiji Matsuura, MD, Kagoshima, Japan, personal communication, 2007). The HTLV-I–specific cytotoxic T cells appear to be preferentially recruited or expanded in the CNS [48]. HLA class I and II expression is up-regulated on endothelial cells and microglia [49]. Proinflammatory and potentially neurotoxic cytokines such as IL-1β, TNF-α, and interferon (INF)-γ are detected in perivascular infiltrating macrophages, microglia, and astrocytes [50,51].

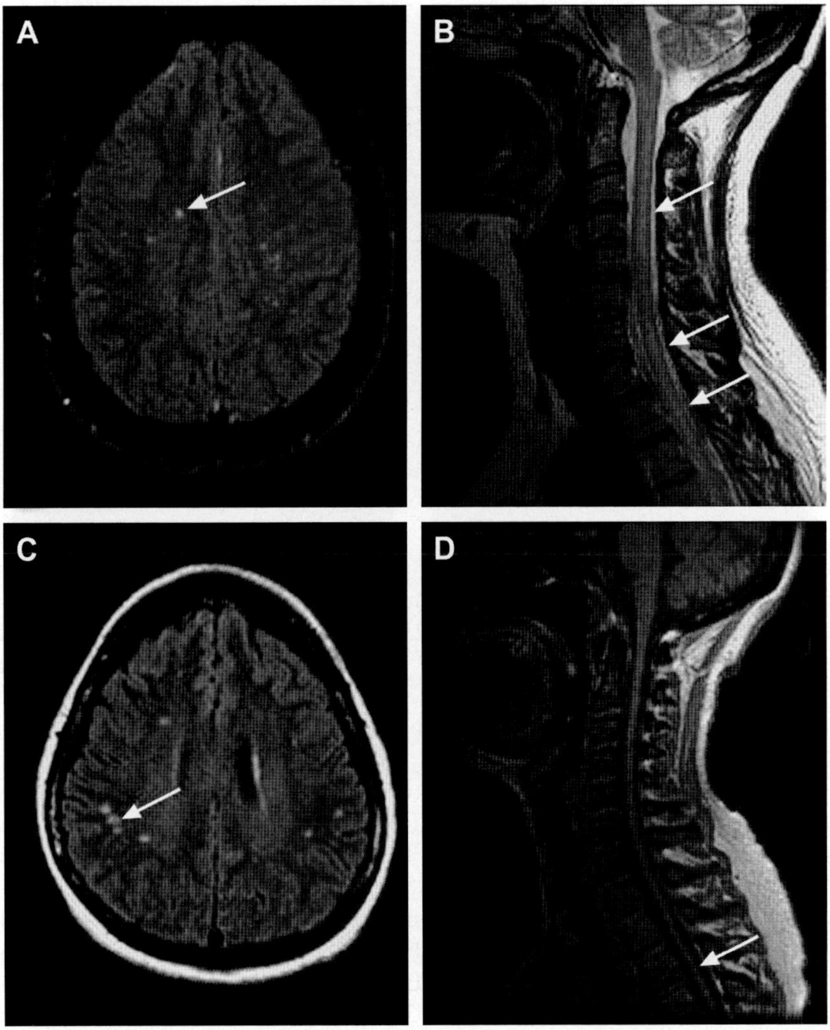

Fig. 1. MRI findings in HAM/TSP. MRI of the brain and spine in a 39-year-old woman with 2 years of gait difficulty and urinary dysfunction showing brain white matter abnormalities (*arrow*) on FLAIR (*A*), and T2 signal changes (*arrows*) in the cervical and upper thoracic cord (*B*). MRI of the brain and spine in a 58-year-old woman with over 20-year history of spastic paraparesis showing brain white matter abnormalities (*arrow*) on FLAIR (*C*) and atrophy of the spinal cord (*D*).

Pathogenesis

The striking CNS inflammatory changes demonstrated in HAM/TSP favor immune-mediated mechanisms of pathogenesis. Several hypotheses have emerged. HTLV infection of glial cells could lead to direct lysis by HTLV-specific cytotoxic CD8+ T cells. Alternatively, the target of immune attack

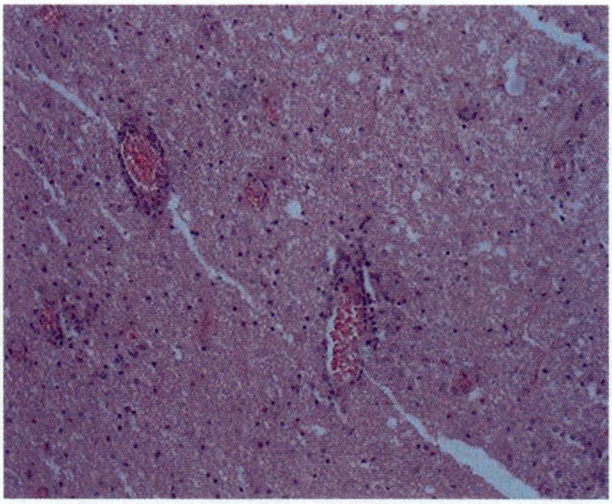

Fig. 2. Pathologic findings in HAM/TSP. Hematoxylin and eosin staining of spinal cord section from a subject with HAM/TSP showing perivascular and parenchymal mononuclear cell infiltration. (*Courtesy of* Eiji Matsuura, PhD, Okayama, Japan.)

may be HTLV-infected activated lymphocytes that have migrated into the CNS. *Bystander damage* [52] is then mediated through the release of neurotoxic inflammatory cytokines during the HTLV-specific cellular immune response. In addition, evidence has been put forth for the presence of anti-HTLV antibodies that, by *molecular mimicry* [53], cross-react with self-antigens (hnRNP-A1) in the CNS, leading to neuronal dysfunction and damage [54].

The sequence of events leading to bystander damage may be as follows [28]. Activation of T cells following HTLV infection allows the activation and migration of CD4+ and CD8+ T cells across the blood-brain barrier from the peripheral blood to the CNS. HTLV-I-specific CD8+ T cells are preferentially recruited or expanded in the CNS, and respond to HTLV-I antigen-expressing cells, most likely HTLV-I–infected CD4+ T cells or possibly HTLV-I–infected glial cells. HTLV-specific immune responses that occur in the CNS then lead to release of neurotoxic cytokines and subsequent CNS damage.

Therapy

No therapy has been conclusively shown to alter long-term disability associated with HAM/TSP. Clinical improvements have been reported for a number of agents in open-label studies including corticosteroids, plasmapheresis, danazol [55], pentoxifylline [56], and interferon. With the exception of interferon-α, however, these drugs lack the quality of evidence required to merit a strong recommendation for their use in HAM/TSP. The role of interferon-α in HAM/TSP also remains in question, as no study has conclusively shown long-term benefit.

Although significant challenges to the development of effective disease-modifying therapy for HAM/TSP remain, therapeutic trials for HAM/TSP, collectively, continue to advance our understanding of this disease, setting the stage for better qualification of biomarkers and refinement of specific therapeutic targets.

HAM/TSP relevant biomarkers

The lack of sensitive measures of disease activity remains a challenge for assessing the effect of therapy in a chronic progressive neurologic disease such as HAM/TSP. Progression of clinical disease in HAM/TSP may be subtle, especially later in disease, and difficult to quantify even over the course of a year. Biomarkers of HAM/TSP that reflect underlying pathogenic mechanisms would hold great potential with respect to screening and evaluation of novel therapeutics in a timely, effective manner.

There are no validated surrogate markers of disease in HAM/TSP. Among the number of potential relevant biomarkers studied, *HTLV-I proviral load* has the most data supporting its relationship with disease course. HTLV-I proviral load is determined as the amount of integrated HTLV provirus, measured by a real-time quantitative PCR assay and typically expressed as the number of copies of HTLV-I proviral DNA per number of peripheral blood mononuclear cells (PBMC). There is little, if any, cell-free virus found in plasma of HTLV-infected individuals [57]. Evidence from clinical studies suggests HTLV-I proviral load is potentially a marker of disease activity. Analysis from observational studies showed a relationship between HTLV-I proviral load and disability [33,40,58,59]. Analysis of interventional studies (interferon-alpha) showed concordance between reduced proviral load and clinical benefit [60]. Full validation of HTLV-I proviral load as a marker of disease activity will require data from prospective controlled trials that demonstrate correlation between change in HTLV-I proviral load and clinical impairment/disability.

A related biomarker is the *HTLV-I Tax mRNA load* measured by real-time quantitative reverse transcriptase (RT) PCR. One study showed correlation between HTLV-I Tax mRNA load and disease severity, and showed higher ratio of Tax mRNA load to the proviral load (mRNA/DNA) in HAM/TPS than in healthy volunteers [61]. The reduction in HTLV-I Tax mRNA load corresponded to the reduction in the frequency of *Tax-specific CD8+ T cells* following interferon-β therapy, suggesting that Tax mRNA load identifies a transcriptionally active fraction of HTLV-I–infected cells that disproportionally influence host cellular immune response [62]. Tax-specific CD8+ T cells have been implicated as a key cellular element contributing to bystander damage in HAM/TSP [28]. The availability of peptide/major histocompatibility complex (MHC), tetrameric and pentameric complexes allows the direct identification of Tax-specific CD8+ T cells for individuals with the HLA A*0201 and HLA A*2404 alleles. The high frequency of Tax-specific CD8+ T cells distinguished HAM/TSP from asymptomatic carrier within an HTLV-I infected cohort [63]. Further characterization of these cells with respect to effector/memory phenotype and cytokine production and function is likely essential to understanding their role in HAM/TSP [64–66].

Spontaneous lymphoproliferation indicates ex vivo proliferation of peripheral blood leukocytes from patients with HAM/TSP in the absence of exogenous stimulation. The observation of spontaneous lymphoproliferation is often interpreted as a marker of persistent immune activation in HAM/TSP. The up-regulation of IL-2 and IL-15 by the HTLV-I–encoded Tax protein underlie the autocrine/paracrine loops driving spontaneous lymphoproliferation in HAM/TSP [67,68]. Thus, inhibition of spontaneous lymphoproliferation reflects the ability to modulate key inflammatory cytokines dysregulated in HAM/TSP [69,70].

Novel biomarkers currently being examined include CD4+CD25+ regulatory T cells and their marker Foxp3, which are impaired in HAM/TSP [71,72], altered transforming growth factor (TGF)-β receptor expression in HAM/TSP [73] and the spontaneous degranulation (CD107a mobilization) demonstrated by HTLV-I-specific CD8+ T cells in HAM/TSP [74].

Corticosteroids

The evidence for supporting the use of corticosteroids in HAM/TSP is based on uncontrolled case series [75]. The immunomodulatory effects of corticosteroids [76] form the rationale for their use in HAM/TSP. One of the earliest reported therapy-producing clinical benefits in HAM/TSP [75], corticosteroids are now generally considered to confer transient clinical improvement [77]. However, because of lack of well-designed controlled trials, many questions remain regarding the usefulness of corticosteroids in HAM/TSP. Anecdotal reports suggest corticosteroids may be most efficacious early in disease, in rapidly progressive disease, and that clinical benefit may be prolonged with prolonged course (eg, monthly pulsed intravenous corticosteroids), but there are no reliable data to guide the clinician with regard to the time of initiation, the dose, or the duration of corticosteroid therapy in HAM/TSP.

Plasmapheresis and intravenous gammaglobulin

The rationale for the use plasmapheresis would be to modulate the contribution of the humoral response to the pathogenesis of HAM/TSP. There is some evidence to suggest that anti-HTLV antibodies could potentially be neuronopathic (see earlier in this article). Limited data from uncontrolled case series showed that plasmapheresis led to transient (2 to 4 weeks) clinical improvement in more than half of the patients treated for HAM/TSP [78]. Anti-HTLV-I-antibody titers were lowered in serum but not in the CSF following plasmapheresis. Future studies analyzing the relationship between changes to the titers of anti-neuronal antibody implicated in HAM/TSP (hnRNP-A1) and clinical response following plasmapheresis may further our understanding of the relative pathogenic contribution of anti-HTLV-I humoral response. A similar transient clinical benefit was reported for intravenous gammaglobulin (IVIG) [79], which has a number of effects on the immune system including neutralization of activated complement, signaling via Fc receptors, and modulation of proinflammatory cytokines [80]. It is yet unknown whether periodic plasmapheresis or IVIG treatment could lead to sustained clinical benefit.

Daclizumab (humanized anti-Tac)

HTLV-mediated up-regulation of IL-2 signaling represents a key component of the immune activation observed in HAM/TSP. HTLV-encoded transactivating factor Tax up-regulates expression of IL-2 as well as CD25, the alpha subunit of the IL-2 receptor in T cells [68]. An expansion of the CD4+CD25+ compartment in the peripheral blood is observed and identifies a major reservoir of HTLV infection [81]. In vitro studies showed that an antibody against CD25 that blocks IL-2 signaling inhibits the characteristic spontaneous proliferation of peripheral blood leukocytes from patients with HAM/TSP [68]. Clinical trials had already demonstrated the potential for anti-CD25 monoclonal antibody (humanized anti-Tac) to reduce CD25+ cells in adult T-cell leukemia/lymphoma [82,83]. Thus, preliminary studies suggested that a monoclonal antibody therapy against CD25 could reduce HTLV-infected cells as well as block HTLV-driven immune activation.

Anti-CD25 monoclonal antibody therapy was tested in a single-center open-label trial using humanized anti-Tac (daclizumab) [84]. Anti-CD25 antibody therapy had the anticipated result of reducing the number of CD4+CD25+ cells in subjects with HAM/TSP. Moreover, the anti-CD25 antibody therapy resulted in a significant concomitant reduction in the HTLV proviral load. Analysis of ex vivo proliferative capacity demonstrated a partial suppression of spontaneous lymphoproliferation during anti-CD25 antibody therapy. Three of the nine treated subjects demonstrated clinical improvements by Expanded Disability Status Scale (EDSS). None had clinical progression of disease during the trial. The reduction in HTLV proviral load likely represents loss of HTLV-infected CD4+CD25+ cells during humanized anti-Tac therapy, and supports the use of this marker (CD25) to target HTLV-infected cells. The reduction in spontaneous lymphoproliferation suggested that the anti-CD25 antibody therapy achieved at least partial inhibition of immune activation in HAM/TSP.

Interferons

Class I interferons (IFN-α and -β), which have cytostatic and antiviral properties, represent an approach for long-term immunomodulatory therapy in patients with HAM/TSP.

Interferon-α, considered a standard of care for HAM/TSP in Japan, demonstrated clinical benefit for subjects with HAM/TSP in a blinded trial [85]. A dose-effect was demonstrated using human lymphoblastoid interferon with subjects on higher dose interferon-α more likely to improve than those on lower dose. The benefit of long-term interferon-α therapy, however, has not been well studied. A small study extending interferon-α treatment to 24 weeks reported sustained clinical response [86].

The main marker of efficacy during interferon-α therapy appears to be a reduction in HTLV-I proviral load [60]. The reduction in HTLV-I proviral load during interferon-α therapy was associated with changes in CD8+ T cells. A reduction in CD8+ T cells of the central memory phenotype was observed with interferon-α treatment. In addition, a reduction in CD8+ T-cell perforin expression was observed, and a concordance between reduced perforin and clinical improvement was suggested by the analysis. These changes in cellular immune response following interferon-α study lend some support to cytotoxic CD8+ T-cell involvement in bystander damage postulated to occur in HAM/TSP.

Following the modest success of interferon-α therapy in HAM/TSP, long-term treatment with interferon-β was tested in a cohort of HAM/TSP patients. In an open-label, single-center trial, 12 subjects who met WHO criteria for HAM/TSP received up to twice-weekly injections of 60 μg interferon-β1a for 28 weeks. No significant clinical changes were observed during the trial in this cohort, whose mean disease duration was 6.5 years (range 2 to 20). Among the number of biomarkers measured during the trial, significant reduction in the frequency of Tax-specific CD8+ T cells was observed along with a reduction in HTLV-I Tax mRNA load and the spontaneous lymphoproliferation. No significant changes were observed in the HTLV proviral loads. These results suggested that interferon-β therapy had an overall immunomodulatory effect in HAM/TSP as evidenced by the relative suppression of spontaneous proliferation, and demonstrated that interferon-β therapy had a profound effect on virus-specific cellular immune responses.

The observation that reduction in Tax-specific CD8+ T-cell frequencies were more closely tied to HTLV-I Tax mRNA load than to the HTLV-I proviral load following interferon-β treatment suggest that a small transcriptionally active fraction of HTLV-infected cells may be largely responsible for the extent of virus specific CD8+ response in HTLV-I infection. Thus, changes in HTLV-I transcriptional activity, through increased HTLV-I protein expression, processing, and presentation as peptide in association with the major histocompatibility complex, may profoundly influence HTLV-I–specific cellular immune response. These results also suggest that one could modify the magnitude of the HTLV-I–specific immune response by the altering the transcriptional activity of the HTLV-I provirus.

Nucleoside analogs

Initial observational studies suggested clinical benefit and reduction of HTLV-I proviral load with zidovudine [87,88] and lamivudine [25], which had demonstrated some in vitro activity against the HTLV-I reverse transcriptase [89,90]. A randomized, double-blind, placebo-controlled study of 6 months of combination therapy with these two nucleoside analogs was conducted in 16 patients [24]. No significant clinical changes were observed as a result of zidovudine/lamivudine combination therapy in the measures tested. No significant change was observed in HTLV-I proviral loads. The contribution of reverse transcriptase activity to maintaining HTLV-I proviral load in vivo remains controversial, despite this well-designed trial targeting HTLV-I reverse transcription.

Emerging therapies

Hu MiK β1: Targeting Interleukin-15

Proinflammatory cytokines implicated in the pathogenesis of immune-mediated inflammatory diseases represent rational targets for therapy. IL-15 is a proinflammatory cytokine that appears to be dysregulated in HAM/TSP, and may play a pathogenic role in this and a number of other immune-mediated inflammatory diseases [91]. The role of IL-15 is particularly relevant in HAM/TSP because the HTLV-I encoded Tax protein promotes the expression of IL-15 and its private receptor (IL-15Rα) by transactivating IL-15 and IL-15Ra gene transcription [67]. The up-regulation of IL-15 contributes to the sustained immune activation in HAM/TSP that can be inhibited in vitro by the administration of monoclonal antibodies that block IL-15 signaling [69]. An important function of IL-15 in the immune system is the maintenance of memory CD8+ T cells. The upregulation of the IL-15 paracrine loop may explain the persistent expansion of HTLV-I–specific CD8+ T cells in HAM/TSP [92]. In short, evidence suggests that IL-15 plays a specific role in the immune activation observed in HAM/TSP.

A phase I/II clinical trial is under way to test the safety and provide preliminary efficacy data of a monoclonal antibody, Hu MiK-β1, that blocks IL-15 signaling. Based on its in vitro effects, we hypothesize that the administration of Hu MiK-β1in subjects with HAM/TSP will lead to inhibition of spontaneous proliferation and reduction in HTLV-I–specific cellular immune responses. The study of targeted therapies such as Hu MiK-β1 has the potential to elucidate the contribution of specific pathways to the pathogenesis of HAM/TSP.

Valproic acid

A large fraction of the HTLV-I–infected cells in the peripheral blood is transcriptionally quiescent in HAM/TSP, escaping cytolysis by HTLV-I–specific cytotoxic CD8+ cells and the effects of reverse transcriptase inhibitors. Recent work in HIV suggests that histone deacetylase inhibitors such as valproic acid (VPA) may induce expression of HIV from previously quiescent HIV-infected CD4+ T cells [93,94], subjecting them to the effects of antiretroviral therapy.

A single-arm, open-label trial tested the benefit of VPA (20 mg/kg/day) in 19 patients with HAM/TSP. Preliminary data suggest VPA administration induced a transient increase then a drop in HTLV-I proviral loads in HAM/TSP patients [95]. Clinical measures were stable overall. The authors hypothesized that VPA led to increased viral expression and subsequent destruction by the host immune system.

Summary

The treatment of HAM/TSP remains a challenge. No agent has shown to significantly modify the long-term disability associated with HAM/TSP. Advances in our understanding of the pathogenesis of HAM/TSP have led to the identification of several biomarkers and therapeutic targets. Clinical trials in HAM/TSP continue to be opportunities for further qualification and refinement of biomarkers and therapeutic targets. The validation of HAM/TSP-relevant biomarkers and the identification of new targets remain key challenges in the development of effective targeted therapy in HAM/TSP.

References

[1] Gessain A, Barin F, Vernant JC, et al. Antibodies to human T-lymphotropic virus type-I in patients with tropical spastic paraparesis. Lancet 1985;2(8452):407–10.

[2] Osame M, Usuku K, Izumo S, et al. HTLV-I associated myelopathy, a new clinical entity. Lancet 1986;1(8488):1031–2.

[3] Poiesz BJ, Ruscetti FW, Reitz MS, et al. Isolation of a new type C retrovirus (HTLV) in primary uncultured cells of a patient with Sezary T-cell leukaemia. Nature 1981;294(5838):268–71.

[4] Poiesz BJ, Ruscetti FW, Gazdar AF, et al. Detection and isolation of type C retrovirus particles from fresh and cultured lymphocytes of a patient with cutaneous T-cell lymphoma. Proc Natl Acad Sci U S A 1980;77(12):7415–9.

[5] Richardson JH, Edwards AJ, Cruickshank JK, et al. In vivo cellular tropism of human T-cell leukemia virus type 1. J Virol 1990;64(11):5682–7.

[6] Nagai M, Brennan MB, Sakai JA, et al. CD8(+) T cells are an in vivo reservoir for human T-cell lymphotropic virus type I. Blood 2001;98(6):1858–61.

[7] Trejo SR, Ratner L. The HTLV receptor is a widely expressed protein. Virology 2000;268(1): 41–8.

[8] Lehky TJ, Fox CH, Koenig S, et al. Detection of human T-lymphotropic virus type I (HTLV-I) tax RNA in the central nervous system of HTLV-I-associated myelopathy/tropical spastic paraparesis patients by in situ hybridization. Ann Neurol 1995;37(2):167–75.

[9] Blattner WA, Gallo RC. Epidemiology of human retroviruses. Leuk Res 1985;9(6):697–8.

[10] Gessain A. Epidemiology of HTLV-I and associated diseases. In: Hollsberg P, Hafler D, editors. Human T-cell lymphotropic virus type I. Chichester (UK): John Wiley and Sons; 1996. p. 33–64.

[11] Kaplan JE, Osame M, Kubota H, et al. The risk of development of HTLV-I-associated myelopathy/tropical spastic paraparesis among persons infected with HTLV-I. J Acquir Immune Defic Syndr 1990;3(11):1096–101.

[12] Tosswill JH, Taylor GP, Tedder RS, et al. HTLV-I/II associated disease in England and Wales, 1993–7: retrospective review of serology requests. BMJ 2000;320(7235):611–2.

[13] Orland JR, Engstrom J, Fridey J, et al. Prevalence and clinical features of HTLV neurologic disease in the HTLV Outcomes Study. Neurology 2003;61(11):1588–94.

[14] Morgan OS, Rodgers-Johnson P, Mora C, et al. HTLV-1 and polymyositis in Jamaica. Lancet 1989;2(8673):1184–7.

[15] Higuchi I, Hashimoto K, Kashio N, et al. Detection of HTLV-I provirus by in situ polymerase chain reaction in mononuclear inflammatory cells in skeletal muscle of viral carriers with polymyositis. Muscle Nerve 1995;18(8):854–8.

[16] Leite AC, Silva MT, Alamy AH, et al. Peripheral neuropathy in HTLV-I infected individuals without tropical spastic paraparesis/HTLV-I-associated myelopathy. J Neurol 2004;251(7): 877–81.

[17] Kiwaki T, Umehara F, Arimura Y, et al. The clinical and pathological features of peripheral neuropathy accompanied with HTLV-I associated myelopathy. J Neurol Sci 2003;206(1):17–21.

[18] Mochizuki M, Watanabe T, Yamaguchi K, et al. HTLV-I uveitis: a distinct clinical entity caused by HTLV-I. Jpn J Cancer Res 1992;83(3):236–9.

[19] Buggage RR. Ocular manifestations of human T-cell lymphotropic virus type 1 infection. Curr Opin Ophthalmol 2003;14(6):420–5.

[20] Nishioka K, Maruyama I, Sato K, et al. Chronic inflammatory arthropathy associated with HTLV-I. Lancet 1989;1(8635):441.

[21] Zucker-Franklin D. Non-HIV retroviral associations with rheumatic disease. Curr Rheumatol Rep 2000;2(2):156–62.

[22] Sugimoto M, Nakashima H, Watanabe S, et al. T-lymphocyte alveolitis in HTLV-I-associated myelopathy. Lancet 1987;2(8569):1220.

[23] Igakura T, Stinchcombe JC, Goon PK, et al. Spread of HTLV-I between lymphocytes by virus-induced polarization of the cytoskeleton. Science 2003;299(5613):1713–6.

[24] Taylor GP, Goon P, Furukawa Y, et al. Zidovudine plus lamivudine in Human T-Lymphotropic Virus type-I-associated myelopathy: a randomised trial. Retrovirology 2006;3:63–71.

[25] Taylor GP, Hall SE, Navarrete S, et al. Effect of lamivudine on human T-cell leukemia virus type 1 (HTLV-1) DNA copy number, T-cell phenotype, and anti-tax cytotoxic T-cell frequency in patients with HTLV-1-associated myelopathy. J Virol 1999;73(12):10289–95.

[26] Grant C, Barmak K, Alefantis T, et al. Human T cell leukemia virus type I and neurologic disease: events in bone marrow, peripheral blood, and central nervous system during normal immune surveillance and neuroinflammation. J Cell Physiol 2002;190(2):133–59.

[27] Jacobson S, Shida H, McFarlin DE, et al. Circulating CD8+ cytotoxic T lymphocytes specific for HTLV-I pX in patients with HTLV-I associated neurological disease. Nature 1990; 348(6298):245–8.

[28] Nagai M, Jacobson S. Immunopathogenesis of human T cell lymphotropic virus type I-associated myelopathy. Curr Opin Neurol 2001;14(3):381–6.

[29] Nagai M, Usuku K, Matsumoto W, et al. Analysis of HTLV-I proviral load in 202 HAM/TSP patients and 243 asymptomatic HTLV-I carriers: high proviral load strongly predisposes to HAM/TSP. J Neurovirol 1998;4(6):586–93.

[30] Manns A, Miley WJ, Wilks RJ, et al. Quantitative proviral DNA and antibody levels in the natural history of HTLV-I infection. J Infect Dis 1999;180(5):1487–93.

[31] Jeffery KJ, Usuku K, Hall SE, et al. HLA alleles determine human T-lymphotropic virus-I (HTLV-I) proviral load and the risk of HTLV-I-associated myelopathy. Proc Natl Acad Sci U S A 1999;96(7):3848–53.

[32] Vine AM, Witkover AD, Lloyd AL, et al. Polygenic control of human T lymphotropic virus type I (HTLV-I) provirus load and the risk of HTLV-I-associated myelopathy/tropical spastic paraparesis. J Infect Dis 2002;186(7):932–9.

[33] Olindo S, Cabre P, Lezin A, et al. Natural history of human T-lymphotropic virus 1-associated myelopathy: a 14-year follow-up study. Arch Neurol 2006;63(11):1560–6.
[34] Report of the Scientific Group on HTLV-I Infections and Associated Diseases. World Health Organization Report (WP) CDS (S) / ICP / PHC / 014. 1988.
[35] De Castro-Costa CM, Araujo AQ, Barreto MM, et al. Proposal for diagnostic criteria of tropical spastic paraparesis/HTLV-I-associated myelopathy (TSP/HAM). AIDS Res Hum Retroviruses 2006;22(10):931–5.
[36] Alcindor F, Valderrama R, Canavaggio M, et al. Imaging of human T-lymphotropic virus type I-associated chronic progressive myeloneuropathies. Neuroradiology 1992;35(1): 69–74.
[37] Kira J, Fujihara K, Itoyama Y, et al. Leukoencephalopathy in HTLV-I-associated myelopathy/tropical spastic paraparesis: MRI analysis and a two year follow-up study after corticosteroid therapy. J Neurol Sci 1991;106(1):41–9.
[38] Kuroda Y, Matsui M, Yukitake M, et al. Assessment of MRI criteria for MS in Japanese MS and HAM/TSP. Neurology 1995;45(1):30–3.
[39] Puccioni-Sohler M, Yamano Y, Rios M, et al. Differentiation of HAM/TSP from patients with multiple sclerosis infected with HTLV-I. Neurology 2007;68(3):206–13.
[40] Takenouchi N, Yamano Y, Usuku K, et al. Usefulness of proviral load measurement for monitoring of disease activity in individual patients with human T-lymphotropic virus type I-associated myelopathy/tropical spastic paraparesis. J Neurovirol 2003;9(1):29–35.
[41] Lezin A, Olindo S, Oliere S, et al. Human T lymphotropic virus type I (HTLV-I) proviral load in cerebrospinal fluid: a new criterion for the diagnosis of HTLV-I-associated myelopathy/tropical spastic paraparesis? J Infect Dis 2005;191(11):1830–4.
[42] Levin MC, Jacobson S. HTLV-I associated myelopathy/tropical spastic paraparesis (HAM/TSP): a chronic progressive neurologic disease associated with immunologically mediated damage to the central nervous system. J Neurovirol 1997;3(2):126–40.
[43] Izumo S, Ijichi T, Higuchi I, et al. Neuropathology of HTLV-I-associated myelopathy—a report of two autopsy cases. Acta Paediatr Jpn 1992;34(3):358–64.
[44] Iwasaki Y. Pathology of chronic myelopathy associated with HTLV-I infection (HAM/TSP). J Neurol Sci 1990;96(1):103–23.
[45] Cartier LM, Cea JG, Vergara C, et al. Clinical and neuropathological study of six patients with spastic paraparesis associated with HTLV-I: an axomyelinic degeneration of the central nervous system. J Neuropathol Exp Neurol 1997;56(4):403–13.
[46] Iwasaki Y, Ohara Y, Kobayashi I, et al. Infiltration of helper/inducer T lymphocytes heralds central nervous system damage in human T-cell leukemia virus infection. Am J Pathol 1992; 140(5):1003–8.
[47] Aye MM, Matsuoka E, Moritoyo T, et al. Histopathological analysis of four autopsy cases of HTLV-I-associated myelopathy/tropical spastic paraparesis: inflammatory changes occur simultaneously in the entire central nervous system. Acta Neuropathol (Berl) 2000;100(3): 245–52.
[48] Nagai M, Yamano Y, Brennan MB, et al. Increased HTLV-I proviral load and preferential expansion of HTLV-I Tax-specific CD8+ T cells in cerebrospinal fluid from patients with HAM/TSP. Ann Neurol 2001;50(6):807–12.
[49] Umehara F, Izumo S, Nakagawa M, et al. Immunocytochemical analysis of the cellular infiltrate in the spinal cord lesions in HTLV-I-associated myelopathy. J Neuropathol Exp Neurol 1993;52(4):424–30.
[50] Umehara F, Izumo S, Ronquillo AT, et al. Cytokine expression in the spinal cord lesions in HTLV-I-associated myelopathy. J Neuropathol Exp Neurol 1994;53(1):72–7.
[51] Umehara F, Okada Y, Fujimoto N, et al. Expression of matrix metalloproteinases and tissue inhibitors of metalloproteinases in HTLV-I-associated myelopathy. J Neuropathol Exp Neurol 1998;57(9):839–49.
[52] Osame M. Pathological mechanisms of human T-cell lymphotropic virus type I-associated myelopathy (HAM/TSP). J Neurovirol 2002;8(5):359–64.

[53] Fujinami RS, Oldstone MB. Amino acid homology between the encephalitogenic site of myelin basic protein and virus: mechanism for autoimmunity. Science 1985;230(4729): 1043–5.

[54] Levin MC, Lee SM, Kalume F, et al. Autoimmunity due to molecular mimicry as a cause of neurological disease. Nat Med 2002;8(5):509–13.

[55] Harrington WJ Jr, Sheremata WA, Snodgrass SR, et al. Tropical spastic paraparesis/HTLV-1-associated myelopathy (TSP/HAM): treatment with an anabolic steroid danazol. AIDS Res Hum Retroviruses 1991;7(12):1031–4.

[56] Shirabe S, Nakamura T, Tsujino A, et al. Successful application of pentoxifylline in the treatment of HTLV-I associated myelopathy. J Neurol Sci 1997;151(1):97–101.

[57] Wodarz D, Nowak MA, Bangham CR. The dynamics of HTLV-I and the CTL response. Immunol Today 1999;20(5):220–7.

[58] Matsuzaki T, Nakagawa M, Nagai M, et al. HTLV-I proviral load correlates with progression of motor disability in HAM/TSP: analysis of 239 HAM/TSP patients including 64 patients followed up for 10 years. J Neurovirol 2001;7(3):228–34.

[59] Olindo S, Lezin A, Cabre P, et al. HTLV-1 proviral load in peripheral blood mononuclear cells quantified in 100 HAM/TSP patients: a marker of disease progression. J Neurol Sci 2005;237(1–2):53–9.

[60] Saito M, Nakagawa M, Kaseda S, et al. Decreased human T lymphotropic virus type I (HTLV-I) provirus load and alteration in T cell phenotype after interferon-alpha therapy for HTLV-I-associated myelopathy/tropical spastic paraparesis. J Infect Dis 2004;189(1): 29–40.

[61] Yamano Y, Nagai M, Brennan M, et al. Correlation of human T-cell lymphotropic virus type 1 (HTLV-1) mRNA with proviral DNA load, virus-specific CD8(+) T cells, and disease severity in HTLV-1-associated myelopathy (HAM/TSP). Blood 2002;99(1):88–94.

[62] Oh U, Yamano Y, Mora CA, et al. Interferon-beta1a therapy in human T-lymphotropic virus type I-associated neurologic disease. Ann Neurol 2005;57(4):526–34.

[63] Kubota R, Nagai M, Kawanishi T, et al. Increased HTLV type 1 tax specific CD8+ cells in HTLV type 1-asociated myelopathy/tropical spastic paraparesis: correlation with HTLV type 1 proviral load. AIDS Res Hum Retroviruses 2000;16(16):1705–9.

[64] Nagai M, Kubota R, Greten TF, et al. Increased activated human T cell lymphotropic virus type I (HTLV-I) Tax11-19-specific memory and effector CD8+ cells in patients with HTLV-I-associated myelopathy/tropical spastic paraparesis: correlation with HTLV-I provirus load. J Infect Dis 2001;183(2):197–205.

[65] Parker CE, Daenke S, Nightingale S, et al. HTLV-1-specific cytotoxic T-lymphocytes are found in healthy seropositives as well as in patients with tropical spastic paraparesis. Virology 1992;188(2):628–36.

[66] Asquith B, Mosley AJ, Barfield A, et al. A functional CD8+ cell assay reveals individual variation in CD8+ cell antiviral efficacy and explains differences in human T-lymphotropic virus type 1 proviral load. J Gen Virol 2005;86(Pt 5):1515–23.

[67] Azimi N, Brown K, Bamford RN, et al. Human T cell lymphotropic virus type I Tax protein trans-activates interleukin 15 gene transcription through an NF-kappaB site. Proc Natl Acad Sci U S A 1998;95(5):2452–7.

[68] Tendler CL, Greenberg SJ, Blattner WA, et al. Transactivation of interleukin 2 and its receptor induces immune activation in human T-cell lymphotropic virus type I-associated myelopathy: pathogenic implications and a rationale for immunotherapy. Proc Natl Acad Sci U S A 1990;87(13):5218–22.

[69] Azimi N, Jacobson S, Leist T, et al. Involvement of IL-15 in the pathogenesis of human T lymphotropic virus type I-associated myelopathy/tropical spastic paraparesis: implications for therapy with a monoclonal antibody directed to the IL-2/15R beta receptor. J Immunol 1999;163(7):4064–72.

[70] Waldmann TA, Dubois S, Tagaya Y. Contrasting roles of IL-2 and IL-15 in the life and death of lymphocytes: implications for immunotherapy. Immunity 2001;14(2):105–10.

[71] Yamano Y, Takenouchi N, Li HC, et al. Virus-induced dysfunction of CD4(+)CD25(+) T cells in patients with HTLV-I-associated neuroimmunological disease. J Clin Invest 2005; 115(5):1361–8.
[72] Oh U, Grant C, Griffith C, et al. Reduced Foxp3 protein expression is associated with inflammatory disease during human t lymphotropic virus type 1 Infection. J Infect Dis 2006; 193(11):1557–66.
[73] Grant C, Oh U, Yao K, et al. Dysregulation of TGF-β signaling and regulatory and effector T cell function in virus-induced neuroinflammatory disease. Manuscript accepted for publication.
[74] Yoshimi Akahta, et al. submitted for publication.
[75] Osame M, Matsumoto M, Usuku K, et al. Chronic progressive myelopathy associated with elevated antibodies to human T-lymphotropic virus type I and adult T-cell leukemialike cells. Ann Neurol 1987;21(2):117–22.
[76] Cupps TR, Fauci AS. Corticosteroid-mediated immunoregulation in man. Immunol Rev 1982;65:133–55.
[77] Gessain A, Gout O. Chronic myelopathy associated with human T-lymphotropic virus type I (HTLV-I). Ann Intern Med 1992;117(11):933–46.
[78] Matsuo H, Nakamura T, Tsujihata M, et al. Plasmapheresis in treatment of human T-lymphotropic virus type-I associated myelopathy. Lancet 1988;2(8620):1109–13.
[79] Kuroda Y, Takashima H, Ikeda A, et al. Treatment of HTLV-I-associated myelopathy with high-dose intravenous gammaglobulin. J Neurol 1991;238(6):309–14.
[80] Gold R, Stangel M, Dalakas MC. Drug Insight: the use of intravenous immunoglobulin in neurology—therapeutic considerations and practical issues. Nat Clin Pract Neurol 2007; 3(1):36–44.
[81] Yamano Y, Cohen CJ, Takenouchi N, et al. Increased expression of human T lymphocyte virus type I (HTLV-I) Tax11-19 peptide-human histocompatibility leukocyte antigen A*201 complexes on CD4+ CD25+ T cells detected by peptide-specific, major histocompatibility complex-restricted antibodies in patients with HTLV-I-associated neurologic disease. J Exp Med 2004;199(10):1367–77.
[82] Waldmann TA, White JD, Goldman CK, et al. The interleukin-2 receptor: a target for monoclonal antibody treatment of human T-cell lymphotrophic virus I-induced adult T-cell leukemia. Blood 1993;82(6):1701–12.
[83] Junghans RP, Waldmann TA, Landolfi NF, et al. Anti-Tac-H, a humanized antibody to the interleukin 2 receptor with new features for immunotherapy in malignant and immune disorders. Cancer Res 1990;50(5):1495–502.
[84] Lehky TJ, Levin MC, Kubota R, et al. Reduction in HTLV-I proviral load and spontaneous lymphoproliferation in HTLV-I-associated myelopathy/tropical spastic paraparesis patients treated with humanized anti-Tac. Ann Neurol 1998;44(6):942–7.
[85] Izumo S, Goto I, Itoyama Y, et al. Interferon-alpha is effective in HTLV-I-associated myelopathy: a multicenter, randomized, double-blind, controlled trial. Neurology 1996;46(4): 1016–21.
[86] Yamasaki K, Kira J, Koyanagi Y, et al. Long-term, high dose interferon-alpha treatment in HTLV-I-associated myelopathy/tropical spastic paraparesis: a combined clinical, virological and immunological study. J Neurol Sci 1997;147(2):135–44.
[87] Gout O, Gessain A, Iba-Zizen M, et al. The effect of zidovudine on chronic myelopathy associated with HTLV-1. J Neurol 1991;238(2):108–9.
[88] Sheremata WA, Benedict D, Squilacote DC, et al. High-dose zidovudine induction in HTLV-I-associated myelopathy: safety and possible efficacy. Neurology 1993;43(10): 2125–9.
[89] Derse D, Hill SA, Lloyd PA, et al. Examining human T-lymphotropic virus type 1 infection and replication by cell-free infection with recombinant virus vectors. J Virol 2001;75(18): 8461–8.

[90] Balestrieri E, Forte G, Matteucci C, et al. Effect of lamivudine on transmission of human T-cell lymphotropic virus type 1 to adult peripheral blood mononuclear cells in vitro. Antimicrobial Agents Chemother 2002;46(9):3080–3.

[91] McInnes IB, Gracie JA. Interleukin-15: a new cytokine target for the treatment of inflammatory diseases. Curr Opin Pharmacol 2004;4(4):392–7.

[92] Azimi N, Nagai M, Jacobson S, et al. IL-15 plays a major role in the persistence of Tax-specific CD8 cells in HAM/TSP patients. Proc Natl Acad Sci U S A 2001;98(25):14559–64.

[93] Ylisastigui L, Archin NM, Lehrman G, et al. Coaxing HIV-1 from resting CD4 T cells: histone deacetylase inhibition allows latent viral expression. AIDS 2004;18(8):1101–8.

[94] Lehrman G, Hogue IB, Palmer S, et al. Depletion of latent HIV-1 infection in vivo: a proof-of-concept study. Lancet 2005;366(9485):549–55.

[95] Lezin A, Gillet N, Olindo S, et al. Histone deacetylase mediated transcriptional activation reduces proviral loads in HTLV-1 associated myelopathy/tropical spastic paraparesis patients. Blood 2007;110:3722–8.

ELSEVIER
SAUNDERS

Neurol Clin 26 (2008) 799–819

NEUROLOGIC CLINICS

HIV Infection of the Central Nervous System: Clinical Features and Neuropathogenesis

Lysa Boissé, MD, MSc, BSc[a], M. John Gill, MBBS[b], Christopher Power, MD[c,*]

[a]*Division of Neurology, Queen's University, 76 Stuart Street, Kingston, ON K7L 2V7, Canada*
[b]*Departments of Medicine and Microbiology and Infectious Diseases, University of Calgary, 213, 906 - 8th Avenue SW, Calgary, AB T2P 1H9, Canada*
[c]*Departments of Medicine (Neurology), Medical Microbiology and Immunology, and Psychiatry, University of Alberta, 6-11 Heritage Medical Research Centre, Edmonton, AB T6G 2S2, T2P 1H9, Canada*

Almost 65 million people worldwide have been infected with HIV since it was first identified in the early 1980s. In North America, 1 million individuals are currently infected with HIV, largely HIV type 1 (HIV-1). The HIV-1 B subtype (or clade) is the predominant virus causing infection in the industrialized world; hence, the neurologic disorders described herein reflect those syndromes caused by HIV-1 subtype B. Neurologic disorders associated with HIV-1 affect between 40% and 70% of infected individuals. They involve the central nervous system (CNS) and the peripheral nervous system (PNS) at any point during the course of infection [1], although most neurologic disorders become evident with advanced immunosuppression during the AIDS phase of disease (Box 1).

There are two principal groups of neurologic disorders that occur in association with HIV-1 infection. One group—the opportunistic infections of the CNS and PNS arising as consequences of HIV-induced immunosuppression (of which some are listed in Box 1; reviewed in Refs. [2,3])—have become less frequently observed with the availability of combination antiretroviral therapy (cART). Conversely, the primary HIV-induced neurologic syndromes, including neurocognitive disorders and peripheral neuropathies, have emerged as the more frequently encountered nervous system disorders in the clinics of industrialized countries (Fig. 1). The primary HIV-induced

* Corresponding author.
E-mail address: chris.power@ualberta.ca (C. Power).

doi:10.1016/j.ncl.2008.04.002 *neurologic.theclinics.com*

Box 1. Definition of AIDS

AIDS is defined as a $CD4^+$ T-lymphocyte count below 200 cells/μL of blood with or without the presence of an AIDS-defining illness, including the following:

- HIV-associated dementia
- Opportunistic infections such as *Pneumocystis jiroveci* pneumonia, Kaposi sarcoma, *Mycobacterium avium*, and tuberculosis; and neurologic disorders including cryptococcal meningitis, toxoplasmic encephalitis, progressive multifocal leukoencephalopathy, primary CNS lymphoma, and cytomegalovirus encephalitis.
- HIV wasting syndrome

Data from Centers for Disease Control and Prevention. 1993 revised classification system for HIV infection and expanded surveillance case definition for AIDS among adolescents and adults. Morb Mortal Wkly Rep Surveill Summ 1992;41(RR-17):1–19.

neurologic syndromes involving the CNS are the subject of this review and represent HIV-1's immediate and deleterious effects on neural cells, causing damage to the brain and spinal cord. The most significant of these primary HIV-associated CNS disorders include HIV-associated neurocognitive disorder (HAND), which comprises HIV-associated dementia (HAD) (also termed AIDS dementia complex or HIV encephalopathy), mild neurocognitive disorder (MND), and asymptomatic neurocognitive impairment [4]. Nonetheless, other CNS disorders complicate the course of HIV-1 infection (Box 2). The neurologic syndrome can often be predicted by the $CD4^+$ T-lymphocyte count in blood or by estimating the time from original infection (Fig. 2). For example, HAD is typically a feature of AIDS, whereas MND may present earlier in the disease course at higher $CD4^+$ T-cell counts.

With the advent of cART in the mid-1990s, many HIV-1 seropositive individuals in industrialized countries now live upwards of 20 years after initial infection. Despite the availability of cART, HIV-related CNS disorders continue to represent a substantial personal, economic, and societal burden [5,6]. This review summarizes the clinical manifestations, diagnosis, treatment, and pathogenesis of the primary HIV-associated CNS disorders.

HIV-1-associated neurocognitive disorders

The development of HAD is among the most devastating consequences of HIV-1 infection because of its unique and progressive clinical manifestations. This syndrome is characterized by neurocognitive impairments (forgetfulness, poor concentration), emotional disturbance (agitation, apathy), and motor

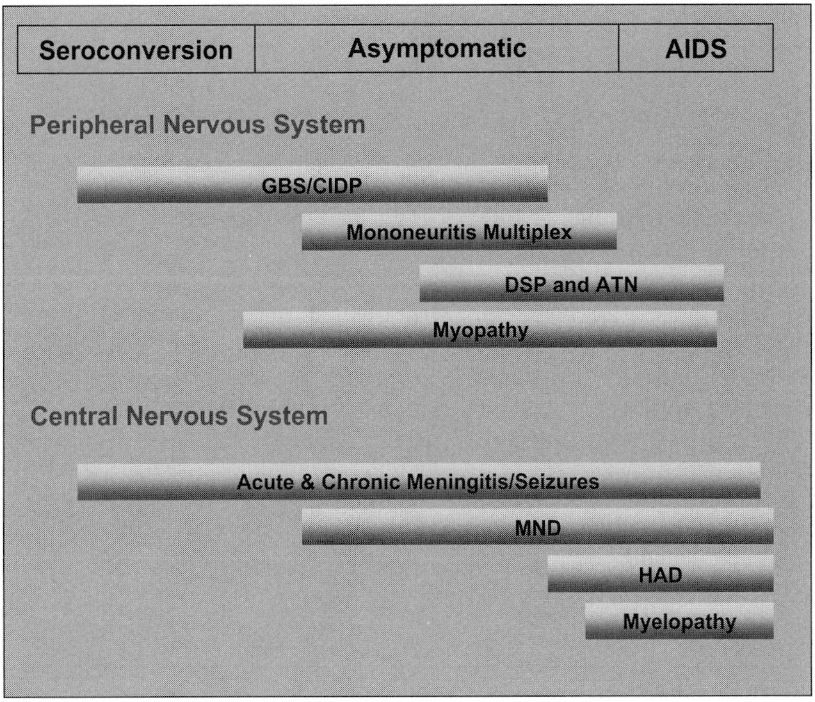

Fig. 1. Primary HIV-induced neurologic syndromes occurring during the course of infection. Individual syndromes arise depending on the degree of immune suppression; that is, acute meningitis and Guillain-Barré Syndrome (GBS) tend to arise earlier in the disease course, whereas HIV-associated dementia (HAD) and distal sensory polyneuropathy (DSP) present later in the disease course. All levels of the nervous system can be affected by HIV. ATN, antiretroviral toxic neuropathy; CIDP, chronic inflammatory demyelinating polyneuropathies; MND, mild neurocognitive disorder. (*Adapted from* Power C, Gill MJ, Johnson RT. The neuropathogenesis of HIV infection: host-virus interaction and the impact of therapy. Can J Neurol Sci 2002;29:19–32; with permission.)

dysfunction (tremor, ataxia, and spasticity) (Box 3). HAD, however, can display remarkable diversity in its clinical phenotype, including movement disorders [7–9] and occasionally mania or psychosis [10,11]. The course of the dementia is highly variable from patient to patient and can present as an abrupt decline over a few weeks or a protracted course over several months [12]. MND is often an antecedent syndrome that can precede the onset of frank dementia but presents with the clinical hallmarks of HAD [13,14], albeit with less-severe signs and symptoms and at higher $CD4^+$ T-cell levels [15]. The diagnosis of MND may be more difficult to make because of comorbidities such as prior substance abuse, head injury, or coinfections such as hepatitis C virus (HCV) infection. The full spectrum of asymptomatic neurocognitive impairment remains undefined in terms of impact and outcomes; however, the diagnosis of HAD or MND heralds a prognosis for diminished survival, regardless of whether the patient is receiving cART [16,17].

Box 2. Other primary neurologic syndromes associated with HIV infection involving the central nervous system

- Aseptic meningitis: presents as part of the constellation of symptoms often associated with primary HIV infection/seroconversion [134]
- Multiple sclerosis–like disorders: largely evident on neuroimaging
- Cerebrovascular diseases: ischemic and hemorrhagic disorders; associated with structural changes in the neurologic vasculature
- Pain disorders (primary HIV-induced headache; neuropathic pain): nonthrobbing headache associated with photophobia with no cause other than HIV-1 infection [135]; treatment with low-dose tricyclic medication can be effective [136]; neuropathic pain secondary to myelopathic syndromes and neuropathy
- Psychiatric disorders: delirium; mood, psychotic, and psychoactive substance–use disorders
- Drug-related symptoms: efavirenz-associated nightmares

HAD is an AIDS-defining illness (see Box 1), and approximately 20% to 40% of untreated AIDS patients acquire the diagnosis. Symptoms typically begin once an individual's $CD4^+$ T-cell count drops below 200 cells/μL of blood, although with the advent of cART, HAD is now presenting at higher $CD4^+$ T-cell levels. Clinical risk factors for HAD include low $CD4^+$ T-cell levels, high viral load in cerebrospinal fluid (CSF) or plasma, anemia, extremes of age, injection drug use, and several host genetic polymorphisms [18]. Radiologic features accompanying HAD include cerebral and basal ganglia atrophy and diffuse periventricular white matter hyperintensities on MRI T2-weighted images (Fig. 3) [19,20], although nondemented HIV/AIDS patients can display similar radiologic changes [21]. Magnetic resonance spectroscopy shows diminished *N*-acetylaspartate, a neuronal metabolite [22] together with increased choline levels indicative of inflammation [23]. Other valuable investigations include CSF analyses to exclude opportunistic infections; high protein and immunoglobulin G levels with an accompanying pleocytosis are found in the CSF of 66% of HAD patients. CSF HIV-1 copy number together with markers of immune dysregulation (neopterin, β-microtubulin, chemokines) are associated with the presence of HAD [1]. Although HIV genome and protein can be detected in the CNS, viral abundance in brain parenchyma does not seem to correlate with the diagnosis or severity of HAD. There is a weak association between disease severity and viral load in human studies and animal models of

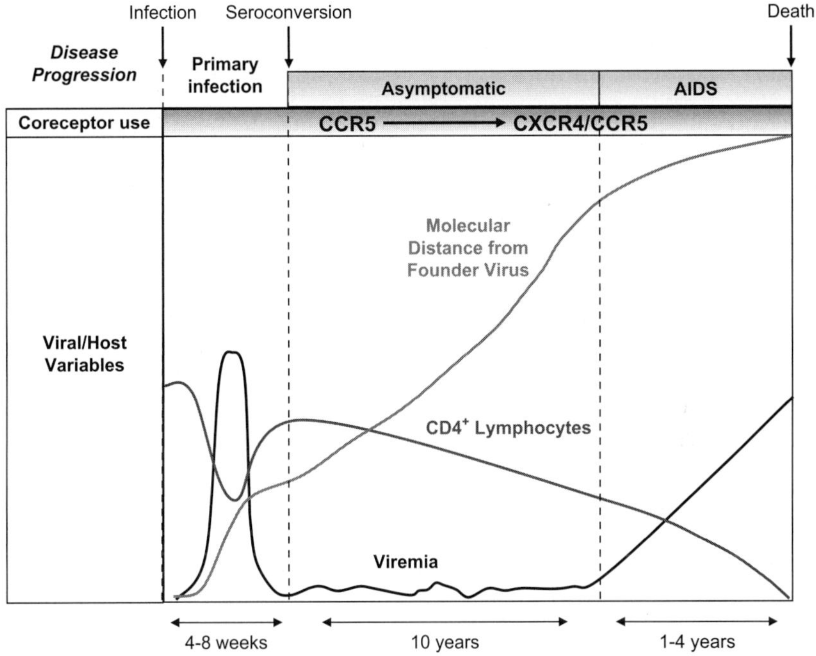

Fig. 2. Natural course of untreated HIV infection. In the first 4 to 8 weeks after infection, there is a dramatic increase in viral load accompanied by a decrease in $CD4^+$ T-cell levels in blood. During the asymptomatic period of infection, viral load remains relatively low. An increase in viral load after approximately 10 years of infection is associated with progression to AIDS. $CD4^+$ T lymphocytes drop after initial infection only to rise again, but their number declines slowly over the course of infection. There is a progressive change from CCR5- to CXCR4-dependent infection in blood during the course of disease.

disease [24], although brain- and blood-derived viral sequences exhibit increased molecular diversity in patients who have HAD (reviewed in Ref. [25]). In contrast, viral load in CSF may be predictive of HAD [1] and is higher in demented than in nondemented HIV patients and in patients who have HIV encephalitis. This latter finding likely reflects blood-derived virus rather than intrinsically brain-derived virus.

Neuropsychologic assessment is a useful tool in confirming the diagnosis of HAD and in evaluating the response to therapy. Given that HIV preferentially infects and affects the basal ganglia and deep white matter, HAD predictably displays the typical features of a "subcortical" dementia. Hence, HAD is not readily detected by the Folstein Mini-Mental Status Exam unless the patient is severely demented. More useful screening tools for HAD include the HIV Dementia Scale [26], the Mental Alteration Test [27], the Executive Interview [28], and the HIV Dementia Assessment [29]. More recently, a derivative of the HIV Dementia Scale has been reported—the International HIV Dementia Scale—which is being used in developing

Box 3. Features of HIV-1-associated neurocognitive disorders (HIV-associated dementia, mild neurocognitive disorder)

- Neurocognitive dysfunction: memory impairment, poor concentration, psychomotor slowing
- Emotional disturbance: apathy and social withdrawal, which can be mistaken for depression; also irritability, mental inflexibility, and decreased sex drive
- Motor abnormalities: weakness, ataxia, clumsy gait, slowing motor skills, tremor, diffuse increase in tone, hyperreflexia, spasticity, abnormal eye movements, and parkinsonism; frontal release signs and myoclonus in advanced stages of disease
- Brain atrophy and abnormal white matter signal on MRI and CT scanning
- Pleocytosis, increased protein, and viral load in cerebospinal fluid
- Abnormal neuropsychologic testing

countries [30]. The Memorial Sloan-Kettering Scale is a widely accepted tool for monitoring the progression and disability of dementia over time [31]. Established neuropsychologic tests including Symbol-Digit, Trail Making A and B, and the Grooved Pegboard tests remain the cornerstone for assessing neuropsychologic performance among HIV/AIDS patients.

The differential diagnoses for HAD/MND are broad, with an immediate need to exclude other treatable infections that are common in HIV/AIDS

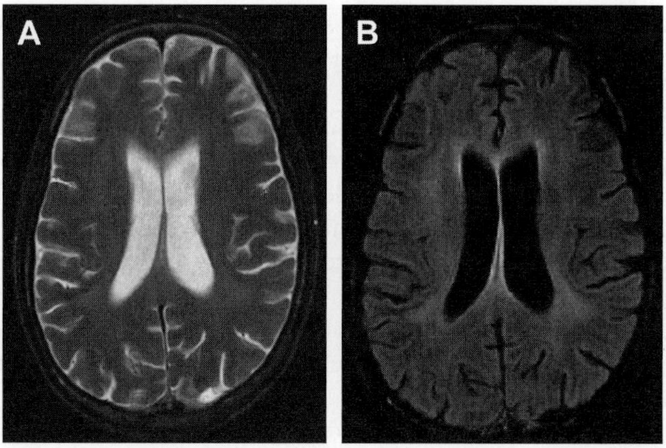

Fig. 3. Brain MRI of a 34-year-old HIV-positive woman who had HAD. (*A*) T2 image shows enlarged ventricles with diffuse white matter hypertensive signal. (*B*) FLAIR image shows white matter changes with cortical atrophy.

patients, including cryptococcal meningitis, toxoplasmic encephalitis, progressive multifocal leukoencephalopathy, and cytomegalovirus encephalitis. The combination of neuroimaging and CSF analyses is usually sufficient to exclude these infections. It is essential to ascertain a history of past or current substance abuse to exclude delirium or residual neurologic deficits, especially evident with long-standing crack/cocaine use. A paradoxic deterioration in neurologic status following initiation of cART, which is termed neurologic immune reconstitution inflammatory syndrome (neuroIRIS; discussed later), requires exclusion based on history and neuroimaging. Other neurodegenerative disorders including Alzheimer's disease and vascular dementia are becoming more prevalent as the HIV/AIDS population ages. Again, the neuropsychologic profile together with neuroimaging and CSF analysis permit distinction between HAND and these disorders. Finally, given the evolving demographics of HIV infection, especially in North America and Europe, it is essential to evaluate other causes of neurocognitive impairment such as head injury, HCV infection, and pre-existing neuropsychiatric disorders.

The most effective therapy for HAD is treatment of the underlying HIV infection with cART. cART usually consists of two nucleoside analog reverse transcriptase inhibitors (NRTIs) plus a protease inhibitor (PI) or a nonnucleoside analog reverse transcriptase inhibitor (NNRTI). Neuropsychologic performance is improved in HIV/AIDS patients treated with two NRTIs and the NNRTI nevirapine or efavirenz [32] or with two NRTIs and a PI [33,34]. Specific antiretroviral drugs have higher CSF penetration than others, including the NRTIs zidovudine (AZT), stavudine (d4T), and abacavir, the NNRTI efavirenz, and the PI indinavir [35]. It is clear that there is a positive relationship between viral load in CSF and CNS penetration of antiretroviral drugs [36]. cART shows some benefit in reversing HAD, although poor drug penetration into the CNS limits the efficacy of cART, and thus other agents have been investigated for their potential neuroprotective properties. Patients who have had substantial cART exposure before developing HAND or MND do not appear to respond as well to a change in cART, raising the possibility of drug-resistant viruses in the CNS and the emergence of a "burnt-out" or plateau phase of the disease. Memantine is a promising treatment for HAD but failed to show benefits in neuropsychologic performance in a recent clinical trial, in part because the study was underpowered [37]. Human growth hormone has been shown to be neuroprotective and may be a component of HAD treatment in the future [38]. Other neurotrophic and neuroprotective factors are currently being developed for the treatment of HAD, including minocycline, valproic acid, and new antiretroviral drugs [1].

Pediatric HIV-associated neurocognitive disorders

Although the risk of perinatal HIV transmission has been reduced to 1% to 2% with cART [39], among children who are congenitally infected with HIV, a progressive encephalopathy can develop as a frank encephalopathy or as neurodevelopmental delays in motor, language, and cognitive milestones [40]. The prevalence of neurologic involvement in HIV-infected children can be as high as 50% depending on the level of immunosuppression and treatment, usually in the first 2 years of life. Neurologic features of congenital HIV infection include microcephaly, delayed developmental milestones, spastic paraparesis, and rarely seizures. Risk factors for the development of neurologic disease in children include high CSF viral load and maternal high viral burden at the time of birth. The neuropathology in children who have HIV infection differs from adults in that there is typically high viral abundance and frequent infection of astrocytes together with basal ganglia calcification and cortical atrophy that is also evident on neuroimaging [41]. The prognosis for congenital HIV infection with encephalopathy is poor, with survival ranging from 6 to 24 months in untreated individuals. With the availability of cART, outcomes have improved, although residual neurocognitive impairment continues to be an ongoing problem, with associated behavioral complications [42].

Seizures/epilepsy

New-onset seizures occur in 8% of adults and up to 20% of children who have HIV infection, although among patients followed regularly in HIV care clinics, the prevalence of seizures is lower (2%–4%). Most seizures are generalized tonic-clonic seizures that can progress to status epilepticus in 8% to 18% of instances [43]. There are many potential causes for seizures in HIV-seropositive patients, including opportunistic infections, medications, substance use or withdrawal, metabolic disturbance, HIV infection, and pre-existing disorders. The opportunistic infections most commonly associated with seizures are toxoplasmic encephalitis, followed by cryptococcal meningitis and primary CNS lymphoma [44]. The seizure may be the only sign that the patient is suffering from an infectious or malignant process, especially in the case of CNS lymphoma. Several medications are associated with decreases in seizure threshold, including selective serotonin reuptake inhibitors, tricyclic antidepressants, ganciclovir, and foscarnet. Cocaine and heroine are also associated with an increased risk of seizures, as is alcohol withdrawal. Disturbances of electrolyte balance can also cause seizures, as in non-HIV infected patients.

In up to 50% of HIV-seropositive individuals, no underlying cause of the seizure is found except HIV itself [45]. Of interest, 25% of patients who have seizures of unknown etiology have features of HAD [35] and have a propensity to develop HAD within 6 months of the seizure [45]. Without treatment, there is a high likelihood that the seizures will recur. Because HIV-seropositive patients inherently have an increased rate of adverse drug reactions, up to a quarter of these patients will develop a rash with phenytoin. There is also a potential for drug interactions to occur between anticonvulsants and cART because these groups of drugs are largely metabolized by the cytochrome p450 system. In addition, PIs competitively bind to albumin, thus altering anticonvulsant drug levels and efficacy. Despite these concerns, carbamazepine is well tolerated when it is coupled with regular monitoring of anticonvulsant levels and viral load. Valproate has been shown to induce viral replication in vitro, although this has not been demonstrated in vivo [46]. The next generation of anticonvulsants such as gabapentin, levetiracetam, topiramate, and tiagabine may be safer [47]. As caveats, topiramate may exacerbate weight loss and should therefore be avoided in individuals who have low body mass indices, and gabapentin and levetiracetam have the benefit of being renally excreted, precluding interactions with hepatically metabolized antiretroviral drugs. Nonetheless, the risks of using "first-line" anticonvulsants (carbamzepine, phenytoin, valproate) appear to be minimal, assuming viral load and anticonvulsant levels are measured regularly as part of ongoing care.

Vacuolar myelopathy

HIV-associated vacuolar myelopathy (VM) affects 5% to 10% of untreated AIDS-defined patients, usually manifesting as gait ataxia, leg weakness, spasticity, and incontinence [48,49]. Impaired proprioception with sensory ataxia may also be present. It can occur independently or in conjunction with HAD or opportunistic infections and malignancies. Progression is insidious over months, with no associated back pain. On physical examination, symmetric spastic paraparesis is evident, with lower-extremity hyperreflexia and extensor plantar responses. Upper-limb signs are less common, although hyperreflexia of the arms is occasionally present. There is usually no defined sensory dermatomal level. Subclinical VM may be evident on examination in an otherwise asymptomatic patient, as indicated by hyperreflexia, spasticity, and extensor plantar reflexes.

The diagnosis is one of exclusion, and other causes of myelopathy should be eliminated before VM is diagnosed. For example, conditions such as human T-lymphotropic virus type I and type II infection, vitamin B_{12} deficiency, multiple sclerosis, and varicella-zoster virus–related myelopathy can present with similar clinical features. It is also essential to exclude syphilitic, tuberculous, or acute transverse myelitis. Tests such as complete blood count, B_{12} levels, CSF analysis, and neuroimaging studies should be performed to exclude these alternate causes of chronic myelopathy. The neuropathologic correlates of VM are axonal injury and intense macrophage infiltration, which is found in over 50% of HIV/AIDS at autopsy. The vacuolar appearance, primarily localized in the lateral and dorsal columns of thoracic spinal cord, may reflect intramyelinic edema [50].

The incidence of VM has dropped with increasing cART availability to a point that it is infrequently seen in HIV clinics except in severely immunosuppressed patients [49]. Although cART appears to reduce the incidence of VM, limited reversal of the signs or symptoms is observed after therapy is implemented. Symptomatic treatment (including baclofen and gabapentin) for painful spasticity, neurogenic bladder, clonus, and tremor is frequently beneficial to patients who have VM.

Neurologic immune reconstitution inflammatory syndrome

The immune reconstitution inflammatory syndrome (IRIS) is a recently recognized consequence of the introduction of cART in very immunosuppressed patients and is manifested as an abrupt clinical deterioration [51]. As immunologic improvement begins with suppression of the viral load in blood and as the patient's $CD4^+$ T cell-count rises, the clinical deterioration occurs, usually within weeks of cART initiation, although IRIS has been recognized months after cART was started. The mechanism of clinical deterioration during IRIS is thought to be sudden activation and increase in $CD4^+$ T cells (specifically CD45RO memory cells), resulting in a heightened immune response and significant host-mediated inflammatory cell damage [52]. The onset of neurologic signs and symptoms or worsening of prior neurologic disabilities is often apparent in patients who have very low $CD4^+$ T-cell counts and concurrent CNS opportunistic infections [53] after cART is introduced; this syndrome is termed neurologic IRIS (neuroIRIS). The underlying opportunistic infection may be active, although in some cases, it is subclinical and does not manifest itself until after cART is initiated [54]. Across most studies, approximately 20% to 25% of patients started on cART develop some form of IRIS (as reviewed in Ref. [55]), whereas the risk of neuroIRIS is substantially lower (<1%). On average, 60% of patients who develop IRIS present within 60 days of initiating cART, although its presentation can be biphasic, with onset within 30 days of initiating cART or after 90 days [56].

NeuroIRIS results in focal CNS deficits including encephalopathy, hemiparesis and seizures, CSF pleocytosis, and white matter and cortical abnormalities in neuroimaging, and is frequently observed in the context of pre-existing progressive multifocal leukoencephalopathy, cryptococcal meningitis, cytomegalovirus retinitis, and occasionally HAD. Neuropathologic studies of neuroIRIS disclose widespread leukocyte (lymphocyte and macrophage) infiltrates in the brain, frequently proximate to the pre-existing infection [57,58]. The extent of neurologic disability varies widely but may respond to treatment with glucocorticoids, although the full extent of neuroIRIS in terms of epidemiology and underlying pathogenesis warrants further study.

HIV-1 neuropathology and neuropathogenesis

All levels of the neural axis (brain, spinal cord, peripheral nerves) are infected by lentiviruses and are susceptible to virus-mediated damage. The CNS is protected from infectious pathogens and circulating toxins by the blood-brain barrier (BBB), or the blood-nerve barrier (BNB) in the case of the PNS (Fig. 4). Lentiviruses such as HIV enter the nervous system through a process termed *neuroinvasion* [59]. HIV-1 infects leukocytes in the periphery by way of engagement of the CD4 molecule coupled with the chemokine coreceptor CXCR4 or CCR5, although CXCR4 is used by more pathogenic HIV-1 and feline immunodeficiency virus (FIV) strains [60]. Of interest, there is a notable absence of CXCR4-dependent HIV-1 strains in the brain [61]. The infected leukocytes subsequently cross the BBB (or BNB) [62,63], thereby transporting the virus into the nervous system much like a Trojan horse. In the nervous system, HIV infects infiltrating macrophages, microglia, and astrocytes, indicative of its *neurotropism*. It does not infect neurons, oligodendrocytes, or Schwann cells [64,65]. This process culminates in cell injury and death involving neurons, astrocytes, and neural stem cells (reviewed in Ref. [66]).

The pathologic hallmarks of advanced HIV infection of the CNS include HIV encephalitis, defined by multinucleated giant cells, viral antigen detection, perivascular cuffs, and diffuse white matter pallor with ensuing neuronal or synaptodendritic loss [67–69]. The *neurovirulence* of a virus is the degree to which it contributes to neurologic injury and, in this case, neuronal cell death and synaptodendritic "pruning" [15]. HIV exerts its neurovirulence through two separate yet indelibly linked processes. The first is a host-derived cascade of inflammatory molecules released by activated or infected macrophages, microglia, and astrocytes (reviewed in Ref. [70]).

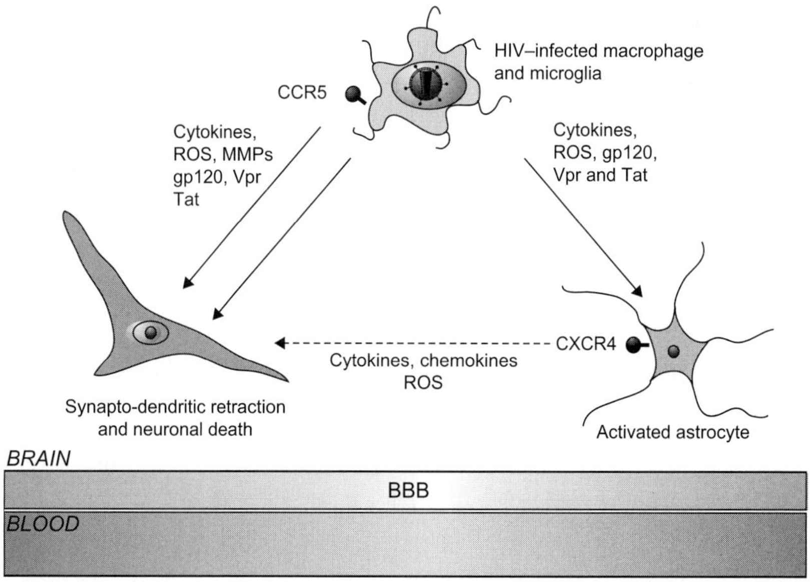

Fig. 4. HIV neuropathogenesis. HIV infects the CNS by HIV-infected leukocytes traversing the BBB and subsequently infects resident microglia, macrophages, and to a lesser extent, astrocytes. Neuronal injury and death occurs through exposure to viral proteins and host neurotoxic molecules. MMPs, matrix metalloproteinases; ROS, reactive oxygen species.

The second process involves the direct neurotoxic effects of HIV-encoded proteins such as gp120, Vpr, and Tat [66]. These mechanisms have reciprocal and complementary effects as inflammatory host cytokines enhance viral replication, whereas viral proteins (such as gp120 and Vpr) can induce aberrant innate immune responses in the CNS and the PNS [71,72]. The mechanisms of neuronal death during lentiviral infection has been shown to be apoptotic and necrotic [73–78]. Furthermore, neuronal vulnerability to injury (or death) is enhanced due to reduced neurotrophin expression during lentivirus infections (reviewed in Refs. [25,66]).

Several cytokines are elevated in brain tissue and CSF following infection by HIV-1, simian immunodeficiency virus (SIV), or FIV [71,79–81]. Tumor necrosis factor alpha (TNF-α) and inteleukin-1β are inflammatory cytokines that have received extensive attention for their potential neurotoxic effects in HIV infection. TNF-α mRNA and protein levels are increased in the brains and CSF of patients who have HIV encephalopathy [82]. Similar observations in SIV and FIV CNS infections have also been made [71,83]. These cytokines not only influence the release of other cytokines and proteases [84,85] but also prevent uptake of glutamate by astrocytes, leading to toxic accumulation of this excitatory neurotransmitter [86]. Excess activation of glutamate (*N*-methyl-D-aspartate or alpha-amino-3-hydroxy-5-methyl-4-isoxaloepropionic acid) receptors results in "excitotoxcity," with ensuing neuronal apoptosis (reviewed in Ref. [66]).

Several chemokines such as CXCL-10 (IP-10) [87] and CXCL-11 (chemokine stromal cell-derived factor-1) [88] appear to have neurotoxic properties. Other host proteins, including the matrix metalloproteinases [89] and proteinase-activated receptors [90], also display neuropathogenic actions. There is also an increasing appreciation for the role of circulating T lymphocytes ($CD4^+$ and $CD8^+$ T cells) in governing the CNS neuroimmune environment and neural cell viability in lentivirus infections [91], which is partly driven by their levels of activation as observed for other neuroinflammatory diseases [92]. Altered permeability of the BBB arising from chronic neuroinflammation is a common feature of HIV infection and is associated with the development of HIV encephalopathy as originally shown by Power and colleagues [93] and confirmed by other groups [94]. This altered permeability likely allows for greater influx of inflammatory cytokines and leukocytes from the periphery to the CNS or PNS, causing a vicious circle of increasing immune activation.

The term *neurosusceptiblity* was coined to define the relative vulnerability of different hosts to HIV-mediated neurologic disease and the associated underlying mechanisms [59]. Although it is true that the consistent host response to HIV infection is neuroinflammation, each host mounts a distinct profile of immune gene activation in the nervous system based on the host's unique genotype and phenotype. Indeed, neurosusceptibility to HIV infection is driven by many factors, including host age and state of immune health, coupled with specific genetic factors [59]. For example, susceptibility to HAD is associated with polymorphisms of the CCR5 and TNF-α genes [95,96], and HAD and DSP are associated with the E4 allele of the APOE gene [97].

Interactions between neurons and HIV-encoded proteins (such as gp120, Vpr, and Tat) directly result in neuronal death in vivo [77,98] and ex vivo [99–101]. In addition, the same proteins can indirectly mediate neuronal death through the activation of proximal glial cells with the ensuing release of neurotoxins [102–105]. Specific brain-derived HIV-1 sequences are associated with enhanced in vivo [106] and ex vivo neurovirulence [12,107]. Molecular sequence diversity is significantly greater in viruses derived from blood and brain in demented patients compared with nondemented AIDS patients [12,106,108], suggesting that diversity contributes significantly to neurotoxicity. Diversity is caused by the virus' error-prone reverse transcriptase, high rates of replication in permissive cells, intense immune selection, and recombination between closely related and disparate viral strains [25]. The consequences of viral diversity include an enhanced ability to evade the host immune system, with ensuing immune "exhaustion" and eventual collapse. Some mutations bestow a "gain of function" that expands the repertoire of receptors the virus recognizes, which leads to increased infectivity of more cell types and the ability to activate a wider spectrum of pathogenic signaling pathways. Of interest, HIV can segregate into organ-specific strains within a single host that act in concert to escalate the disease course; for example, immunosuppressive and neurovirulent viral strains cooperating within a single host [12,109–111].

The contributions of individual HIV strains as determinants of disease has remained controversial, largely because of the difficulty defining specific viral sequences associated with disease due to their immense molecular diversity. Specific HIV-1 clades are associated with an accelerated disease course compared with socioeconomically and geographically matched patients infected with other HIV-1 clades [112]. Individual strains that coinfect the same host can exacerbate and accelerate the disease course [113–115].

Despite viral load suppression in the peripheral circulation with cART and the subsequent improvement in immunity, tissue penetration of cART remains a major obstacle to suppressing viral replication in the nervous system. This remains an issue even for drugs such as zidovudine, stavudine, and efavirenz, which exhibit greater brain penetration and efficacy [15]. Impaired cART penetration to the CNS and PNS may create a viral reservoir [116] by which HIV can escape exposure to circulating antiviral regimens. These circumstances might convert the nervous system into a virus reservoir, allowing the virus to seed the entire body unhindered from the relative safety of the BBB and BNB [117,118]. The occurrence of drug-resistant HIV-1 strains, which are suppressed but then re-emerge depending on the therapeutic regimen, is well recognized [119]. Small studies indicate that these drug-resistant viruses can emerge in the CNS [108,118]. Clinical benefits from primary neurologic disease can be derived from cART (reviewed in Ref. [120]), although persistent viral replication in the brain likely contributes to the ongoing high prevalence of nervous system disease among persons who have HIV/AIDS.

Future prospects

Given the rising prevalence of HIV/AIDS in the industrialized and developing worlds, it is likely that new HIV-associated neurologic syndromes will emerge as time progresses. There is intense interest in the appearance and transmission of drug-resistant HIV-1 strains whose associated neurologic sequelae have not been well described [58]. Although there is wide experience with the neurologic disorders caused by the HIV-1 subtype B, a gap exists in understanding the effects of non–type B HIV-1 and HIV-2 infections on the nervous system. Likewise, concurrent HCV infection has become a rising concern among HIV/AIDS patients. HCV appears to infect the nervous system and is neuropathogenic, but its impact on the already vulnerable HIV-infected neural tissue is unclear [121].

There are multiple agents that may prove to be valuable neuroprotectants [66]. For example, several neurosteroids such as dehydroepiandrosterone or allopregnenolone demonstrate neuroprotective effects (reviewed in Ref. [122]) and antiviral effects [123–126], although their clinical efficacy has not yet been proved. Human growth hormone is also neuroprotective [38,127] and shows promise in clinical usage. There is also substantial interest in using anticonvulsants such as valproic acid as neuroprotective agents [128]. Likewise, antibiotics including minocycline, cephalosporins, and derivatives thereof are promising [129,130]. Neurotrophins including brain-derived neurotrophic factor, insulin-like growth factor I, and others might be beneficial, especially in patients who do not improve neurologically after the introduction of cART.

Finally, as life expectancies among HIV/AIDS patients increase and as the population ages, neurologic disorders associated with aging are becoming more prevalent. Indeed, the risk of stroke [131] and diabetes mellitus [132] is increasing among in HIV/AIDS patients, due in part to the marked metabolic abnormalities such as hyperglycemia and hyperlipidemia associated with different cART regimens coupled with advancing age. As congenitally infected children with HIV survive to adulthood, the developmental and cognitive consequences of NeuroAIDS will require further study [133]. Similarly, the risk of neurodegenerative disease associated with aging, including Alzheimer's disease, Parkinson's disease, and vascular dementia, is likely to become more evident as survival among HIV/AIDS patients continues to improve and assumes the status of a chronic and manageable disease.

Acknowledgments

The authors thank Leah DeBlock and Stephanie Skinner for their assistance with manuscript preparation.

References

[1] McArthur JC, Brew BJ, Nath A. Neurological complications of HIV infection. Lancet Neurol 2005;4:543–55.
[2] Mamidi A, DeSimone JA, Pomerantz RJ. Central nervous system infections in individuals with HIV-1 infection. J Neurovirol 2002;8:158–67.
[3] Roullet E. Opportunistic infections of the central nervous system during HIV-1 infection (emphasis on cytomegalovirus disease). J Neurol 1999;246:237–43.
[4] Antinori A, Arendt G, Becker JT, et al. Updated research nosology for HIV-associated neurocognitive disorders. Neurology 2007;69:1789–99.
[5] Pandya R, Krentz HB, Gill MJ, et al. HIV-related neurological syndromes reduce health-related quality of life. Can J Neurol Sci 2005;32:201–4.
[6] Yeung H, Krentz HB, Gill MJ, et al. Neuropsychiatric disorders in HIV infection: impact of diagnosis on economic costs of care. AIDS 2006;20:2005–9.
[7] Maher J, Choudhri S, Halliday W, et al. AIDS dementia complex with generalized myoclonus. Mov Disord 1997;12:593–7.
[8] Mirsattari SM, Power C, Nath A. Parkinsonism with HIV infection. Mov Disord 1998;3:684–9.
[9] Navia BA, Jordan BD, Price RW. The AIDS dementia complex: I. Clinical features. Ann Neurol 1986;19:517–24.
[10] Alciati A, Fusi A, D'Arminio Monforte A, et al. New-onset delusions and hallucinations in patients infected with HIV. J Psychiatry Neurosci 2001;26:229–34.
[11] Koutsilieri E, Scheller C, Sopper S, et al. Psychiatric complications in human immunodeficiency virus infection. J Neurovirol 2002;(8) Suppl 2:129–33.
[12] Power C, McArthur JC, Johnson RT, et al. Demented and nondemented patients with AIDS differ in brain-derived human immunodeficiency virus type 1 envelope sequences. J Virol 1994;68:4643–9.
[13] A randomized, double-blind, placebo-controlled trial of deprenyl and thioctic acid in human immunodeficiency virus-associated cognitive impairment. Dana Consortium on the Therapy of HIV Dementia and Related Cognitive Disorders [see comments]. Neurology 1998;50:645–51.
[14] Ellis RJ, Deutsch R, Heaton RK, et al. Neurocognitive impairment is an independent risk factor for death in HIV infection. San Diego HIV Neurobehavioral Research Center Group. Arch Neurol 1997;54:416–24.
[15] Ellis R, Langford D, Masliah E. HIV and antiretroviral therapy in the brain: neuronal injury and repair. Nat Rev Neurosci 2007;8:33–44.
[16] Grant I. Neurocognitive disturbances in HIV. Int Rev Psychiatry 2008;20:33–47.
[17] McArthur JC, Hoover DR, Bacellar H, et al. Dementia in AIDS patients: incidence and risk factors. Multicenter AIDS Cohort Study. Neurology 1993;43:2245–52.
[18] Brew BJ. Evidence for a change in AIDS dementia complex in the era of highly active antiretroviral therapy and the possibility of new forms of AIDS dementia complex. AIDS 2004;(18 Suppl 1):S75–8.
[19] Dal Pan GJ, McArthur JH, Aylward E, et al. Patterns of cerebral atrophy in HIV-1-infected individuals: results of a quantitative MRI analysis. Neurology 1992;42:2125–30.
[20] Simpson DM, Tagliati M. Neurologic manifestations of HIV infection [published erratum appears in Ann Intern Med 1995;122(4):317]. Ann Intern Med 1994;121:769–85.
[21] McArthur JC, Cohen BA, Farzedegan H, et al. Cerebrospinal fluid abnormalities in homosexual men with and without neuropsychiatric findings. Ann Neurol 1988;23:S34–7.
[22] Sacktor N, Skolasky RL, Ernst T, et al. A multicenter study of two magnetic resonance spectroscopy techniques in individuals with HIV dementia. J Magn Reson Imaging 2005;21:325–33.
[23] Ernst T, Chang L, Arnold S. Increased glial metabolites predict increased working memory network activation in HIV brain injury. Neuroimage 2003;19:1686–93.

[24] Power C, Buist R, Johnston JB, et al. Neurovirulence in feline immunodeficiency virus-infected neonatal cats is viral strain specific and dependent on systemic immune suppression. J Virol 1998;72:9109–15.
[25] van Marle G, Power C. Human immunodeficiency virus type 1 genetic diversity in the nervous system: evolutionary epiphenomenon or disease determinant? J Neurovirol 2005; 11:107–28.
[26] Power C, Selnes OA, Grim JA, et al. HIV Dementia Scale: a rapid screening test. J Acquir Immune Defic Syndr Hum Retrovirol 1995;8:273–8.
[27] Jones BN, Teng EL, Folstein MF, et al. A new bedside test of cognition for patients with HIV infection. Ann Intern Med 1993;119:1001–4.
[28] Berghuis JP, Uldall KK, Lalonde B. Validity of two scales in identifying HIV-associated dementia. J Acquir Immune Defic Syndr 1999;21:134–40.
[29] Grassi MP, Perin C, Borella M, et al. Assessment of cognitive function in asymptomatic HIV-positive subjects. Eur Neurol 1999;42:225–9.
[30] Sacktor NC, Wong M, Nakasujja N, et al. The international HIV Dementia Scale: a new rapid screening test for HIV dementia. AIDS 2005;19:1367–74.
[31] Price RW, Brew BJ. The AIDS dementia complex. J Infect Dis 1988;158:1079–83.
[32] Price RW, Yiannoutos T, Clifford DB, et al. Neurological outcomes in late HIV infection: adverse impact of neurological survival and protection effect of antiretroviral therapy. AIDS 1999;13:1677–85.
[33] Sacktor NC, Lyles RH, Skolasky RL, et al. Combination antiretroviral therapy improves psychomotor speed performance in HIV-seropositive homosexual men. Multicenter AIDS Cohort Study (MACS). Neurology 1999;52:1640–7.
[34] von Giesen HJ, Koller H, Theisen A, et al. Therapeutic effects of nonnucleoside reverse transcriptase inhibitors on the central nervous system in HIV-1-infected patients. J Acquir Immune Defic Syndr 2002;29:363–7.
[35] Nath A, Berger J. HIV dementia. Curr Treat Options Neurol 2004;6:139–51.
[36] Letendre S, Marquie-Beck J, Capparelli E, et al. Validation of the CNS penetration-effectiveness rank for quantifying antiretroviral penetration into the central nervous system. Arch Neurol 2008;65:65–70.
[37] Schifitto G, Navia BA, Yiannoutsos CT, et al. Memantine and HIV-associated cognitive impairment: a neuropsychological and proton magnetic resonance spectroscopy study. AIDS 2007;21:1877–86.
[38] Silva C, Zhang K, Tsutsui S, et al. Growth hormone prevents human immunodeficiency virus-induced neuronal p53 expression. Ann Neurol 2003;54:605–14.
[39] Fowler MG, Lampe MA, Jamieson DJ, et al. Reducing the risk of mother-to-child human immunodeficiency virus transmission: past successes, current progress and challenges, and future directions. Am J Obstet Gynecol 2007;197:S3–9.
[40] Van Rie A, Harrington PR, Dow A, et al. Neurologic and neurodevelopmental manifestations of pediatric HIV/AIDS: a global perspective. Eur J Paediatr Neurol 2007;11:1–9.
[41] Tardieu M, Le Chenadec J, Persoz A, et al. HIV-1-related encephalopathy in infants compared with children and adults. French Pediatric HIV Infection Study and the SEROCO Group. Neurology 2000;54:1089–95.
[42] Shanbhag MC, Rutstein RM, Zaoutis T, et al. Neurocognitive functioning in pediatric human immunodeficiency virus infection: effects of combined therapy. Arch Pediatr Adolesc Med 2005;159:651–6.
[43] Kellinghaus C, Engbring C, Kovac S, et al. Frequency of seizures and epilepsy in neurological HIV-infected patients. Seizure 2008;17(1):27–33.
[44] Garg RK. HIV infection and seizures. Postgrad Med J 1999;75:387–90.
[45] Dore GJ, Law MG, Brew BJ. Prospective analysis of seizures occurring in human immunodeficiency virus type-1 infection. J Neuro AIDS 1996;1:59–69.
[46] Romanelli F, Jennings HR, Nath A, et al. Therapeutic dilemma: the use of anticonvulsants in HIV-positive individuals. Neurology 2000;54:1404–7.

[47] Romanelli F, Ryan M. Seizures in HIV-seropositive individuals: epidemiology and treatment. CNS Drugs 2002;16:91–8.
[48] Dal Pan GJ, Berger JR. Spinal cord disease in human immunodeficiency virus infection. In: Berger JR, Levy RM, editors. AIDS and the nervous system. 2nd edition. Philadelphia: Lippincott-Raven Publishers; 1997. p. 173–87.
[49] DiRocco L, Dalton T, Liang D, et al. Nonallelism for the audiogenic seizure prone (Asp1) and the aryl hydrocarbon receptor (Ahr) loci in mice. J Neurogenet 1998;12:191–203.
[50] Eyer-Silva WA, Auto I, Pinto JF, et al. Myelopathy in a previously asymptomatic HIV-1-infected patient. Infection 2001;29:99–102.
[51] Riedel DJ, Pardo CA, McArthur J, et al. Therapy insight: CNS manifestations of HIV-associated immune reconstitution inflammatory syndrome. Nat Clin Pract Neurol 2006;2: 557–65.
[52] Lipman M, Breen R. Immune reconstitution inflammatory syndrome in HIV. Curr Opin Infect Dis 2006;19:20–5.
[53] Venkataramana A, Pardo CA, McArthur JC, et al. Immune reconstitution inflammatory syndrome in the CNS of HIV-infected patients. Neurology 2006;67:383–8.
[54] Subsai K, Kanoksri S, Siwaporn C, et al. Neurological complications in AIDS patients receiving HAART: a 2-year retrospective study. Eur J Neurol 2006;13:233–9.
[55] Murdoch DM, Venter WD, Van Rie A, et al. Immune reconstitution inflammatory syndrome (IRIS): review of common infectious manifestations and treatment options. AIDS Res Ther 2007;4:9.
[56] Shelburne SA, Visnegarwala F, Darcourt J, et al. Incidence and risk factors for immune reconstitution inflammatory syndrome during highly active antiretroviral therapy. AIDS 2005;19:399–406.
[57] Miller RF, Isaacson PG, Hall-Craggs M, et al. Cerebral CD8+ lymphocytosis in HIV-1 infected patients with immune restoration induced by HAART. Acta Neuropathol (Berl) 2004;108:17–23.
[58] Vendrely A, Bienvenu B, Gasnault J, et al. Fulminant inflammatory leukoencephalopathy associated with HAART-induced immune restoration in AIDS-related progressive multifocal leukoencephalopathy. Acta Neuropathol (Berl) 2005;109(4):449–55.
[59] Patrick MK, Johnston JB, Power C. Lentiviral neuropathogenesis: comparative neuroinvasion, neurotropism, neurovirulence and host neurosusceptibility. J Virol 2002;76:7923–31.
[60] Kimata JT, Kuller L, Anderson DB, et al. Emerging cytopathic and antigenic simian immunodeficiency virus variants influence AIDS progression. Nat Med 1999;5:535–41.
[61] Klasse PJ, Moore JP. Is there enough gp120 in the body fluids of HIV-1-infected individuals to have biologically significant effects? Virology 2004;323:1–8.
[62] Haase AT. Pathogenesis of lentivirus infections. Nature 1986;322:130–6.
[63] Power C. Retroviral diseases of the nervous system: pathogenic host response or viral gene-mediated neurovirulence? Trends Neurosci 2001;24:162–9.
[64] Brew, BJ. HIV Neurology. Oxford University; 2001. p. 8–12.
[65] Trillo-Pazos G, Diamanturos A, Rislove L, et al. Detection of HIV-1 DNA in microglia/macrophages, astrocytes and neurons isolated from brain tissue with HIV-1 encephalitis by laser capture microdissection. Brain Pathol 2003;13:144–54.
[66] Jones G, Power C. Regulation of neural cell survival by HIV-1 infection. Neurobiol Dis 2006;21:1–17.
[67] Navia BA, Cho ES, Petito CK, et al. The AIDS dementia complex: II. Neuropathology. Ann Neurol 1986;19:525–35.
[68] Sharer LR. Pathology of HIV-1 infection of the central nervous system. A review. J Neuropathol Exp Neurol 1992;51:3–11.
[69] Vinters HV, Anders KH. Neuropathology of AIDS. Boca Raton (FL): CRC Press, Inc.; 1990.
[70] Persidsky Y, Gendelman HE. Mononuclear phagocyte immunity and the neuropathogenesis of HIV-1 infection. J Leukoc Biol 2003;74:691–701.

[71] Johnston JB, Jiang Y, van Marle G, et al. Lentiviral infection in the brain induces matrix metalloproteinase expression: the role of envelope diversity. J Virol 2000;74:7211–20.
[72] Zhu Y, Jones G, Tsutsui S, et al. Lentivirus infection causes neuroinflammation and neuronal injury in dorsal root ganglia: pathogenic effects of STAT-1 and inducible nitric oxide synthase. J Immunol 2005;175:1118–26.
[73] Adle-Biassette H, Chretien F, Wingertsmann L, et al. Neuronal apoptosis does not correlate with dementia in HIV infection but is related to microglial activation and axonal damage. Neuropathol Appl Neurobiol 1999;25:123–33.
[74] Adle-Biassette H, Levy Y, Colombel M, et al. Neuronal apoptosis in HIV infection in adults. Neuropathol Appl Neurobiol 1995;21:218–27.
[75] Chen W, Sulcove J, Frank I, et al. Development of a human neuronal cell model for human immunodeficiency virus (HIV)-infected macrophage-induced neurotoxicity: apoptosis induced by HIV type 1 primary isolates and evidence for involvement of the Bcl-2/Bcl-xL-sensitive intrinsic apoptosis pathway. J Virol 2002;76:9407–19.
[76] Gelbard HA, James HJ, Sharer LR, et al. Apoptotic neurons in brains from paediatric patients with HIV-1 encephalitis and progressive encephalopathy. Neuropathol Appl Neurobiol 1995;21:208–17.
[77] Jones GJ, Barsby NL, Cohen EA, et al. HIV–1 Vpr causes neuronal apoptosis and in vivo neurodegeneration. J Neurosci 2007;27(14):3703–11.
[78] Zhang K, Rana F, Silva C, et al. Human immunodeficiency virus type 1 envelope-mediated neuronal death: uncoupling of viral replication and neurotoxicity. J Virol 2003;77:6899–912.
[79] Mintz M, Rapaport R, Oleske JM, et al. Elevated serum levels of tumor necrosis factor are associated with progressive encephalopathy in children with acquired immunodeficiency syndrome. Am J Dis Child 1989;143:771–4.
[80] Wesselingh SL, Power C, Glass JD, et al. Intracerebral cytokine messenger RNA expression in acquired immunodeficiency syndrome dementia. Ann Neurol 1993;33:576–82.
[81] Zink MC, Amedee AM, Mankowski JL, et al. Pathogenesis of SIV encephalitis. Selection and replication of neurovirulent SIV. Am J Pathol 1997;151:793–803.
[82] Brew BJ, Pemberton L, Cunningham P, et al. Levels of human immunodeficiency virus type 1 RNA in cerebrospinal fluid correlate with AIDS dementia stage. J Infect Dis 1997;175:963–6.
[83] Berman NE, Marcario JK, Yong C, et al. Microglial activation and neurological symptoms in the SIV model of NeuroAIDS: association of MHC-II and MMP-9 expression with behavioral deficits and evoked potential changes. Neurobiol Dis 1999;6:486–98.
[84] Benveniste EN. Cytokine circuits in brain: implications for AIDS dementia complex. In: Price RW, Perry SW, editors. HIV, AIDS, and the brain. New York: Raven Press; 1994. p. 71–88.
[85] Wilt SG, Milward E, Zhou JM, et al. In vitro evidence for a dual role of tumor necrosis factor-alpha in human immunodeficiency virus type 1 encephalopathy. Ann Neurol 1995;37:381–94.
[86] Fine SM, Angel RA, Perry SW, et al. Tumor necrosis factor alpha inhibits glutamate uptake by primary human astrocytes. Implications for pathogenesis of HIV-1 dementia. J Biol Chem 1996;271:15303–6.
[87] van Marle G, Henry S, Todoruk T, et al. Human immunodeficiency virus type 1 Nef protein mediates neural cell death: a neurotoxic role for IP-10. Virology 2004;329:302–18.
[88] Zhang K, McQuibban GA, Silva C, et al. HIV-induced metalloproteinase processing of the chemokine stromal cell derived factor-1 causes neurodegeneration. Nat Neurosci 2003;6:1064–71.
[89] Yong VW, Power C, Forsyth P, et al. Metalloproteinases in the biology and pathology of the nervous system. Nat Rev Neurosci 2001;2:502–11.
[90] Boven LA, Vergnolle N, Henry SD, et al. Upregulation of proteinase-activated receptor 1 expression in astrocytes during HIV encephalitis. J Immunol 2003;170:2638–46.

[91] Persidsky Y, Poluektova L. Immune privilege and HIV-1 persistence in the CNS. Immunol Rev 2006;213:180–94.
[92] Sospedra M, Martin R. Immunology of multiple sclerosis. Annu Rev Immunol 2005;23: 683–747.
[93] Power C, Kong PA, Crawford TO, et al. Cerebral white matter changes in acquired immunodeficiency syndrome dementia: alterations of the blood-brain barrier. Ann Neurol 1993; 34:339–50.
[94] Berger JR, Nath A, Greenberg RN, et al. Cerebrovascular changes in the basal ganglia with HIV dementia. Neurology 2000;54:921–6.
[95] Quasney MW, Zhang Q, Sargent S, et al. Increased frequency of the tumor necrosis factor-alpha-308 A allele in adults with human immunodeficiency virus dementia. Ann Neurol 2001;50:157–62.
[96] Van Rij RP, Portegies P, Hallaby T, et al. Reduced prevalence of the CCR5 delta32 heterozygous genotype in human immunodeficiency virus-infected individuals with AIDS dementia complex. J Infect Dis 1999;180:854–7.
[97] Corder EH, Robertson K, Lannfelt L, et al. HIV-infected subjects with the E4 allele for APOE have excess dementia and peripheral neuropathy. Nat Med 1998;4:1182–4.
[98] Johnston JB, Zhang K, Silva C, et al. HIV-1 Tat neurotoxicity is prevented by matrix metalloproteinase inhibitors. Ann Neurol 2001;49:230–41.
[99] Adamson DC, Wildemann B, Sasaki M, et al. Immunologic NO synthase: elevation in severe AIDS dementia and induction by HIV-1 gp41. Science 1996;274:1917–21.
[100] Nath A, Psooy K, Martin C, et al. Identification of a human immunodeficiency virus type 1 Tat epitope that is neuroexcitatory and neurotoxic. J Virol 1996;70:1475–80.
[101] Overholser ED, Coleman GD, Bennett JL, et al. Expression of simian immunodeficiency virus (SIV) nef in astrocytes during acute and terminal infection and requirement of nef for optimal replication of neurovirulent SIV in vitro. J Virol 2003;77:6855–66.
[102] Barks JD, Sun R, Malinak C, et al. gp120, an HIV-1 protein, increases susceptibility to hypoglycemic and ischemic brain injury in perinatal rats. Exp Neurol 1995;132: 123–33.
[103] Gruol DL, Yu N, Parsons KL, et al. Neurotoxic effects of feline immunodeficiency virus, FIV-PPR. J Neurovirol 1998;4:415–25.
[104] Kaiser PK, Offermann JT, Lipton SA. Neuronal injury due to HIV-1 envelope protein is blocked by anti-gp120 antibodies but not by anti-CD4 antibodies. Neurology 1990;40: 1757–61.
[105] Meeker RB, Robertson K, Barry T, et al. Neurotoxicity of CSF from HIV-infected humans. J Neurovirol 1999;5:507–18.
[106] van Marle G, Rourke SB, Zhang K, et al. HIV dementia patients exhibit reduced viral neutralization and increased envelope sequence diversity in blood and brain. AIDS 2002; 16:1905–14.
[107] Power C, McArthur JC, Nath A, et al. Neuronal death induced by brain-derived human immunodeficiency virus type 1 envelope genes differs between demented and nondemented AIDS patients. J Virol 1998;72:9045–53.
[108] Bratanich AC, Liu C, McArthur JC, et al. Brain-derived HIV-1 tat sequences from AIDS patients with dementia show increased molecular heterogeneity. J Neurovirol 1998;4: 387–93.
[109] Hughes ES, Bell JE, Simmonds P. Investigation of the dynamics of the spread of human immunodeficiency virus to brain and other tissues by evolutionary analysis of sequences from the p17gag and env genes. J Virol 1997;71:1272–80.
[110] Iwasa Y, Michor F, Nowak MA. Virus evolution within patients increases pathogenicity. J Theor Biol 2005;232:17–26.
[111] Liu P, Hudson LC, Tompkins MB, et al. Compartmentalization and evolution of feline immunodeficiency virus between the central nervous system and periphery following intracerebroventricular or systemic inoculation. J Neurovirol 2006;12:307–21.

[112] Kanki PJ, Hamel DJ, Sankale JL, et al. Human immunodeficiency virus type 1 subtypes differ in disease progression. J Infect Dis 1999;179:68–73.
[113] Gottlieb GS, Nickle DC, Jensen MA, et al. Dual HIV-1 infection associated with rapid disease progression. Lancet 2004;363:619–22.
[114] Grobler J, Gray CM, Rademeyer C, et al. Incidence of HIV-1 dual infection and its association with increased viral load set point in a cohort of HIV-1 subtype C-infected female sex workers. J Infect Dis 2004;190:1355–9.
[115] Koelsch KK, Smith DM, Little SJ, et al. Clade B HIV-1 superinfection with wild-type virus after primary infection with drug-resistant clade B virus. AIDS 2003;17:F11–6.
[116] Nath A, Sacktor N. Influence of highly active antiretroviral therapy on persistence of HIV in the central nervous system. Curr Opin Neurol 2006;19:358–61.
[117] Clements JE, Babas T, Mankowski JL, et al. The central nervous system as a reservoir for simian immunodeficiency virus (SIV): steady-state levels of SIV DNA in brain from acute through asymptomatic infection. J Infect Dis 2002;186:905–13.
[118] Smit TK, Brew BJ, Tourtellotte W, et al. Independent evolution of human immunodeficiency virus (HIV) drug resistance mutations in diverse areas of the brain in HIV-infected patients, with and without dementia, on antiretroviral treatment. J Virol 2004;78: 10133–48.
[119] Imamichi T. Action of anti-HIV drugs and resistance: reverse transcriptase inhibitors and protease inhibitors. Curr Pharm Des 2004;10:4039–53.
[120] Power C, Gill MJ, Johnson RT. The neuropathogenesis of HIV infection: host-virus interaction and the impact of therapy. Can J Neurol Sci 2002;29:19–32.
[121] Letendre S, Paulino AD, Rockenstein E, et al. Pathogenesis of hepatitis C virus coinfection in the brains of patients infected with HIV. J Infect Dis 2007;196:361–70.
[122] Mackenzie EM, Odontiadis J, Le Melledo JM, et al. The relevance of neuroactive steroids in schizophrenia, depression, and anxiety disorders. Cell Mol Neurobiol 2007;27(5):541–74.
[123] Bradley WG, Kraus LA, Good RA, et al. Dehydroepiandrosterone inhibits replication of feline immunodeficiency virus in chronically infected cells. Vet Immunol Immunopathol 1995;46:159–68.
[124] Padgett DA, Loria RM, Sheridan JF. Endocrine regulation of the immune response to influenza virus infection with a metabolite of DHEA-androstenediol. J Neuroimmunol 1997;78:203–11.
[125] Pedersen NC, North TW, Rigg R, et al. 16alpha-Bromo-epiandrosterone therapy modulates experimental feline immunodeficiency virus viremia: initial enhancement leading to long-term suppression. Vet Immunol Immunopathol 2003;94:133–48.
[126] Yang JY, Schwartz A, Henderson EE. Inhibition of HIV-1 latency reactivation by dehydroepiandrosterone (DHEA) and an analog of DHEA. AIDS Res Hum Retroviruses 1993;9:747–54.
[127] van Marle G, Antony JM, Silva C, et al. Aberrant cortical neurogenesis in a pediatric neuroAIDS model: neurotrophic effects of growth hormone. AIDS 2005;19:1781–91.
[128] Schifitto G, Peterson DR, Zhong J, et al. Valproic acid adjunctive therapy for HIV-associated cognitive impairment: a first report. Neurology 2006;66:919–21.
[129] Rumbaugh JA, Li G, Rothstein J, et al. Ceftriaxone protects against the neurotoxicity of human immunodeficiency virus proteins. J Neurovirol 2007;13:168–72.
[130] Zink MC, Uhrlaub J, DeWitt J, et al. Neuroprotective and anti-human immunodeficiency virus activity of minocycline. JAMA 2005;293:2003–11.
[131] Ortiz G, Koch S, Romano JG, et al. Mechanisms of ischemic stroke in HIV-infected patients. Neurology 2007;68:1257–61.
[132] Ledergerber B, Furrer H, Rickenbach M, et al. Factors associated with the incidence of type 2 diabetes mellitus in HIV-infected participants in the Swiss HIV Cohort Study. Clin Infect Dis 2007;45:111–9.
[133] Jeremy RJ, Kim S, Nozyce M, et al. Neuropsychological functioning and viral load in stable antiretroviral therapy-experienced HIV-infected children. Pediatrics 2005;115:380–7.

[134] Cooper DA, Gold J, Maclean P, et al. Acute AIDS retrovirus infection. Definition of a clinical illness associated with seroconversion. Lancet 1985;1:537–40.
[135] Mirsattari SM, Power C, Nath A. Primary headaches in HIV-infected patients. Headache 1999;39:3–10.
[136] Brew BJ, Miller J. Human immunodeficiency virus-related headache. Neurology 1993;43:1098–100.

ELSEVIER
SAUNDERS

Neurol Clin 26 (2008) 821–832

NEUROLOGIC
CLINICS

Diagnosis and Management of HIV-Associated Neuropathy

Alejandra Gonzalez-Duarte, MD, Jessica Robinson-Papp, MD, David M. Simpson, MD*

The Mount Sinai School of Medicine, One Gustave L. Levy Place, Box 1187, New York, NY 10029-6574, USA

Significant advances have been made in the management of HIV infection. Longer life spans, however, have led to an increased prevalence of certain neurologic disorders due to longer duration of infection and toxicity associated with the use of multiple drugs. Of these disorders, distal sensory polyneuropathy (DSP) is the most common neurologic complication and a major cause of morbidity in HIV-infected patients. DSP may occur secondary to HIV (HIV-DSP) or be due to antiretroviral (ARV) drug toxicity (ARV-DSP). The clinical features of HIV-DSP and ARV-DSP are identical, with pain and paresthesias being the principal symptoms.

It is imperative to recognize the symptoms and signs of DSP in patients who have HIV. Timely detection may allow for the reversal of the toxic effects of ARVs and for the initiation of symptomatic treatment. It is unfortunate that DSP is frequently misdiagnosed or delayed in diagnosis, leading to adverse effects on overall health and quality of life.

The pathogenic mechanism of HIV-DSP is uncertain but likely multifactorial. HIV-DSP tends to appear in advanced stages of HIV infection. It is thought that HIV disease progression leads to dysregulation of macrophages and overproduction of proinflammatory cytokines and chemokines, which in turn leads to DSP. ARV-DSP is associated with the use of the nucleoside analog reverse transcriptase inhibitors didanosine (ddI), zalcitarabine (ddC), and stavudine (d4T). More recently, some protease inhibitors (PIs), including indinavir, saquinavir, and ritonavir, have also been implicated. These ARVs appear to have adverse effects on metabolism, with increased risk for mitochondrial toxicity. In many cases, ARV toxicity

* Corresponding author. Clinical Neurophysiology Laboratories, Annenberg 2nd Floor, Box 1052, New York, NY 10029.

E-mail address: david.simpson@mssm.edu (D.M. Simpson).

doi:10.1016/j.ncl.2008.04.001 **neurologic.theclinics.com**

may unmask a previously asymptomatic HIV-DSP or may act synergistically with the pathogenic effect of HIV itself.

Before the introduction of highly active ARV treatment (HAART), age, high plasma viral load, and low CD4 cell count were the most common predisposing factors to the development of DSP.

As with other forms of DSP, the pathologic hallmark is distal axonal degeneration. Small-diameter nociceptive sensory axons and their respective soma in the dorsal root ganglion are the principal cellular structures affected.

Restorative therapies for DSP are not currently available; however, recent advances have led to novel symptomatic therapies. Although the strength of the clinical trial data varies, symptomatic agents commonly used in clinical practice include antidepressants such as amitryptiline and duloxetine; anticonvulsants, particularly gabapentin, lamotrigine, and pregabalin; and topical therapies such as lidocaine and capsaicin.

Incidence, prevalence, and risk factors

The reported incidence and prevalence of DSP in the HIV-positive population vary among studies, partly due to evolution in the nature of HIV disease and to differences in diagnostic criteria for DSP, including clinical, electrophysiologic, and pathologic features. In addition, most epidemiologic studies do not distinguish between HIV-DSP and ARV-DSP or they have differing definitions for the diagnosis of ARV-DSP. In the pre-ARV era, studies suggested incidence rates of DSP as high as 89% [1]. Most prospective studies after the introduction of HAART estimate the incidence of neuropathy of 30% [1,2] and with a prevalence of 53% [3]; however, the histologic evidence of neuropathy in autopsy series is close to 100% [1].

Before the introduction of HAART, the main risks factors for DSP were those related to advanced disease, such as low $CD4^+$ T-cell count and high plasma HIV viral load [4]. In the current HAART era, however, the correlation of DSP with most of these risk factors is less clear, which may be the result of acquisition bias in the cohort studies from which this information is derived [3,5,6]. Recent studies have shown HIV-DSP at higher $CD4^+$ T-cell levels than previously observed and in the context of undetectable plasma viral load [7]. Several risk factors are associated with a heightened risk for symptomatic DSP. The most common risk factors in most of the prospective post-HAART cohorts are older age, nadir CD4 counts of less than 50 cells/mm^3, nutritional deficiency, coexisting diabetes mellitus, use of neurotoxic drugs, and alcohol exposure [8]. Low baseline epidermal nerve fiber density (<11 fibers/mm) in skin biopsy specimens correlates with the development of DSP [8]. There may also be differences between risk factors for HIV-DSP and ARV-DSP. HIV-DSP patients appear to be older and have higher peak plasma viral loads than patients who have ARV-DSP [9]. Additional studies are needed to determine the prevalence of neuropathy across cohorts that have differences in race and ethnicity.

Pathogenesis

Although clinically similar, the pathophysiologies of HIV-DSP and ARV-DSP are distinct. Several hypotheses have been postulated, but it is likely that multiple mechanisms interact to produce the axonal injury associated with HIV-DSP. HIV does not infect axons or Schwann cells, the primary cell types of peripheral nerves. HIV-related peripheral and central nervous system toxicity may be due to cytokine-mediated effects. There is evidence that an HIV-associated protein (gp120) may play a key role in this process, perhaps by way of ligation of chemokine receptors on glial cells and neurons [1] or by acting on chemokine receptors related to Schwann cell–neuron interaction [10]. The resulting inflammatory reaction in the nerve and surrounding tissues leads to axonal damage and the characteristic pain associated with DSP. This hypothesis is supported by animal studies that show that gp120 induces allodynia when injected into rat paws [11]. Neuropathic pain may also be produced by the spontaneous activity of uninjured type C nociceptive fibers after injury to neighboring peripheral nerve fibers. Finally, gp120-responsive neurons can express the capsaicin receptor VR1/TRKV1 or induce proinflammatory cytokines such as interleukin (IL)-1 and tumor necrosis factor alpha (TNF-α) [1]. Central remodeling within the dorsal horn of the spinal cord is thought to play an important role in the maintenance of neuropathic pain [12].

ARV-DSP primarily occurs in patients treated with the nucleoside analog reverse transcriptase inhibitors commonly known as "d-drugs" (ie, ddI, ddC, and d4T). Peripheral neuropathy in this context occurs mostly within the first year of treatment or in patients who have a pre-existing diagnosis of peripheral neuropathy [13]. The mechanism of ARV-DSP is unknown but may be related to mitochondrial dysfunction. Increased numbers of abnormal mitochondria in neuronal axons and Schwann cells and depletion of mitochondrial DNA have been demonstrated in ddC-associated DSP [14]. Mitochondrial dysfunction may be related to the inhibition of mitochondrial DNA polymerase-gamma, resulting in energy failure due to loss of the normal electrical potential difference across the inner mitochondrial membrane. Genetic variations in polymerase-gamma could explain the enhanced susceptibility for toxic neuropathy among some populations [8]. Toxicity also seems to depend on the agent and the dose used. In cell culture assays, the d-drug ARVs most frequently clinically associated with DSP (ddI, ddC, and d4T) result in the greatest degree of mitochondrial DNA inhibition [12,15]. Efforts to demonstrate mitochondrial toxicity in vivo have produced mixed results. Elevated serum lactate levels may discriminate between d4T-induced DSP and HIV-DSP [16], but other studies have not replicated these results. Deficiency in serum levels of acetyl-L-carnitine, a critical substrate for normal mitochondrial function, has not been useful in discriminating between DSP due to ARV and HIV-DSP [17,18].

The association between specific PIs and the development of ARV-DSP is being studied. Patients treated with indinavir, saquinavir, and ritonavir may exhibit more symptomatic DSP than those treated with other PIs. This observation may be related to more efficient penetration of neural compartments by these PIs [9]. In addition, HIV-infected dorsal root ganglion cultures exposed to indinavir or ddI showed significant neuronal atrophy, neurite retraction, and process loss compared with controls. Recent cohort studies have not confirmed the clinical association between PI exposure and DSP [19].

Pathology

The pathologic hallmark of DSP is distal axonal degeneration, affecting predominantly small, unmyelinated fibers. Axonal loss in myelinated fibers, and less commonly demyelination and remyelination, may also be seen [1]. Changes also occur more proximally. Macrophage activation, inflammatory infiltrates of lymphocytes, and proinflammatory cytokine production, including TNF-α, IL-6, and nitric oxide, have been identified within the dorsal root ganglion and may be related to central remodeling and sensitization. Increased macrophage activation is correlated with symptomatic HIV-DSP [2]. Skin-punch biopsies have been used effectively to show the epidermal denervation characteristic of small-fiber neuropathies.

Prominent mitochondrial abnormalities have been shown in association with exposure to nucleoside reverse transcriptase inhibitors. In vitro observation of a graded inhibition of DNA polymerase-gamma by specific dideoxynucleosides (ddC, ddI, and d4T) on the mitochondria is seen as a reduction in the copy numbers of mitochondrial DNA and metabolic abnormalities. In addition, discontinuation of these neurotoxin medications normalizes these changes.

Clinical features

The symptoms of DSP are dominated by pain and uncomfortable sensory phenomena. Distal painful dysesthesias, allodynia, severe burning pain, pins and needles sensations, and numbness are the main symptoms. Symptoms usually begin in the feet, often on the soles, and progress up the legs. In more severe cases, symptoms may reach the knees or involve the hands, assuming the classic "stocking and glove distribution" (Fig. 1). The symptoms are typically bilateral as the condition progresses but can be asymptomatic at the outset.

Physical examination confirms the distal loss of sensory function. Sensory dysfunction may be present in all modalities but is most marked in those related to small-fiber function, namely pain and temperature. Ankle reflexes are almost universally lost or diminished in comparison to the knee reflexes. In some patients, brisk knee reflexes with normal ankle reflexes may indicate combined myelopathy and DSP. Joint position sense and strength are

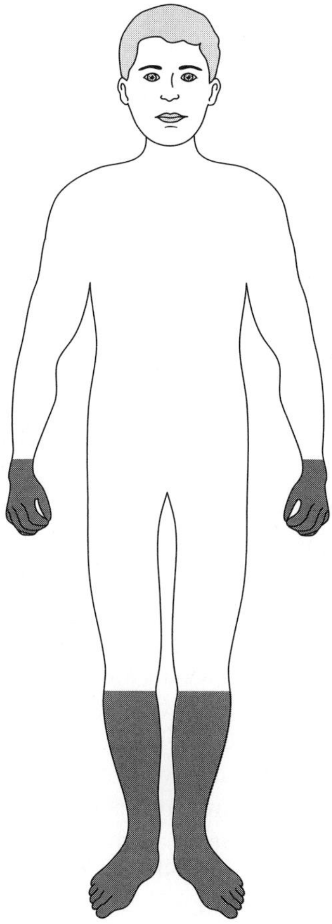

Fig. 1. DSP "stocking and glove" distribution.

relatively preserved. Subclinical neuropathy is common in HIV patients, and clinical signs of DSP are often found in asymptomatic patients.

DSP due to ARV use cannot be differentiated clinically from HIV-DSP and is a diagnosis made largely on historical grounds. ARV-DSP usually bears a temporal relationship to initiation of the drug. A history of symptom onset soon after the initiation of a potentially neurotoxic ARV, often within weeks, supports the diagnosis of ARV-DSP. Most important, symptoms improve or resolve within 8 weeks following dose reduction or drug cessation, although in some patients, a "coasting phenomenon" of increased symptoms may follow drug discontinuation before clinical improvement occurs. Incomplete resolution of symptoms and signs following cessation of neurotoxic drugs may be due to coexistent HIV-DSP.

Patients who have any systemic polyneuropathy, including HIV- and ARV-DSP, may be at higher risk for superimposed entrapment neuropathies such as carpal tunnel syndrome. Asymmetric findings, focal pain, sensory loss, parasthesias, or weakness should alert the physician to the possibility of an entrapment neuropathy. Symptoms of carpal tunnel syndrome include pain, numbness, and paresthesias in the four lateral fingers and hand.

Diagnostic studies

DSP remains primarily a clinical diagnosis, and the optimal combination of diagnostic studies has not been defined. Blood work should be performed to exclude other causes of neuropathy, such as diabetes mellitus, impaired glucose tolerance, vitamin B_{12} deficiency, renal or liver impairment, thyroid dysfunction, and syphilis. Although elevated serum lactate levels and decreased serum acetyl-L-carnitine levels have been demonstrated in some studies of ARV-DSP, these findings are not consistent enough to be clinically useful [6,20].

Nerve conduction studies (NCS) are not routinely required for the clinical diagnosis of DSP but when performed, generally show an axonal, length-dependent, predominantly sensory polyneuropathy. NCS can also be normal due to the relatively low sensitivity for the detection of small-fiber neuropathy. Needle electromyography (EMG) is usually normal but may demonstrate chronic partial denervation and reinnervation typically confined to muscles of the foot and distal leg. NCS and EMG are useful in the diagnosis of superimposed entrapment neuropathies or more complex forms of asymmetrical neuropathies with motor involvement.

Nerve biopsy is rarely necessary and is considered only when there are unusual features such as progressive asymmetric distribution or associated weakness that call the diagnosis of DSP into question. When a biopsy is performed, HIV DNA, RNA, or viral proteins are not consistently detected in the nerve biopsy samples.

Skin biopsy is a validated technique for the detection of small-fiber neuropathy. A punch biopsy is performed at a distal (calf) and proximal (thigh) site in the lower extremity, and intraepidermal nerve fiber density is measured [18]. This technique is valuable for the diagnosis of DSP, for monitoring its progression over time, and for predicting the likelihood of the condition developing in asymptomatic patients. Epidermal nerve fiber density is a useful quantitative marker in clinical trials of neuroregenerative agents [21]. Low epidermal nerve fiber density correlates with the severity of painful symptoms [6]. In asymptomatic patients, low epidermal nerve fiber density (<11 fibers/mm) increases the likelihood of developing symptomatic DSP in the following 6 to 12 months.

Quantitative sensory testing is useful in research settings in which standard measurements for establishing the diagnosis of DSP and tracking its severity are needed. The vibratory threshold has the best discriminatory ability for symptomatic HIV-DSP in comparison to the usual findings of impairment of threshold for heat, pain, and cooling in diabetic neuropathy.

Differential diagnosis

Other forms of neuropathies are much less common than DSP in HIV patients. Mononeuritis multiplex may occur at any stage of HIV. It manifests as multifocal motor and sensory deficits in an asymmetric distribution. The most common presentations include involvement of the common peroneal nerve (foot drop), lateral femoral cutaneous nerve (meralgia paresthetica), facial nerve, and phrenic nerve (diaphragmatic paralysis) [22].

Inflammatory demyelinating polyneuropathies (IDP), including the acute and chronic forms, may be seen in the context of HIV infection. Motor findings usually predominate, with lesser sensory symptoms. Patients exhibit progressive weakness of the extremities, and neurologic examination reveals diffuse hyporeflexia. Electrophysiologic studies reveal demyelinating features. Protein level is elevated in the cerebrospinal fluid (CSF), although in contrast to IDP in seronegative patients, mild pleocytosis is usually present in HIV.

Progressive polyradiculoneuropathy (PP) is usually characterized by a cauda equina syndrome [23]. Symptoms include low back pain with radiation into the legs, asymmetric leg weakness, lower-extremity and saddle distribution sensory disturbance, and bowel and bladder dysfunction. PP can progress to flaccid paraplegia and, in late stages, may involve the upper extremities and cranial nerves. CSF reveals polymorphonuclear pleocytosis, hypoglycorrachia, and elevated protein. Cytomegalovirus is the most common cause of PP in HIV infection. CSF assay for cytomegalovirus by way of polymerase chain reaction is the diagnostic assay of choice. Lumbar MRI with contrast may reveal nerve root enhancement.

HIV-associated neuromuscular weakness syndrome is commonly associated with the use of d4T. It is characterized by progressive weakness that develops over days to weeks and may progress to respiratory failure and death [24]. Lactic acidosis or hyperlactatemia is often present. Electrophysiologic studies reveal axonal neuropathy or myopathy. Other common peripheral neuropathies associated with comorbidities also need to be excluded, including diabetes mellitus, hypothyroidism, nutritional deficiency, excess ethanol consumption, hepatitis B or C virus or syphilis infections, and other drug-related conditions.

Treatment

The approach to treatment of DSP begins with eliminating neurotoxic ARVs when possible. Any coexisting risk factors identified during the patient's evaluation, such as other neurotoxic medications, diabetes mellitus or impaired glucose tolerance, vitamin B_{12} deficiency, renal or liver impairment, thyroid dysfunction, or syphilis, should be addressed.

No neuroregenerative therapies are currently available for the treatment of any form of DSP. There are also no Food and Drug Administration

(FDA)-approved treatments for the painful symptoms associated with HIV-DSP; however, several agents are used in an off-label fashion. Table 1 summarizes the results of several clinical trials of treatment for HIV-DSP. Clinical trials of peptide T [21,25] 6 failed to demonstrate efficacy. Although recombinant human nerve growth factor produced reduction of pain in HIV-DSP, it did not result in evidence of neuroregeneration, and further development of the compound was halted [26]. The hematopoietic growth factor erythropoietin prevented axonal degeneration in cell cultures previously exposed to HIV gp120 protein or the ARV stavudine (d4T) [27]. Erythropoietin is currently being studied in clinical trials as a neuroregenerative agent in patients who have HIV-DSP.

Table 1
Relevant clinical trials for HIV DSP treatment

Drug	Clinical trial
Peptide T	In a randomized double-blind multicentered study, intranasal peptide T was ineffective in the treatment of HIV-DSP [28].
Acetyl-L-carnitine	In small open-label cohort studies, oral and intramuscular acetyl-L-carnitine improved symptoms and caused peripheral nerve regeneration in HIV and ARV-DSP patients [29–31].
Recombinant human nerve growth factor	Open-label and double-blind studies demonstrated pain reduction but no evidence of nerve regeneration [26,31].
Erythropoetin	Under evaluation. Erythropoetin was demonstrated to be neuroprotective in in vitro models of HIV-DSP [27].
Amitriptyline	Used off-label for symptomatic treatment of DSP. Amitriptyline was not superior to placebo in reducing pain in HIV-DSP in a randomized double-blind trial of 145 patients [32].
Duloxetine	Approved for the treatment of diabetic neuropathy [33] and under study in HIV-DSP.
Gabapentin	Reduced pain related to HIV-DSP in a single, small placebo-controlled trial [34].
Pregabalin	A randomized double-blind placebo-controlled multicenter trial of pregabalin versus placebo for the treatment of painful HIV peripheral neuropathy is underway.
Lamotrigine	In a double-blind placebo-controlled multicenter study of 237 patients, lamotrigine was well tolerated and effective for DSP associated with d-drug ARV therapy [35].
Topical lidocaine gel	In a randomized controlled trial, 5% lidocaine gel was ineffective in the treatment of HIV-DSP [36].
High-dose capsaicin patch	A double-blind randomized controlled design showed efficacy of a high-concentration capsaicin patch in painful HIV-DSP [37,38].

There is extensive experience in the symptomatic treatment of HIV-DSP. Most of the agents used have demonstrated efficacy in other forms of painful neuropathy, particularly diabetic peripheral neuropathic pain (DPNP) and postherpetic neuralgia. These agents typically fall into one of three categories: antidepressants, antiepileptic drugs (AEDs), and topical agents. Traditional analgesics and opioids, although not specific treatments for neuropathic pain, may also be of use. Many clinicians find that successful treatment of pain requires combinations of therapies that have different mechanisms of action, an approach termed *rational polypharmacy* [39].

Tricyclic antidepressants (TCAs) and selective serotonin-norepinepherine reuptake inhibitors (SNRIs) are the two major classes of antidepressants used in the treatment of neuropathic pain. TCAs and SNRIs inhibit the reuptake of serotonin and norepinepherine, neurotransmitters responsible for modulation of descending pain inhibition pathways. TCAs are relatively nonselective catecholamine reuptake inhibitors and have been available since the 1950s. Agents of this class, such as amitriptyline and nortriptyline, are still commonly used off-label for the treatment of pain associated with diabetic neuropathy and HIV-DSP. Because TCAs may cause sedation and other anticholinergic side effects, they are typically started as a single low evening dose, such as amitriptyline (10 to 25 mg), and titrated to efficacy as tolerated.

Two agents of the SNRI class are available: duloxetine and venlafaxine. Duloxetine is approved by the FDA for the treatment of DPNP and is under study in an HIV-DSP. Duloxetine is generally well tolerated. The most common side effect of duloxetine (nausea) can be ameliorated by starting treatment at 30 mg daily and increasing to the recommended dose of 60 mg daily up to 120 mg daily after approximately 1 week.

Several AEDs are commonly used in the treatment of HIV-DSP. Gabapentin and the related compound pregabalin are among the most popular. Both agents produce analgesia by way of similar mechanisms involving the inhibition of α-2-Δ presynaptic voltage-gated calcium channels [40]. Both agents have the advantage of renal metabolism, which reduces the potential for interaction with hepatically metabolized drugs, particularly ARVs. There is extensive clinical experience with gabapentin, which was shown to be effective in reducing pain related to HIV-DSP in one small placebo-controlled trial [34]. Although large doses of gabapentin, up to 3600 mg daily, are often tolerated without significant side effects, clinicians frequently begin treatment with gabapentin at lower doses, such as 100 mg three times a day, with subsequent escalation to an effective dose. Pregabalin is a newer agent with a similar mechanism of action and is FDA approved for the treatment of DPNP and postherpetic neuralgia [41]. Pregabalin is under study in clinical trials for the treatment of HIV-DSP. Some clinicians have found pregabalin to have a more rapid effect on painful symptoms of neuropathy than gabapentin, although head-to-head comparisons have not been done. Among the other AEDs, lamotrigine has been found to be effective in HIV-DSP [35]. Although used in the treatment of other pain syndromes such as migraine, the role for topiramate and other AEDs in the treatment of HIV-DSP is unclear and associated with substantial side effects (weight loss, leukopenia, and so forth).

There has been significant interest in the use of topical therapies for the treatment of HIV-DSP. The advantages of a topical therapy include reduced potential for systemic side effects and drug–drug interactions. In addition, the limited distribution of pain in HIV-DSP, typically the soles of the feet, lends itself to a topical treatment. Topical lidocaine, in gel and patch form, has been tried with limited success [36].

The vanilloid receptor agonist capsaicin interacts with C-fiber polymodal nociceptors. Epidermal nerve fibers have been shown to be capsaicin-sensitive nociceptors. A 30- to 90-minute application of high-dose topical capsaicin patch, approximately once every 3 months, has shown significant pain relief in HIV-DSP [37,38]. Side effects include mild irritation, erythema, or burning pain at the application site.

Nonsteroidal anti-inflammatory drugs (NSAIDs) such as ibuprofen and naprosyn do not have proven benefit in the treatment of painful DSP. Along with acetaminophen, however, they are commonly prescribed and may be helpful to some patients. Opioids may play a role in intractable or severe chronic neuropathic pain and in breakthrough pain. Clinicians may find the approach to pain management as described by the World Health Organization helpful in regard to the use of opioids [42]. In this model, mild opiates such as low-dose codeine, hydrocodone, and propoxyphene, usually in combination with NSAIDs, are used for moderate pain; more potent opioids including morphine, oxycodone, fentanyl, and methadone are used for severe pain. There has been recent interest in more rapid–onset treatments for breakthrough pain in chronic pain syndromes such as HIV-DSP. Recent data show efficacy for an oral effervescent form of fentanyl in several forms of neuropathic pain [43].

Summary

As advances in the treatment of HIV lead to improved longevity, managing the neurologic sequelae of HIV infection becomes an increasingly important part of comprehensive HIV care. DSP is the most common neurologic complication of HIV. Although neuroregenerative therapies are not yet available, early recognition of this condition can lead to a treatment strategy to identify modifiable risk factors, to control symptoms, and to improve patients' quality of life.

References

[1] Hoke A, Cornblath DR. Peripheral neuropathies in human immunodeficiency virus infection. Suppl Clin Neurophysiol 2004;57:195–210.

[2] McArthur JC, Brew BJ, Nath A. Neurological complications of HIV infection. Lancet Neurol 2005;4:543–55.
[3] Morgello S, Estanislao L, Simpson D, et al. HIV-associated distal sensory polyneuropathy in the era of highly active antiretroviral therapy: The Manhattan HIV Brain Bank. Arch Neurol 2004;61:546–51.
[4] Childs EA, Lyles RH, Selnes OA, et al. Plasma viral load and CD4 lymphocytes predict HIV-associated dementia and sensory neuropathy. Neurology 1999;52:607–13.
[5] Schifitto G, McDermott MP, McArthur JC, et al. Markers of immune activation and viral load in HIV-associated sensory neuropathy. Neurology 2005;64:842–8.
[6] Simpson DM, Kitch D, Evans SR, et al. HIV neuropathy natural history cohort study: assessment measures and risk factors. Neurology 2006;66:1679–87.
[7] Verma S, Estanislao L, Mintz L, et al. Controlling neuropathic pain in HIV. Curr HIV/AIDS Rep 2004;1:136–41.
[8] Cornblath DR, McArthur JC. Predominantly sensory neuropathy in patients with AIDS and AIDS-related complex. Neurology 1988;38:794–6.
[9] Pettersen JA, Jones G, Worthington C, et al. Sensory neuropathy in human immunodeficiency virus/acquired immunodeficiency syndrome patients: protease inhibitor-mediated neurotoxicity. Ann Neurol 2006;59:816–24.
[10] Keswani SC, Polley M, Pardo CA, et al. Schwann cell chemokine receptors mediate HIV-1 gp120 toxicity to sensory neurons. Ann Neurol 2003;54:287–96.
[11] Oh SB, Tran PB, Gillard SE, et al. Chemokines and glycoprotein120 produce pain hypersensitivity by directly exciting primary nociceptive neurons. J Neurosci 2001;21:5027–35.
[12] Keswani SC, Pardo CA, Cherry CL, et al. HIV-associated sensory neuropathies. AIDS 2002;16:2105–17.
[13] Lichtenstein KA, Armon C, Baron A, et al. Modification of the incidence of drug-associated symmetrical peripheral neuropathy by host and disease factors in the HIV outpatient study cohort. Clin Infect Dis 2005;40:148–57.
[14] Dalakas MC, Semino-Mora C, Leon-Monzon M. Mitochondrial alterations with mitochondrial DNA depletion in the nerves of AIDS patients with peripheral neuropathy induced by 2'3'-dideoxycytidine (ddC). Lab Invest 2001;81:1537–44.
[15] Estanislao L, Thomas D, Simpson D. HIV neuromuscular disease and mitochondrial function. Mitochondrion 2004;4:131–9.
[16] Brew BJ, Tisch S, Law M. Lactate concentrations distinguish between nucleoside neuropathy and HIV neuropathy. AIDS 2003;17:1094–6.
[17] Simpson DM, Katzenstein D, Haidich B, et al. Plasma carnitine in HIV-associated neuropathy. AIDS 2001;15:2207–8.
[18] Luciano CA, Pardo CA, McArthur JC. Recent developments in the HIV neuropathies. Curr Opin Neurol 2003;16:403–9.
[19] Ellis R, Delaney P, Alexander T, et al. HIV protease inhibitors and risk of peripheral neuropathy [abstract]. Presented at the 15th Conference on Retroviruses and Opportunistic Infections (CROI 2008). Boston (MA), February 3–6, 2008.
[20] Famularo G, Moretti S, Marcellini S, et al. Acetyl-carnitine deficiency in AIDS patients with neurotoxicity on treatment with antiretroviral nucleoside analogues. AIDS 1997;11:185–90.
[21] Polydefkis M, Yiannoutsos CT, Cohen BA, et al. Reduced intraepidermal nerve fiber density in HIV-associated sensory neuropathy. Neurology 2002;58:115–9.
[22] Piliero P, Estanislao L, Simpson D. Diaphragmatic paralysis due to isolated phrenic neuropathy in an HIV-infected man. Neurology 2004;62(1):154–5.
[23] Anders H, Goebel F. Cytomegalovirus polyradiculopathy in patients with AIDS. Clin Infect Dis 1998;27:345–52.
[24] HIV Neuromuscular Syndrome Study Group. HIV-associated neuromuscular weakness syndrome. AIDS 2004;18:1403–12.
[25] Evans S, Simpson D, Kitch D, the Prosaptide Study Team. Phase II trial of Prosaptide™ prosaposin-derived peptides for HIV-associated sensory neuropathies: first use of electronic

diary to record HIV-associated neuropathic pain. Presented at the 13th Conference on Retroviruses and Opportunist Infections. Denver (CO), February 6, 2006.

[26] McArthur JC, Yiannoutsos C, Simpson DM, et al. A phase II trial of nerve growth factor for sensory neuropathy associated with HIV infection. AIDS Clinical Trials Group Team 291. Neurology 2000;54:1080–8.

[27] Keswani SC, Leitz GJ, Hoke A. Erythropoietin is neuroprotective in models of HIV sensory neuropathy. Neurosci Lett 2004;371:102–5.

[28] Simpson DM, Dorfman D, Olney RK, et al. Peptide T in the treatment of painful distal neuropathy associated with AIDS: results of a placebo-controlled trial. The Peptide T Neuropathy Study Group. Neurology 1996;47:1254–9.

[29] Hart AM, Wilson AD, Montovani C, et al. Acetyl-l-carnitine: a pathogenesis based treatment for HIV-associated antiretroviral toxic neuropathy. AIDS 2004;18:1549–60.

[30] Scarpini E, Sacilotto G, Baron P, et al. Effect of acetyl-L-carnitine in the treatment of painful peripheral neuropathies in HIV+ patients. J Peripher Nerv Syst 1997;2:250–2.

[31] Schifitto G, Yiannoutsos C, Simpson DM, et al. Long-term treatment with recombinant nerve growth factor for HIV-associated sensory neuropathy. Neurology 2001;57:1313–6.

[32] Kieburtz K, Simpson D, Yiannoutsos C, et al. A randomized trial of amitriptyline and mexiletine for painful neuropathy in HIV infection. AIDS clinical trial group 242 protocol team. Neurology 1998;51:1682–8.

[33] Goldstein DJ, Lu Y, Detke MJ, et al. Duloxetine vs. placebo in patients with painful diabetic neuropathy. Pain 2005;116:109–18.

[34] Hahn K, Arendt G, Braun JS, et al. A placebo-controlled trial of gabapentin for painful HIV-associated sensory neuropathies. J Neurol 2004;251:1260–6.

[35] Simpson DM, McArthur JC, Olney R, et al. Lamotrigine for HIV-associated painful sensory neuropathies: a placebo-controlled trial. Neurology 2003;60:1508–14.

[36] Estanislao L, Carter K, McArthur J, et al, Neuropathy Group. A randomized controlled trial of 5% lidocaine gel for HIV-associated distal symmetric polyneuropathy. J Acquir Immune Defic Syndr 2004;37:1584–6.

[37] Simpson D, Brown S, Sampson J. Novel high-concentration capsaicin patch for the treatment of painful HIV-associated distal symmetrical polyneuropathy: results of an open label trial. Program Abstr Conf Retrovir Oppor Infect 11th 2004 San Francisco, CA. February 8-11, 2004;(11): abstract no. 490.

[38] Simpson DM, Brown SJ, Chang S, NGX-400 Study Group. Controlled study of high-concentration capsaicin patch for painful HIV-associated distal sensory polyneuropathy. Presented at the 13th Conference on Retroviruses and Opportunist Infections. Denver (CO), February 6, 2006.

[39] Gilron I, Bailey JM, Tu D, et al. Morphine, gabapentin, or their combination for neuropathic pain. N Engl J Med 2005;352:1324–34.

[40] Lesser H, Sharma U, LaMoreaux L, et al. Pregabalin relieves symptoms of painful diabetic neuropathy: a randomized controlled trial. Neurology 2004;63:2104–10.

[41] Freynhagen R, Strojek K, Griesing T, et al. Efficacy of pregabalin in neuropathic pain evaluated in a 12-week, randomised, double-blind, multicentre, placebo-controlled trial of flexible- and fixed-dose regimens. Pain 2005;115:254–63.

[42] Cancer pain relief and palliative care. Report of a WHO expert committee. World Health Organ Tech Rep Ser 1990;804:1–75.

[43] Simpson D, Messina J, Xie F, et al. Management of breakthrough pain in opioid-tolerant patients with chronic low back pain or neuropathic pain: randomized, double-blind studies of efficacy and tolerability of fentanyl buccal tablet (FBT). Presented at the 26th Annual Meeting of the American Pain Society. Washington, DC, May 4, 2007.

ELSEVIER
SAUNDERS

Neurol Clin 26 (2008) 833–854

NEUROLOGIC
CLINICS

Progressive Multifocal Leukoencephalopathy

Thomas Weber, MD

Neurologische Klinik, Marienkrankenhaus Hamburg, Academic Teaching Hospital, University of Hamburg, Alfredstr. 9, D-22087 Hamburg, Germany

Progressive multifocal leukoencephalopathy (PML) is the only known human viral demyelinating disorder [1]. PML is caused by the polyomavirus JC (JCV). It is named after the initials of the first patient (John Cunningham) of whose brain it could first be isolated [2]. The disease was first described by the German neuropathologist Hallervorden who failed to identify PML as a distinct entity but subsummized the disease under "unclassifiable" [3]. Åström and colleagues [4] coined the term PML, after identifying chronic and immunosupressive diseases such as Hodgkin's disease, non-Hodgkin's lymphoma and chronic lymphatic leukemias as underlying illnesses of a possible viral infection. Cavanagh and colleagues identified intranuclear inclusion bodies in 1959 as further evidence of a viral cause of PML [5]. In 1965, Zu Rhein and colleagues [6] demonstrated intranuclear inclusions resembling papovavirus-like particles. Seven years later, Billie Padgett at the University of Wisconsin successfully cultured virus from a patient's brain [2]. In 1984, Frisque and colleagues [7] published the first full-length sequence of this isolate, named Mad-1. Incidentally, the growing AIDS pandemic led to a considerable increase of PML cases [8]. Still today, about 85% of all patients suffer from AIDS as underlying illness [9]. Between 3% and 5% of all HIV-infected individuals develop PML [10]. In 2005, PML became a topic of concern, because of its sudden and unprecedented occurrence in two patients with multiple sclerosis and one patient with Crohn's disease treated with a monoclonal antibody (natalizumab, trade name Tysabri) directed toward the adhesion molecule a4/b1 integrin [11–13]. Further proof for the role of lymphocytes in the dissemination and propagation of PML is provided by at least 25 cases of PML in patients treated with rituximab, a monoclonal antibody directed toward CD20 [14]. The productive infection by JCV leads to multiple, small foci of demyelination that tend to coalesce. The lesions are mainly in the white matter but

E-mail address: dr.thomas.weber@hanse.net

doi:10.1016/j.ncl.2008.03.007 *neurologic.theclinics.com*

can also affect gray matter [15]. At least one third of cases show perivascular round cell infiltrates as indication of an immunologic reaction toward an agent, which was considered for a long time as a "slow virus" typically characterized by the absence of a virus specific immune response [15,16]. The introduction of highly active antiretroviral therapy (HAART) led to a change in the pathology of PML, leading to a severe immune response in patients with AIDS and PML after institution of therapy, termed immune reconstitution (inflammatory) response (IRS or IRIS) [17,18]. By MRI, focal inflammatory lesions appear as gadolinium (Gd) enhancing, which had so far been reported in at most 15% of PML cases [17,19,20]. By MRI, multiple, pinpoint sized to confluent large lesions are typical. These lesions may affect an entire hemisphere and appears hyperintense in T_2-weighted imaging. Rarely, the lesions are space occupying. The most frequently affected regions are the periventricular and subcortical regions of the parietooccipital and frontal lobes [21]. The clinical signs and symptoms of PML are multifold and nonspecific. Mono- or hermiparesis predominate, followed by various forms of aphasia and disturbed gait and stance [22,23]. There is no proven, effective therapy of PML in either AIDS patients or patients with other immunosuppressive diseases [24].

Epidemiology

Before the AIDS pandemic the incidence of PML was at 0.15 cases per million. With AIDS, the incidence rose to 0.6 cases per million [25]. By far the most common immunosuppressive drugs related to the development of PML are corticosteroids, which appear to account for about 40% while chemotherapy accounts or about 16%; in almost half of all patients no causative drug can be identified [9]. Second to AIDS, lympho- and myeloproliferative disorders, solid tumors, and congenital immunodeficiency disorders account for about 4.5% of cases each [9]. Rarely, chronic granulomatous disease such as tuberculosis and sarcoidosis cause PML.

As approximately 85% of all PML cases occur on the basis of an HIV infection, most data are derived from studies of HIV patients with PML [26–35]. As may be anticipated, successful reduction of HIV viral load and resulting immunosuppression is the most important factor responsible for the significantly longer survival times reported for PML patients treated with HAART as compared with historical controls before the AIDS pandemic [23,36–46]. Since the introduction of HAART, the incidence of PML in AIDS patients in the United States and Europe has remained constant [10,47]. The introduction of HAART has led to a significant prolonged median survival time, which has been reported to be 4.5 years in AIDS patients with CD4 cell counts higher than 100/μL while the median survival time for those with counts below 100/μL was 3.4 years [23]. There is some evidence for a decline in mortality and delay of onset of PML in HIV-negative patients with lymphoproliferative disorders such as Hodgkin's disease and non-Hodgkin lymphoma [48]. Furthermore, newer antineoplastic therapies such as treatment with purine analogs and high-dose therapy with hematopoietic stem cell transplantation appears to reduce the incidence of PML and leads to a delay in the onset of PML as does the treatment with rituximab, a monoclonal antibody to CD20 (Table 1) [48]. Another monoclonal antibody, natalizumab, appears to be associated with a risk of developing PML at a rate of 0.1% or at about 1 patient in 1000 patient treated for a period of 17.9 months [49].

JCV is found worldwide. About 80% of the adult population is antibody positive and has undergone an asymptomatic infection [1]. Serology allows for a distinction between JCV and BKV, as there is no serologic cross reactivity between the major capsid protein VP1 of JCV and BKV [50]. Molecular cloning of both VP1 of JCV and BKV permits a simple, inexpensive, and rapid screening by enzyme-linked immunosorbent assay (ELISA) [50,51]. At least 14 genotypes of JCV have been identified. Genotyping of JCV has been used to study human evolution, as the codivergence of host (humans) and virus had been assumed to change at the same rate [52]. Current evidence no longer supports this hypothesis and suggests a rate of divergence of JCV that is about two orders of magnitude faster than that of the human genome [53].

Clinical signs and symptoms

Clinical symptoms and signs of PML are nonspecific. In about 25% of cases, PML is the initial AIDS-defining illness [19]. Due to the small number of series and the retrospective character of all studies published to date, a broad range of symptoms and signs has been variously described as initial or presenting manifestation of PML (Tables 2 and 3) [19,22,23,54–58]. The most common presenting symptom is limb weakness in 52% of cases. In approximately 45%, cognitive deficits lead to the diagnosis of PML. Roughly 30% presented with either disorders of speech and/or visual deficits. About 20% of patients had limb incoordination. Approximately one eighth of all patients presented with either seizures or headache, whereas 10% of patients had sensory symptoms (see Table 1). Clinical signs have been retrospectively collected in a recent Spanish multicenter study and are compared with the older findings of Berenguer and colleagues in Table 2 [19,23]. Again paresis or even paralysis and ataxia/gait abnormalities emerge as the most common first signs leading to the diagnosis of PML.

Diagnosis

Until recently, PML had never been described in the course of or as differential diagnosis to multiple sclerosis (MS), but the unexpected development of PML in two patients treated for MS has led to the need

Table 1
Drugs and treatments associated with PML

Treatment	Drug(s)	Target
Oral glucocortioids	All	Multiple mechanisms
Alkylating agents	Cyclophosphamide, camstine, dacarbazine	DNA
Purine analogs	Fludaribine, cladribine, azathioprine	DNA
Antimetabolate	Methotrexate	Inhibits folic acid metabolism
Monoclonal antibody	Rituximab	CD20
Monoclonal antibody	Infliximab	TNF-α
Monoclonal antibody	Etanercept	TNF-α
Monoclonal antibody	Natalizumab	VLA-4
Monoclonal antibody	Basiliximab	IL-2Ra
Monoclonal antibody	Daclizumab	IL-2R
Monoclonal antibody	Muromonab-cd3	CD3
Immunosuppressant	Cydosporin	cyclophilin
Immunosuppressant	Tacrolimus	calcineurin
Immunosuppressant	Sirolimus	mTOR
Immunosuppressant	Mycophenolate	Inosine monophosphate

to establish criteria and means of detecting PML in MS patients treated with natalizumab, a monoclonal antibody to VLA-4, an a4/b1 integrin [12,13,49]. Seizures, speech disorders or frank aphasia are rare new symptoms in MS patients with a frequency of 2.3% to 4% and about 0.8%, respectively but quite commonly indicative of PML with about 20% to 30% (see Table 1). By far the best noninvasive method of diagnosing PML in an immunosuppressed patient is MRI [21,59]. In about 90% of patients not undergoing immune reconstitution, lesions do not take up gadolinium and are not space occupying (Fig. 1). Predominantly, areas affected include periventricular and subcortical frontal and parieto-occipital white matter followed in order of descending likelihood brainstem, cerebellum, thalamus, basal ganglia,

Table 2
Frequency of symptoms in PML patients as compared with patients with multiple sclerosis

Author	[22]	[170]	[57]	[171]	[56]	[73]	[23]	All PML	MS
Symptom								Median	[172]
Sensory	6	8	7	7	18	4	19	11	50
Seizure	6		**20**	**14**	11	**22**	**13**	**14**	**4**
Headache	7	16	7	—	23	11	—	16	17
Limb incoordination	13	28	40	11	26	11	44	22	47
Visual deficits	33	24	20	36	30	36	20	36	26
Speech deficits	17	16	**53**	7	**31**	**36**	**47**	**31**	**0.8**
Mono- or hemiparesis	33	48	60	46	67	54	70	52	80
Cognitive deficits	36	24	40	54	66	61	—	45	16

In the first seven columns, symptoms as found in retrospective series of PML cases are shown. The last column shows the frequency of initial symptoms in multiple sclerosis (MS). Note the much higher incidence of seizures and speech deficits (all forms of aphasia as symptoms in PML as compared with multiple sclerosis). Dashes, data not given.

Table 3
Frequency (%) of clinical signs in AIDS patients with PML [19,23]

Paresis or paralysis	54	67
Gait abnormalities	28	64
Dysarthria	24	—
Aphasia	19	—
Cranial nerve palsies	—	31
Visual impairment	17	20
Ataxia	—	44

Dashes, data not given.

corpus callosum, and rarely cervical or thoracic spinal chord [21,60]. Signs of a space-occupying lesion are found in about 10% of cases and herald a grave prognosis [61]. Diffusion-weighted imaging (DWI) and diffusion tensor imaging (DTI) revealing areas of diffusion abnormalities correlates with disease progression. By proton magnetic resonance spectroscopy (1HMRS), lesions show a reduced N-acetyl-Aspartate- (NAA) and creatine- (CR) content, while choline-containing substances (CHO) and lactate are elevated [60,62]. The sensitivity of 1HMRS is high, but its low specificity and positive predictive value are insufficient to differentiate PML from other lesions such as demyelinating areas of MS, gliomas, lymphomas, or brain abscess [63]. The detailed description of various MRI characteristics of PML and their differentiation from those of new MS lesions put forward by Yousry and colleagues [49] can be summarized as follows. By MRI, new lesions in PML are mainly diffuse and subcortical, while in MS the lesions are focal and in the white matter. The borders in PML are ill defined and irregularly shaped, while in MS they have sharp edges and may appear finger-like (Dawson finger). In PML, lesions grow asymmetrically and diffuse, are confined to white matter tracts, and spare the cortex, while in MS the lesions are focal and enlarge within days to at most weeks and decrease over time. On T_2-weighted imaging, acute MS lesions show a hyperintense center and an isointense ring with a slight hyperintensity outside the ring structure while the chronic lesion is hyperintense without ring structure. In PML, however, acute lesions tend to appear diffusely hyperintensive and may show only a slightly to moderately increased intensity of newly involved areas. In Gd-enhanced T_1-weighted images, acute MS lesions show a homogeneous enhancement and sharp well-defined edges, while subacute lesions show ring enhancement. Up to very recently, it could be reasonably argued that PML lesions usually show no enhancement, while in patients receiving HAART for newly diagnosed AIDS there may be peripheral or even dense enhancement (Fig. 2) [49]. These criteria have been questioned by an elegant study of Huang and colleagues [64], comparing MRI findings in PML patients with AIDS as underlying cause of immunosuppression to those with other causes of reduced immunity. Surprisingly, two of five HIV-negative patients also showed enhancing lesions as compared with two of four HIV-positive patients.

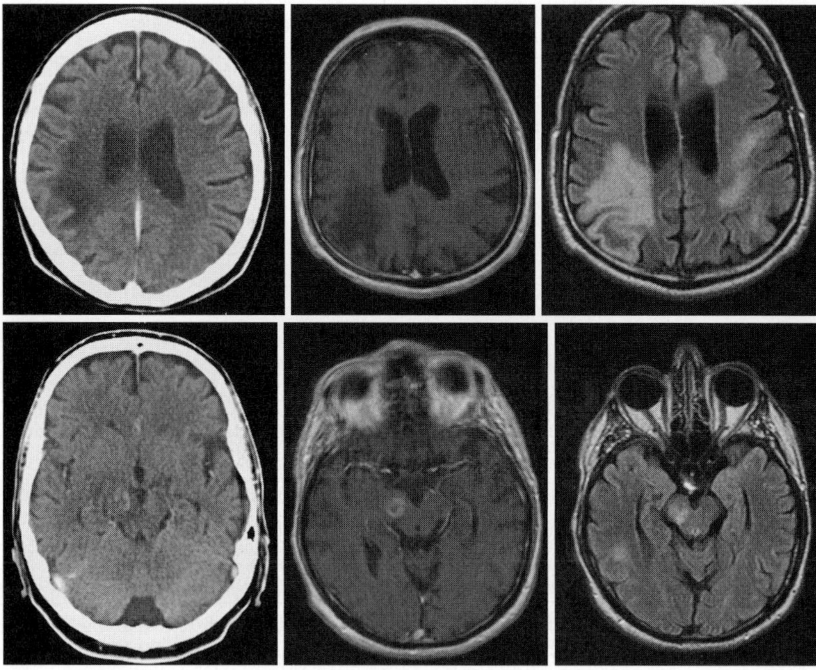

Fig. 1. Comparison of CT and MRI in an AIDS patient with autopsy-proven PML and cerebral toxoplasmosis . In the upper row on the left, CT shows fewer lesions and reveals a smaller affected area than MRI. In he middle column, a large, hypointense lesion can be seen in the region of the right occipital lobe (angular gyrus and intraparietal sulcus) in a T1-weighted image. On the right, a T2-weighted image reveals the typical, multifocal affection of PML with additional lesions in the white matter of the left frontal temporal and occipital lobe. Note sparing of gray matter. In the lower row, CT shows contrast enhancement in the right cerebral peduncles as does MRI with gadolinium in the middle column. This contrast-enhancing lesion corresponded to cerebral toxoplasmosis at autopsy. On the right T2-weighted image, an additional hyperintense lesion due to PML can be seen in the white matter in the inferior posterior temporal gyrus on the right.

The advent of natalizumab and the unexpected development of PML in MS patients treated with this monoclonal antibodies has spurred a large, retrospective analysis of treated patients (91%), which suggests a risk of PML of about 1 in 1000 patients treated for 17.9 months [49]. In those two of three PML patients for whom cerebrospinal fluid (CSF) samples were available, JCV DNA was detected in CSF but not in 34 patients receiving natalizumab [49]. The mechanisms by which natalizumab could lead to PML are manifold [65,66]. Natalizumab causes a long-lasting depletion of lymphocytes, both CD4- and CD8-positive T cells in CSF. In addition, the migratory capacity of white blood cells is reduced and the expression profile of distinct immune cells is differentially regulated with unique profiles in individual patients and the activation threshold of immune cells is modulated. Viremia, or more precisely, detection of JCV DNA in plasma, occurs

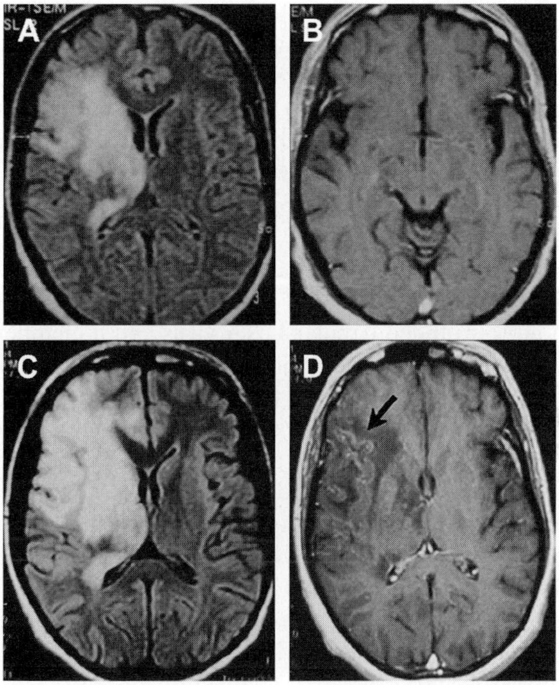

Fig. 2. Inflammatory PML. (*A*) Large demyelinating lesion in the right fronto-parietal lobe (*upper row, left*) revealed by FLAIR imaging (*B*) no gadolinium enhancement seen in T1-weighted imaging. (*C*) Enlarged lesion after 5 weeks of treatment with HAART revealed by FLAIR and (*D*) showing punctuate gadolinium enhancement (*arrow*). (*From* Tulius M, Pacheco MCN, Vaz B. Inflammatory progressive multifocal leukoencephalopathy after antiretroviral treatment. AIDS 2006;20(3):469–71; with permission.)

in about 1.3% of healthy individuals and may increase to 10% to 15% in HIV-positive individuals, while it may be found in up to 33% to 50% of all HIV-positive patients with PML [67,68]. Judging from the rather limited experience in two patients with PML while receiving natalizumab, JCV viremia may either precede or coincide with symptoms and signs of PML [11,69]. All available evidence from studies in HIV-infected patients suggests at best a 50% chance of diagnosing PML by serial plasma JCV load measurements [67,68]. Given this low prior probability of plasma JC virus load, CSF analysis by PCR has for more than a decade been the method of choice for non- or minimally invasive diagnosis of PML [67,68,70–74].

By CSF analysis, pleocytosis is found in only about 14% while intrathecal IgG synthesis is found in about 29% [75]. In non-AIDS patients, PCR for JCV still appears to have a sensitivity of about 80% and a specificity of 95% for the diagnosis of PML [76]. Recent evidence, however, suggests a much lower sensitivity ranging from 58% to 81% in patients under HAART leading to a reduced negative predictive value of about 89% [77,78]. A comparable sensitivity of 78% and specificity of 96.8% can be obtained by the demonstration of intrathecally synthesized IgG antibodies to recombinant JCV-Virus protein 1 (VP1) [75,79]. Over the course of time, however, the percentage of patients positive for intrathecally synthesized IgG antibodies to recombinant VP1 appears to increase, thus making this ELISA-based assay a more reliable method for the confirmatory diagnosis of PML, remaining positive, even when by PCR JCV can no longer be amplified after successful clearance of virus from the brain [80].

Biology

JCV is a double-stranded circular DNA virus and a member of the polyomaviridae. A virus particle has a diameter of about 45 nm and forms an icosaheder. The viral capsid contains three proteins designated as virus particle (VP) VP1, VP2, and VP3. VP1 accounts for about 75% of the capsid protein, the remaining two proteins are designated as VP2 and VP3. VP1 forms virus-like particles (VLPs) independently of VP2 and VP3 [81,82]. VP1 is an efficient transporter or drug delivery system and may be used as an antigen for the detection of JCV-specific humoral and/or cellular immune responses [81,82]. The agnoprotein is a polypeptide with 71 amino acids inhibiting cell growth and arresting the cell in the G2/M phase [83]. In addition, it suppresses T protein–induced viral DNA replication and interacts with capsid assembly [84]. In vitro, coinfection of astrocytes with HIV-1 and JCV causes a slightly higher expression of JCV due to the association of the N-terminal domain of the agnoprotein with Tat and consecutive suppression of HIV replication [85].

The first JCV isolate sequenced is the prototypical strain Mad with a size of 5130 base pairs [7]. The circular genome is read by cellular DNA polymerase in both directions, "clockwise" and "counterclockwise" (Fig. 3) [1,86]. The "early" gene large T (T) and small t (t) are read from a single DNA strand at the origin of replication (ORI). The T-antigen (T-Ag) has long been known to induce brain tumors [87,88]. T-Ag is a multifunctional protein and transforms cells by the interaction with various growth regulatory genes [89,90]. The "late" genes for the structural proteins VP1, VP2, and VP3 and the agnoprotein are read from the opposite strand [7]. The agnoprotein is a polypeptide with 71 amino acids inhibiting cell growth and cell cycle in the G2/M-phase [83]. In addition, it suppresses T protein–mediated viral DNA replication and promotes the assembly of the viral capsid [84].

Lytically infected cells harbor further early proteins designated as T'135, T'136 and T'165. These are derived from T by alternative splicing from the early viral precursor mRNA. By binding to p107 and p130, they alter the phosphorylation status of these proteins and inactivate cell cycle regulators to promote JCV-induced cell transformation [91]. These proteins participate at the replication of JCV-DNA and regulate functions of the T protein [92]. T interacts with the tumor suppressor protein p53 and the cell cycle

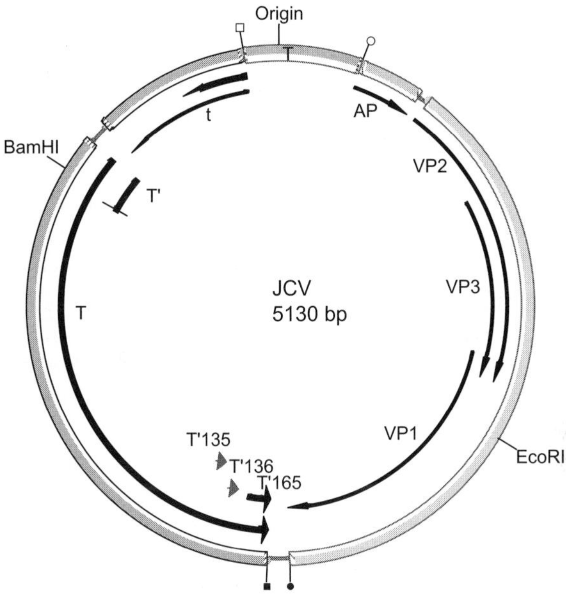

Fig. 3. Circular map of the JCV genome. The origin is flanked by the early genes to the left and the late genes to the right. The early genes consist of t, T, and at least three different splice forms called T', designated according to the number of amino acids as T'_{165}, T'_{136}, and T'_{135}. ⚲ denotes the start codon for t and T, ⫯ gives the stop codon for T. ⚲ gives the start codon for the late genes; AP: agnoprotein, VP1: virus protein 1, VP2: virus protein 2, VP3 virus protein 3. ⫯ indicates the stop codon for the late gene VP1.

regulator protein pRb [93,94]. Depending on certain DNA polymorphisms, p53 either inhibits JC virus replication or does not interact with its replication [95]. Interactions with these regulatory proteins may lead to cell destruction, malignant transformation of cells, and unregulated growth of these cells.

DNA replication and transcription are mediated by a cell-specific complex interaction of a multitude of DNA binding proteins and factors [96–98]. The most important groups of proteins are NF-kappa B and NF-1 [99–101]. The permissiveness of JCV replication appears to depend on the cellular expression profile of the NF-1 transcription factor family and is linked to NF-1X (also called NF-1D) [100]. This serves to explain the extremely restricted expression pattern of JCV [102]. A member of the AP-1 family of DNA-binding protein, c-Jun has recently been demonstrated to regulate JCV activity by binding to NF-1 [103].

In animal models, JCV has been shown to induce a variety of brain tumors such as glioblastoma multiforme (GBM) in owl monkeys, hamsters, and rats, and medulloblastoma/primitive neuroectodermal tumors (PNET) in mice transgenic for the early region of JCV [89,104–106]. By PCR and other DNA-amplification methods, JCV DNA has been detected in human colon cell lines, stroma cells, and lymphocytes in tonsils [107,108]. These and other findings have led to hypotheses giving JCV a role in the development of such diverse tumors as colon carcinoma, glioblastoma multiforme, oligoastrozytoma, ganglioglioma, medulloblastoma, mesothelioma, and non-Hodgkin's lymphomas [109]. So far, however, no conclusive evidence for a definite role in the development of any of these or other tumors by JCV has been presented. These and other findings have led to hypotheses giving JCV a role in the development of such diverse tumors as colon carcinoma, glioblastoma multiforme, oligoastrozytoma, ganglioglioma, medulloblastoma, mesothelioma, and non-Hodgkin's lymphomas [109]. So far, however, no conclusive evidence for a definite role of JCV in the development of any of these or other tumors has been presented [110,111]. Although the odds ratio for JCV VP1 IgG antibody–positive patients to develop a tumor is slightly higher than 1, there is no definitive evidence for an association of JCV with human astroglial tumors [112]. Seroepidemiological data, however, appear to suggest an increased risk of subsequent development of non-Hodgkin's lymphoma with a rising JCV antibody titer [113]. Whether this is related to a direct role of JCV in the development of Non Hodgkin's Lymphoma remains to be shown. Furthermore, a polymorphism at codon 72 in exon 4 of p53 appears to support the development of PML in HIV/AIDS patients [114].

JCV binds to at least two receptors on the cell surface. One carries terminal alpha-2,6–linked sialic acid residues [115]. These residues are found in sialoglycoproteins such as acidic alpha-1 glycoprotein, fetuin, and the transferrin-receptor [116]. In normal human brain tissue, receptors bearing these residues are present on oligodendrocytes and astrocytes but not cortical neurons and are expressed more highly on B- than on T-lymphocytes [117]. The higher levels of expression on B cells as compared with T lymphocytes in normal human spleen and tonsils, further substantiates the hypothesis of viral uptake by the respiratory route with persistence in lymphoid tissue and dissemination to the brain via B-cells [117,118]. Unlike SV-40, JC virions enter glial cells by receptor-mediated clathrin-dependent endocytosis [119]. The 5-hydroxytryptamine 2A (5-HT2A) subtype of the serotonin receptor, which is highly expressed in areas of the brain where the blood-brain-barrier is absent, ie, in the plexus chorioideus and area postrema, has been identified as a receptor for JCV [120]. The 5-HT2A receptor is exclusively found on normal astroglial cells and is thought to regulate either specific glial functions, such as astrocyte proliferation and maturation and development of central serotonergic pathways or influence the microcirculaton by activation of astroglial receptors and release of vasoactive signaling molecules [61]. Interestingly, caveolin-1, a scaffolding protein enriched in caveolae complexes with 5-HT2A and is involved in endocytic trafficking of JCV [121]. All in all, there appears to be a striking correlation between expression of the JCV-receptor-type sialic acid on cells and their susceptibility to JCV infection [117].

Pathogenesis

The pathogenesis of PML may be divided into three phases. The first phase is that of primary, inapparent infection. Possible routes of infection are by respiratory and/or fecal-oral route [122]. The second phase is latency in the human host. The long-known urinary excretion of JCV points to the kidney as one place of latency [123–125]. The genomic structure of the form of JCV most frequently found in the kidney has been designated as archetype [124]. Other likely places of viral latency are bone marrow and spleen [126]. The third and final phase is probably induced by immunologic and molecular alterations of the viral regulatory region resulting in hematogenous dissemination of JCV to the brain [1].

Humoral immunity

Infection with JCV induces a humoral IgG response, predominantly directed toward the major capsid protein VP1 [75,112,127]. It appears that there is also an IgM response toward JCV. Its role is still unknown and could either be an indication of active JCV antigenic stimulation in healthy adults or be the result of nonspecific serologic crossreactivity [128]. All patients with PML investigated so far had detectable serum IgG antibodies [75,129,130]. Antibodies are not protective in PML. By quantitative ELISA using recombinant VP1 of JCV as antigen, an overall seropositivity of 84.5% was found [75]. Although IgG titers are not significantly different among healthy controls, patients with MS, and those with PML, intrathecal synthesis of JCV VP1–specific IgG antibodies can be used to confirm the diagnosis of PML [68,75]. As in other viral infections of the central nervous system, the intrathecal antibody response to JCV develops over time and exceeds the upper limit of normal of 1.5 of the antibody index (AI) 2 to 3 weeks after onset of disease [130]. The diagnosis of PML by the demonstration of intrathecally synthesized IgG antibodies can be achieved with a sensitivity of 76% and a specificity of 96.8% [75].

Cellular immunology

Richardson [131] speculated in 1961 on an altered or suppressed immunity as cause of PML. Substantiation of his idea came from case reports showing anergy to delayed type hypersensitivity reactions to antigens such as tetanus toxoid and 2,4-dinitrofluorobenzene in PML patients [132,133]. In 1980, Willoughby and colleagues [134] could demonstrate a reduced proliferation of lymphocytes in the presence of mitogens such as phytohemagglutinin or concanavalin A in 7 PML patients. Further, clinical evidence for the importance of cellular immunoreactivity is derived from the preferential occurrence of PML in diseases with known impaired cellular immune responses such as in patients with lymphoma or immunosuppressed organ recipients. PML in patients with hypo- or agammaglobulinemia, however, is rare. In HIV-infected patients, the likelihood of development of PML is inversely correlated with the CD4 cell count [19]. Frye and colleagues [135] could show a reduced lymphocyte proliferation in PML patients by using purified JCV virions. The molecular cloning and recombinant expression of VP1 provided sufficient amounts of stable antigen to analyze the cellular and humoral immune response [81]. In PML patients with and those without HIV infection we found a reduced proliferation of lymphocytes and a reduced production of the T_{H1} cytokine interferon-γ after stimulation with VP1. Production of the T_{H2} cytokine IL-10 in HIV-infected PML patients was elevated. These findings point to a suppression of the T-helper function of the T_{H1} type [136]. During the defense of viral infections, CD4+ T lymphocytes recognize extracellular virus proteins that were digested exogenously to peptides and presented by major histocompatibility complex class II molecules (MHC-II molecules) on the surface of antigen-presenting cells. CD8+ cytotoxic T-lymphocytes (CTLs) recognize intracellular synthesized viral peptides, which are degraded endogenously and are presented together with MHC class I molecules on the surface of virus-infected cells. Patients who demonstrate an MHC class I specific cytotoxic CD8+ immune response in association with the MHC class I antigen HLA 0201 against a JCV-VP1-specific epitope (p100), show a clearly prolonged survival time [137,138]. In a recent prospective study, Koralnik and colleagues [139], could further substantiate a protective role of JCV-specific CTLs. While 13 of 15 PML patients with JCV-specific CTLs developed "chronic" or quiescent PML, 9 of 11 patients lacking JCV-specific CTLs had a rapid course with short survival. Further evidence for a protective role of CD8+-CTLs against the development of PML was observed in healthy individuals in whom 73% showed a CD8+-T-cell response to JCV as compared with only 20% with a specific response to cytomegalovirus [140]. Early detection of JCV-specific CTLs in PML appears to have a good positive predictive value for the control of PML [141]. The precursor frequency of JCV A*0201 epitope-specific cytotoxic T cells is not significantly different among healthy individuals, HIV-positive patients, and HIV-positive patients with PML and cannot be used to diagnose or monitor PML [138].

Therapy

The first drug initially thought to be effective but later proven in a well-designed clinical trial to lack any efficiency in PML was cytarabine or ARA-C [32,142]. Clinical benefit in patients with genital warts treated with either interferon alpha, beta, or gamma initiated a pilot study with alpha 2a interferon in 13 PML patients without any relevant benefit [143,144]. These findings were disputed by some case reports and one retrospective analysis but weighted evidence points to the absence of any effect of interferon alpha2a on the course of PML [43,145–147]. Although case reports suggest a favorable response of PML to treatment with interleukin-2 (IL-2), sufficient clinical evidence and experimental data are lacking to support these findings [148,149].

Preliminary in vitro data suggested an antiviral effect of (S)-1-(3-hydroxy-2-phosphonylmethoxypropyl) cytosine (HPMC or cidofovir) [150]. Some case reports, small series, or retrospective analyses suggest an at least partial response of PML to and prolonged survival by cidofovir treatment [35,37,151–153]. In contrast, a larger retrospective analysis of 35 cases with PML even showed a significantly shorter median survival in patients treated with HAART and cidofovir as compared with those receiving HAART only [154]. A prospective phase 1 study in 17 HIV-positive patients with PML failed to show any effect of cidofovir on survival [29]. In a second prospective study of 24 patients, survival was found not to be associated with treatment modalities, ie, HAART versus HAART + cidofovir but with HIV-1 RNA plasma levels ≤ than 500 copies/mL and CD4 T-cell count [28]. Comparable data have been reported in a large single-center observational study of 46 patients, comparing 22 patients with HAART only to 24 with HAART and cidofovir [155]. Taken together, the currently available data suggest an association of prolonged survival with immunologic (CD4 T-cell count) and virological (HIV-1 RNA plasma level, JC viral load) parameters rather than with treatment modalities. Few reports of a positve effect of cidofovir treatment on PML in patients with hematological malignancies [156] or the absence of any efficiency [157–161] strongly support an observational bias (over-reporting of positive and under-reporting of negative results) as the cause of a presumed beneficial effect of cidofovir in PML. Data for camptothecin, a topoisomerase I inhibitor found to be effective in repressing JCV replication in a cell culture model, are sparse [162]. A case report suggests prolongation of survival in a patient with systemic lupus erythematosus taken off steroids and treated with camptothecin [163]. A small series of nine patients receiving topotecin also seemed to show a slight benefit but was associated with serious hematological toxicities [27]. Again, on balanced evidence, there is no sign of meaningful efficiency of treatment with either drug.

By far the best therapy of PML in AIDS patients is an optimal antiretroviral therapy, significantly lowering HIV-1 RNA viral load in plasma and possibly in CSF, as well as improving cellular immune function [23,35,37,45,80,164].

Prognosis

A CD4 count of 100/mm^3 or higher, a low JC viral load in CSF of 100 copies/μL or less, and the presence of JCV-specific cytotoxic T cells in HLA-A2+ HIV-1–positive patients have all been shown to be associated with a prolonged survival [71,139,164,165,166]. Up to 10% of patients with PML and newly treated with HAART for HIV develop an immune restitution inflammatory syndrome (IRIS) [17,167,168]. It is characterized by rapid clinical deterioriation in conjunction with a massive inflammatory reaction in the brain parenchyma [23,169]. Recent quantitative immunostaining of brain tissue showed a significantly enhanced number of infiltrating $CD3^+$ and mainly $CD8^+$ T cells in gadolinium-enhancing lesions as compared with nonenhancing lesions [64].

References

[1] Weber T, Major EO. Progressive multifocal leukoencephalopathy: molecular biology, pathogenesis and clinical impact. Intervirology 1997;40(2–3):98–111.

[2] Padgett BL, Walker DL, Zu Rhein GM, et al. Cultivation of papova-like virus from human brain with progressive multifocal leucoencephalopathy. Lancet 1971;297(7712):1257–60.

[3] Hallervorden J. Eigenartige und nicht rubrizierbare prozesse. In: Bumke O, editor. Handbuch der Geisteskrankheiten Die Anatomie der Psychosen. Berlin: Springer; 1930. p. 1036–107.

[4] Åström K-E, Mancall EL, Richardson EP. Progressive multifocal leuko-encephalopathy. A hitherto unrecognized complication of chronic lymphatic leukaemia and Hodgkin's disease. Brain 1958;81:93–111.

[5] Cavanagh JB, Greenbaum D, Marshall AHE, et al. Cerebral demyelination associated with disorders of the reticuloendothelial system. Lancet 1959;274(7102):524–9.

[6] Zu Rhein G, Padgett BL, Walker DL, et al. Pituitary function after removal of microadenomas for Cushing's disease [letter]. N Engl J Med 1978;299(5):256–7.

[7] Frisque RJ, Bream GL, Cannella MT. Human polyomavirus JC virus genome. J Virol 1984; 51(2):458–69.

[8] Miller JR, Barrett RE, Britton CB, et al. Progressive multifocal leukoencephalopathy in a male homosexual with T-cell immune deficiency. N Engl J Med 1982;307(23):1436–8.

[9] Eng PM, Turnbull BR, Cook SF, et al. Characteristics and antecedents of progressive multifocal leukoencephalopathy in an insured population. Neurology 2006;67(5):884–6.

[10] San-Andres FJ, Rubio R, Castilla J, et al. Incidence of acquired immunodeficiency syndrome-associated opportunistic diseases and the effect of treatment on a cohort of 1115 patients infected with human immunodeficiency virus, 1989–1997. Clin Infect Dis 2003;36(9): 1177–85.

[11] Sandborn WJ, Colombel JF, Enns R, et al. Natalizumab induction and maintenance therapy for Crohn's disease. N Engl J Med 2005;353(18):1912–25.

[12] Langer-Gould A, Atlas SW, Green AJ, et al. Progressive multifocal leukoencephalopathy in a patient treated with natalizumab. N Engl J Med 2005;353(4):375–81.

[13] Kleinschmidt-DeMasters BK, Tyler KL. Progressive multifocal leukoencephalopathy complicating treatment with natalizumab and interferon beta-1a for multiple sclerosis. N Engl J Med 2005;353(4):369–74.

[14] Important drug warning updated safety information. Available at: Genentech biogenidec 2006. Available at: www.rituxan.com/ra/content/Rituxan_DHCP letter. Accessed July 1, 2005.

[15] Jellinger KA, Setinek U, Drlicek M, et al. Neuropathology and general autopsy findings in AIDS during the last 15 years [In Process Citation]. Acta Neuropathol (Berl) 2000;100(2): 213–20.

[16] Neuenburg JK, Brodt HR, Herndier BG, et al. HIV-related neuropathology, 1985 to 1999: rising prevalence of HIV encephalopathy in the era of highly active antiretroviral therapy. J Acquir Immune Defic Syndr 2002;31(2):171–7.

[17] Venkataramana A, Pardo CA, McArthur JC, et al. Immune reconstitution inflammatory syndrome in the CNS of HIV-infected patients. Neurology 2006;67(3):383–8.

[18] Hoffmann C, Horst HA, Albrecht H, et al. Progressive multifocal leucoencephalopathy with unusual inflammatory response during antiretroviral treatment. J Neurol Neurosurg Psychiatr 2003;74(8):1142–4.

[19] Berger JR, Pall L, Lanska D, et al. Progressive multifocal leukoencephalopathy in patients with HIV infection. J Neurovirol 1998;4(1):59–68.

[20] Lima MA, Hanto DW, Curry MP, et al. Atypical radiological presentation of progressive multifocal leukoencephalopathy following liver transplantation. J Neurovirol 2005;11(1): 46–50.

[21] Whiteman ML, Post MJ, Berger JR, et al. Progressive multifocal leukoencephalopathy in 47 HIV-seropositive patients: neuroimaging with clinical and pathologic correlation. Radiology 1993;187:233–40.

[22] Brooks BR, Walker DL. Progressive multifocal leukoencephalopathy. Neurol Clin 1984; 2(2):299–313.

[23] Berenguer J, Miralles P, Arrizabalaga J, et al. Clinical course and prognostic factors of progressive multifocal leukoencephalopathy in patients treated with highly active antiretroviral therapy. Clin Infect Dis 2003;36(8):1047–52.

[24] Roskopf J, Trofe J, Stratta RJ, et al. Pharmacotherapeutic options for the management of human polyomaviruses. Adv Exp Med Biol 2006;577:228–54.

[25] Holman RC, Janssen RS, Buehler JW, et al. Epidemiology of progressive multifocal leukoencephalopathy in the United States: analysis of national mortality and AIDS surveillance data. Neurology 1991;41(11):1733–6.

[26] Skromne E, Rivera VM, Ontaneda D, et al. Progression of progressive multifocal leukoencephalopathy despite treatment with beta-interferon. Neurology 2006;66(11):1787–8, author reply -8.

[27] Royal W III, Dupont B, McGuire D, et al. Topotecan in the treatment of acquired immunodeficiency syndrome-related progressive multifocal leukoencephalopathy. J Neurovirol 2003;9(3):411–9.

[28] Marra CM, Rajicic N, Barker DE, Cohen BA, et al. A pilot study of cidofovir for progressive multifocal leukoencephalopathy in AIDS. AIDS 2002;16(13):1791–7.

[29] Marra CM, Rajicic N, Barker DE, et al. A pilot study of cidofovir for progressive multifocal leukoencephalopathy in AIDS. AIDS 2002;16(13):1791–7.

[30] Levy RM, Major E, Ali MJ, et al. Convection-enhanced intraparenchymal delivery (CEID) of cytosine arabinoside (AraC) for the treatment of HIV-related progressive multifocal leukoencephalopathy (PML). J Neurovirol 2001;7(4):382–5.

[31] Dodge RT. A case study: the use of cidofovir for the management of progressive multifocal leukoencephalopathy. J Assoc Nurses AIDS Care 1999;10(4):70–4.

[32] Hall CD, Dafni U, Simpson D, et al. Failure of cytarabine in progressive multifocal leukoencephalopathy associated with human immunodeficiency virus infection. AIDS Clinical Trials Group 243 Team. N Engl J Med 1998;338(19):1345–51.

[33] Steiger MJ, Tarnesby G, Gabe S, et al. Successful outcome of progressive multifocal leukoencephalopathy with cytarabine and interferon. Ann Neurol 1993;33(4):407–11.

[34] Tashiro K, Doi S, Moriwaka F, et al. Progressive multifocal leucoencephalopathy with magnetic resonance imaging verification and therapeutic trials with interferon. J Neurol 1987;234(6):427–9.

[35] De Luca A, Giancola ML, Ammassari A, et al. Potent anti-retroviral therapy with or without cidofovir for AIDS- associated progressive multifocal leukoencephalopathy: extended follow- up of an observational study. J Neurovirol 2001;7(4):364–8.

[36] Silva MT, Araujo A. Highly active antiretroviral therapy access and neurological complications of human immunodeficiency virus infection: impact versus resources in Brazil. J Neurovirol 2005;11(Suppl 3):11–5.

[37] Antinori A, Cingolani A, Lorenzini P, et al. Clinical epidemiology and survival of progressive multifocal leukoencephalopathy in the era of highly active antiretroviral therapy: data from the Italian Registry Investigative Neuro AIDS (IRINA). J Neurovirol 2003; 9(Suppl 1):47–53.
[38] Cinque P, Pierotti C, Vigano MG, et al. The good and evil of HAART in HIV-related progressive multifocal leukoencephalopathy. J Neurovirol 2001;7(4):358–63.
[39] Tassie JM, Gasnault J, Bentata M, et al. Survival improvement of AIDS-related progressive multifocal leukoencephalopathy in the era of protease inhibitors. Clinical Epidemiology Group. French Hospital Database on HIV. AIDS 1999;13(14):1881–7.
[40] Gasnault J, Taoufik Y, Goujard C, et al. Prolonged survival without neurological improvement in patients with AIDS-related progressive multifocal leukoencephalopathy on potent combined antiretroviral therapy. J Neurovirol 1999;5(4):421–9.
[41] Dworkin MS, Wan PC, Hanson DL, et al. Progressive multifocal leukoencephalopathy: improved survival of human immunodeficiency virus-infected patients in the protease inhibitor era. J Infect Dis 1999;180(3):621–5.
[42] Clifford DB, Yiannoutsos C, Glicksman M, et al. HAART improves prognosis in HIV-associated progressive multifocal leukoencephalopathy. Neurology 1999;52(3):623–5.
[43] Huang SS, Skolasky RL, Dal Pan GJ, et al. Survival prolongation in HIV-associated progressive multifocal leukoencephalopathy treated with alpha-interferon: an observational study. J Neurovirol 1998;4(3):324–32.
[44] Berger JR, Levy RM, Flomenhoft D, et al. Predictive factors for prolonged survival in acquired immunodeficiency syndrome-associated progressive multifocal leukoencephalopathy. Ann Neurol 1998;44(3):341–9.
[45] De Luca A, Giancola ML, Ammassari A, et al. The effect of potent antiretroviral therapy and JC virus load in cerebrospinal fluid on clinical outcome of patients with AIDS-associated progressive multifocal leukoencephalopathy. J Infect Dis 2000;182(4):1077–83.
[46] Elliot B, Aromin I, Gold R, et al. 2.5 year remission of AIDS-associated progressive multifocal leukoencephalopathy with combined antiretroviral therapy. Lancet 1997; 349:850.
[47] Sacktor N, Lyles RH, Skolasky R, et al. HIV-associated neurologic disease incidence changes: multicenter AIDS Cohort Study, 1990–1998. Neurology 2001;56(2):257–60.
[48] Garcia-Suarez J, de Miguel D, Krsnik I, et al. Changes in the natural history of progressive multifocal leukoencephalopathy in HIV-negative lymphoproliferative disorders: impact of novel therapies. Am J Hematol 2005;80(4):271–81.
[49] Yousry TA, Major EO, Ryschkewitsch C, et al. Evaluation of patients treated with natalizumab for progressive multifocal leukoencephalopathy. N Engl J Med 2006;354(9): 924–33.
[50] Hamilton RS, Gravell M, Major EO. Comparison of antibody titers determined by hemagglutination inhibition and enzyme immunoassay for JC virus and BK virus. J Clin Microbiol 2000;38(1):105–9.
[51] Weber T, Beck R, Stark E, et al. Comparative analysis of intrathecal antibody synthesis and DNA amplification for the diagnosis of cytomegalovirus infection of the central nervous system in AIDS patients. J Neurol 1994;241(7):407–14.
[52] Agostini HT, Yanagihara R, Davis V, et al. Asian genotypes of JC virus in Native Americans and in a Pacific Island population: markers of viral evolution and human migration. Proc Natl Acad Sci U S A 1997;94(26):14542–6.
[53] Shackelton LA, Rambaut A, Pybus OG, et al. JC virus evolution and its association with human populations. J Virol 2006;80(20):9928–33.
[54] Berger JR, Concha M. Progressive multifocal leukoencephalopathy: the evolution of a disease once considered rare. J Nephrol 1995;1(1):5–18.
[55] Fong IW, Toma E. The natural history of progressive multifocal leukoencephalopathy in patients with AIDS. Canadian PML Study Group. Clin Infect Dis 1995;20(5):1305–10.

[56] Gillespie SM, Chang Y, Lemp G, et al. Progressive multifocal leukoencephalopathy in persons infected with human immunodeficiency virus, San Francisco, 1981–1989. Ann Neurol 1991;30(4):597–604.
[57] von Einsiedel RW, Fife TD, Aksamit AJ, et al. Progressive multifocal leukoencephalopathy in AIDS: a clinicopathologic study and review of the literature. J Neurol 1993;240:391–406.
[58] Weber T, Turner RW, Frye S, et al. Progressive multifocal leukoencephalopathy diagnosed by amplification of JC virus-specific DNA from cerebrospinal fluid. AIDS 1994;8:49–57.
[59] Thurnher MM, Thurnher SA, Muhlbauer B, et al. Progressive multifocal leukoencephalopathy in AIDS: initial and follow-up CT and MRI. Neuroradiology 1997;39(9):611–8.
[60] Chang L, Ernst T, Tornatore C, et al. Metabolite abnormalities in progressive multifocal lenkoencephalopathy by proton magnetic resonance spectroscopy. Neurology 1997;48: 836–45.
[61] Post MJ, Yiannoutsos C, Simpson D, et al. Progressive multifocal leukoencephalopathy in AIDS: are there any MR findings useful to patient management and predictive of patient survival? AIDS Clinical Trials Group, 243 Team. AJNR Am J Neuroradiol 1999;20(10): 1896–906.
[62] Iranzo A, Moreno A, Pujol J, et al. Proton magnetic resonance spectroscopy pattern of progressive multifocal leukoencephalopathy in AIDS. J Neurol Neurosurg Psychiatr 1999; 66(4):520–3.
[63] Simone IL, Federico F, Tortorella C, et al. Localised 1H-MR spectroscopy for metabolic characterisation of diffuse and focal brain lesions in patients infected with HIV. J Neurol Neurosurg Psychiatr 1998;64(4):516–23.
[64] Huang D, Cossoy M, Li M, et al. Inflammatory progressive multifocal leukoencephalopathy in human immunodeficiency virus-negative patients. Ann Neurol 2007;62(1):34–9.
[65] Stuve O, Marra CM, Bar-Or A, et al. Altered CD4+/CD8+ T-cell ratios in cerebrospinal fluid of natalizumab-treated patients with multiple sclerosis. Arch Neurol 2006;63(10): 1383–7.
[66] Niino M, Bodner C, Simard ML, et al. Natalizumab effects on immune cell responses in multiple sclerosis. Ann Neurol 2006;59(5):748–54.
[67] Dubois V, Moret H, Lafon ME, et al. Prevalence of JC virus viraemia in HIV-infected patients with or without neurological disorders: a prospective study. J Neurovirol 1998;5(5): 539–44.
[68] Weber T. Laboratory findings in progressive multifocal leukoencephalopathy. In: Kahalili K, Stoner GL, editors. Human polyomaviruses molecular and clinical perspectives. New York: Wiley-Lis; 2001. p. 337–58.
[69] Langer-Gould A, Steinman L. Progressive multifocal leukoencephalopathy and multiple sclerosis: lessons from natalizumab. Curr Neurol Neurosci Rep 2006;6(3):253–8.
[70] Moret H, Brodard V, Barranger C, et al. New commercially available PCR and microplate hybridization assay for detection and differentiation of human polyomaviruses JC and BK in cerebrospinal fluid, serum, and urine samples. J Clin Microbiol 2006;44(4):1305–9.
[71] Koralnik IJ, Boden D, Mai VX, et al. JC virus DNA load in patients with and without progressive multifocal leukoencephalopathy. Neurology 1999;52(2):253–60.
[72] Antinori A, Ammassari A, De Luca A, et al. Diagnosis of AIDS-related focal brain lesions: a decision-making analysis based on clinical and neuroradiologic characteristics combined with polymerase chain reaction assays in CSF. Neurology 1997;48(3):687–94.
[73] Weber T, Turner RW, Frye S, et al. Specific diagnosis of progressive multifocal leukoencephalopathy by polymerase chain reaction. J Infect Dis 1994;169:1138–41.
[74] Cinque P, Brytting M, Vago L, et al. Epstein-Barr virus DNA in cerebrospinal fluid from patients with AIDS-related primary lymphoma of the central nervous system. Lancet 1993; 342(8868):398–401.
[75] Weber T, Trebst C, Frye S, et al. Analysis of the systemic and intrathecal humoral immune response in progressive multifocal leukoencephalopathy. J Infect Dis 1997;176(1):250–4.

[76] Weber T. Cerebrospinal fluid analysis for the diagnosis of human immunodeficiency virus-related neurologic diseases. Semin Neurol 1999;19(2):223–33.
[77] Bossolasco S, Calori G, Moretti F, et al. Prognostic significance of JC virus DNA levels in cerebrospinal fluid of patients with HIV-associated progressive multifocal leukoencephalopathy. Clin Infect Dis 2005;40(5):738–44.
[78] Marzocchetti A, Di Giambenedetto S, Cingolani A, et al. Reduced rate of diagnostic positive detection of JC virus DNA in cerebrospinal fluid in cases of suspected progressive multifocal leukoencephalopathy in the era of potent antiretroviral therapy. J Clin Microbiol 2005;43(8):4175–7.
[79] Guillaume B, Sindic CJ, Weber T. Progressive multifocal leukoencephalopathy: simultaneous detection of JCV DNA and anti-JCV antibodies in the cerebrospinal fluid. Eur J Neurol 2000;7(1):101–6.
[80] Giudici B, Vaz B, Bossolasco S, et al. Highly active antiretroviral therapy and progressive multifocal leukoencephalopathy: effects on cerebrospinal fluid markers of JC virus replication and immune response. Clin Infect Dis 2000;30(1):95–9.
[81] Goldmann C, Petry H, Frye S, et al. Molecular cloning and expression of major structural protein VP1 of the human polyomavirus JC virus: formation of virus-like particles useful for immunological and therapeutic studies. J Virol 1999;73(5):4465–9.
[82] Goldmann C, Stolte N, Nisslein T, et al. Packaging of small molecules into VP1-virus-like particles of the human polyomavirus JC virus. J Virol Methods 2000;90(1):85–90.
[83] Darbinyan A, Darbinian N, Safak M, et al. Evidence for dysregulation of cell cycle by human polyomavirus, JCV, late auxiliary protein. Oncogene 2002;21(36):5574–81.
[84] Safak M, Barrucco R, Darbinyan A, et al. Interaction of JC virus agno protein with T antigen modulates transcription and replication of the viral genome in glial cells. J Virol 2001; 75(3):1476–86.
[85] Kaniowska D, Kaminski R, Amini S, et al. Cross-interaction between JC virus agnoprotein and human immunodeficiency virus type 1 (HIV-1) Tat modulates transcription of the HIV-1 long terminal repeat in glial cells. J Virol 2006;80(18):9288–99.
[86] Okada Y, Endo S, Takahashi H, et al. Distribution and function of JCV agnoprotein. J Neurovirol 2001;7(4):302–6.
[87] Bollag B, Chuke WF, Frisque RJ. Hybrid genomes of the polyomaviruses JC virus, BK virus, and simian virus 40: identification of sequences important for efficient transformation. J Virol 1989;63(2):863–72.
[88] Dörries K. New aspects in the pathogenesis of polyomavirus-induced disease. Adv Virus Res 1997;48:205–61.
[89] Khalili K. Human neurotropic JC virus and its association with brain tumors. Dis Markers 2001;17(3):143–7.
[90] White MK, Gordon J, Reiss K, et al. Human polyomaviruses and brain tumors. Brain Res Brain Res Rev 2005;50(1):69–85.
[91] Bollag B, Kilpatrick LH, Tyagarajan SK, et al. JC virus T'135, T'136 and T'165 proteins interact with cellular p107 and p130 in vivo and influence viral transformation potential. J Neurovirol 2006;12(6):428–42.
[92] Frisque RJ. Structure and function of JC virus T' proteins. J Neurovirol 2001;7(4):293–7.
[93] Henson JW, Schnitker BL, Lee TS, et al. Cell-specific activation of the glial-specific JC virus early promoter by large T antigen. J Biol Chem 1995;270(22):13240–5.
[94] Bollag B, Prins C, Snyder EL, et al. Purified JC virus T and T' proteins differentially interact with the retinoblastoma family of tumor suppressor proteins. Virology 2000;274(1):165–78.
[95] Staib C, Pesch J, Gerwig R, et al. p53 inhibits JC virus DNA replication in vivo and interacts with JC virus large T-antigen. Virology 1996;219(1):237–46.
[96] Raj GV, Khalili K. Transcriptional regulation: lessons from the human neurotropic polyomavirus, JCV. Virology 1995;213(2):283–91.
[97] Sadowska B, Barrucco R, Khalili K, et al. Regulation of human polyomavirus JC virus gene transcription by AP-1 in glial cells. J Virol 2003;77(1):665–72.

[98] Safak M, Khalili K. Physical and functional interaction between viral and cellular proteins modulate JCV gene transcription. J Neurovirol 2001;7(4):288–92.
[99] Devireddy LR, Kumar KU, Pater MM, et al. BAG-1, a novel Bcl-2-interacting protein, activates expression of human JC virus. J Gen Virol 2000;81(Pt 2):351–7.
[100] Monaco MC, Sabath BF, Durham LC, et al. JC virus multiplication in human hematopoietic progenitor cells requires the NF-1 class D transcription factor. PG—9687–95. J Virol 2001;75(20).
[101] Sumner C, Shinohara T, Durham L, et al. Expression of multiple classes of the Nuclear Factor-1 family in the developing human brain: differential expression of two classes of NF-1 genes. J Neurovirol 1996;2:87–100.
[102] Messam CA, Hou J, Gronostajski RM, et al. Lineage pathway of human brain progenitor cells identified by JC virus susceptibility. Ann Neurol 2003;53(5):636–46.
[103] Ravichandran V, Sabath BF, Jensen PN, et al. Interactions between c-Jun, Nuclear Factor-1 and JC virus promoter sequences: implications for viral tropism. J Virol 2006;80(21): 10506–13.
[104] Ohsumi S, Motoi M, Ogawa K. Induction of undifferentiated tumors by JC virus in the cerebrum of rats. Acta Pathol Jpn 1986;36(6):815–25.
[105] Major EO, Mourrain P, Cummins C. JC virus-induced owl monkey glioblastoma cells in culture: biological properties associated with the viral early gene product. Virology 1984; 136(2):359–67.
[106] Krynska B, Otte J, Franks R, et al. Human ubiquitous JCV(CY) T-antigen gene induces brain tumors in experimental animals. Oncogene 1999;18(1):39–46.
[107] Ricciardiello L, Chang DK, Laghi L, et al. Mad-1 is the exclusive JC virus strain present in the human colon, and its transcriptional control region has a deleted 98-base-pair sequence in colon cancer tissues. J Virol 2001;75(4):1996–2001.
[108] Monaco MCG, Jensen PN, Hou J, et al. Detection of JC virus DNA in human tonsil tissue: evidence for site of initial viral infection [in Process Citation]. J Virol 1998;72(12):9918–23.
[109] Croul S, Otte J, Khalili K. Brain tumors and polyomaviruses. J Neurovirol 2003;9(2): 173–82.
[110] Newcomb PA, Bush AC, Stoner GL, et al. No evidence of an association of JC virus and colon neoplasia. Cancer Epidemiol Biomarkers Prev 2004;13(4):662–6.
[111] Gillison ML, Chen R, Goshu E, et al. Human retinoblastoma is not caused by known pRb-inactivating human DNA tumor viruses. Int J Cancer 2007;120(7):1482–90.
[112] Rollison DE, Helzlsouer KJ, Alberg AJ, et al. Serum antibodies to JC virus, BK virus, simian virus 40, and the risk of incident adult astrocytic brain tumors. Cancer Epidemiol Biomarkers Prev 2003;12(5):460–3.
[113] Rollison DE, Engels EA, Halsey NA, et al. Prediagnostic circulating antibodies to JC and BK human polyomaviruses and risk of non-Hodgkin lymphoma. Cancer Epidemiol Biomarkers Prev 2006;15(3):543–50.
[114] Power C, Gladden JG, Halliday W, et al. AIDS- and non-AIDS-related PML association with distinct p53 polymorphism. Neurology 2000;54(3):743–6.
[115] Liu CK, Wei G, Atwood WJ. Infection of glial cells by the human polyomavirus JC is mediated by an N-linked glycoprotein containing terminal alpha(2–6)-linked sialic acids. J Virol 1998;72(6):4643–9.
[116] Komagome R, Sawa H, Suzuki T, et al. Oligosaccharides as receptors for JC virus. J Virol 2002;76(24):12992–3000.
[117] Eash S, Tavares R, Stopa EG, et al. Differential distribution of the JC virus receptor-type sialic acid in normal human tissues. Am J Pathol 2004;164(2):419–28.
[118] Sabath BF, Major EO. Traffic of JC virus from sites of initial infection to the brain: the path to progressive multifocal leukoencephalopathy. J Infect Dis 2002;186(Suppl 2): S180–6.
[119] Pho MT, Ashok A, Atwood WJ. JC virus enters human glial cells by clathrin-dependent receptor-mediated endocytosis. J Virol 2000;74(5):2288–92.

[120] Elphick GF, Querbes W, Jordan JA, et al. The human polyomavirus, JCV, uses serotonin receptors to infect cells. Science 2004;306(5700):1380–3.
[121] Bhatnagar A, Sheffler DJ, Kroeze WK, et al. Caveolin-1 interacts with 5-HT2A serotonin receptors and profoundly modulates the signaling of selected Galphaq-coupled protein receptors. J Biol Chem 2004;279(33):34614–23.
[122] Zheng HY, Kitamura T, Takasaka T, et al. Unambiguous identification of JC polyomavirus strains transmitted from parents to children. Arch Virol 2004;149(2):261–73.
[123] Hogan TF, Borden EC, McBain JA, et al. Human polyomavirus infections with JC virus and BK virus in renal transplant patients. Ann Intern Med 1980;92(3):373–8.
[124] Kitamura T, Aso Y, Kuniyoshi N, et al. High incidence of urinary JC virus excretion in nonimmunosuppressed older patients. J Infect Dis 1990;161(6):1128–33.
[125] Chang H, Wang M, Tsai RT, et al. High incidence of JC viruria in JC-seropositive older individuals. J Neurovirol 2002;8(5):447–51.
[126] Houff SA, Major EO, Katz DA, et al. Involvement of JC virus-infected mononuclear cells from the bone marrow and spleen in the pathogenesis of progressive multifocal leukoencephalopathy. N Engl J Med 1988;318(5):301–5.
[127] Knowles WA, Sasnauskas K. Comparison of cell culture-grown JC virus (primary human fetal glial cells and the JCI cell line) and recombinant JCV VP1 as antigen for the detection of anti-JCV antibody by haemagglutination inhibition. J Virol Methods 2003;109(1):47–54.
[128] Knowles WA, Gibson PE, Hand JF, et al. An M-antibody capture radioimmunoassay (MACRIA) for detection of JC virus-specific IgM. J Virol Methods 1992;40(1):95–105.
[129] Knowles WA, Luxton RW, Hand JF, et al. The JC virus antibody response in serum and cerebrospinal fluid in progressive multifocal leucoencephalopathy. Clin Diagn Virol 1995;4:183–94.
[130] Sindic CJ, Trebst C, Van Antwerpen MP, et al. Detection of CSF-specific oligoclonal antibodies to recombinant JC virus VP1 in patients with progressive multifocal leukoencephalopathy. J Neuroimmunol 1997;76(1–2):100–4.
[131] Richardson EPJ. Progressive multifocal leukoencephalopathy. N Engl J Med 1961;265: 815–23.
[132] Ellison GW. Progressive multifocal leukoencephalopathy (PML). I. Investigation of the immunologic status of a patient with lymphosarcoma and PML. J Neuropathol Exp Neurol 1969;28(3):501–6.
[133] Knight A, O'Brien P, Osoba D. "Spontaneous" progressive multifocal leukoencephalopathy. Immunologic aspects. Ann Intern Med 1972;77(2):229–33.
[134] Willoughby E, Price RW, Padgett BL, et al. Progressive multifocal leukoencephalopathy (PML): in vitro cell-mediated immune responses to mitogens and JC virus. Neurology 1980;30(3):256–62.
[135] Frye S, Trebst C, Dittmer U, et al. Efficient production of JC virus in SVG cells and the use of purified viral antigens for analysis of specific humoral and cellular immune response. J Virol Methods 1997;63(1–2):81–92.
[136] Weber T, Weber F, Petry H, et al. Immune response in progressive multifocal leukoencephalopathy: An overview. J Neurovirol 2001;7(4):311–7.
[137] Koralnik IJ. Overview of the cellular immunity against JC virus in progressive multifocal leukoencephalopathy. J Neurovirol 2002;8(Suppl 2):59–65.
[138] Lima MA, Marzocchetti A, Autissier P, et al. Frequency and phenotype of JC virus-specific CD8+ T lymphocytes in the peripheral blood of patients with progressive multifocal leukoencephalopathy. J Virol 2007;81(7):3361–8.
[139] Koralnik IJ, Du Pasquier RA, Kuroda MJ, et al. Association of prolonged survival in HLA-A2+ progressive multifocal leukoencephalopathy patients with a CTL response specific for a commonly recognized JC virus epitope.PG - 499-504. J Immunol 2002;168(1):499–504.
[140] Du Pasquier RA, Schmitz JE, Jean-Jacques J, et al. Detection of JC virus-specific cytotoxic T lymphocytes in healthy individuals. J Virol 2004;78(18):10206–10.

[141] Du Pasquier RA, Kuroda MJ, Zheng Y, et al. A prospective study demonstrates an association between JC virus-specific cytotoxic T lymphocytes and the early control of progressive multifocal leukoencephalopathy. Brain 2004;127:1970–8.
[142] Bauer WR, Turel APJ, Johnson KP. Progressive multifocal leukoencephalopathy and cytarabine. Remission with treatment. JAMA 1973;226(2):174–6.
[143] Berger JR, Pall L, McArthur J, et al. A pilot study of recombinant alpha 2a interferon in the treatment of AIDS-related progressive multifocal leukoencephalopathy. Neurology 1992; 42(Suppl 3):257.
[144] Weck PK, Buddin DA, Whisnant JK. Interferons in the treatment of genital human papillomavirus infections. Am J Med 1988;85(2A):159–64.
[145] Geschwind M, Skolasky R, Royal W, et al. The relative contributions of HAART and alpha-interferon for therapy of progressive multifocal leukoencephalopathy in AIDS. J Neurovirol 2001;7(4):353–7.
[146] Heide W, Kömpf D, Reusche E, et al. Failure of cytarabine/interferon therapy in progressive multifocal leukoencephalopathy. Ann Neurol 1995;37(3):412–3.
[147] Re D, Bamborschke S, Feiden W, et al. Progressive multifocal leukoencephalopathy after autologous bone marrow transplantation and alpha-interferon immunotherapy. Bone Marrow Transplant 1999;23(3):295–8.
[148] Jaeckle KA, Przepiorka D, Birdwell RR, et al. Favorable response of progressive multifocal leukoencephalopathy (PML) to treatment with interleukin-2. Neurology 1997;48(Suppl 2): A26.
[149] Kunschner L, Scott TF. Sustained recovery of progressive multifocal leukoencephalopathy after treatment with IL-2. Neurology 2005;65(9):1510.
[150] Andrei G, Snoeck R, Vandeputte M, et al. Activities of various compounds against murine and primate polyomaviruses. Antimicrobial Agents Chemother 1997;41(3):587–93.
[151] Blick G, Whiteside M, Griegor P, et al. Successful resolution of progressive multifocal leukoencephalopathy after combination therapy with cidofovir and cytosine arabinoside. Clin Infect Dis 1998;26(1):191–2.
[152] Garvey L, Thomson EC, Taylor GP. Progressive multifocal leukoencephalopathy: prolonged survival in patients treated with protease inhibitors and cidofovir: a case series. AIDS 2006;20(5):791–3.
[153] de Luca A, Fantoni M, Tartaglione T, et al. Response to cidofovir after failure of antiretroviral therapy alone in AIDS-associated progressive multifocal leukoencephalopathy. Neurology 1999;52(4):891–2.
[154] Wyen C, Hoffmann C, Schmeisser N, et al. Progressive multifocal leukencephalopathy in patients on highly active antiretroviral therapy: survival and risk factors of death. J Acquir Immune Defic Syndr 2004;37(2):1263–8.
[155] Gasnault J, Kousignian P, Kahraman M, et al. Cidofovir in AIDS-associated progressive multifocal leukoencephalopathy: a monocenter observational study with clinical and JC virus load monitoring. J Neurovirol 2001;7(4):375–81.
[156] Viallard JF, Lazaro E, Lafon ME, et al. Successful cidofovir therapy of progressive multifocal leukoencephalopathy preceding angioimmunoblastic T-cell lymphoma. Leuk Lymphoma 2005;46(11):1659–62.
[157] Houston S, Roberts N, Mashinter L. Failure of cidofovir therapy in progressive multifocal leukoencephalopathy unrelated to human immunodeficiency virus. Clin Infect Dis 2001; 32(1):150–2.
[158] Salmaggi A, Maccagnano E, Castagna A, et al. Reversal of CSF positivity for JC virus genome by cidofovir in a patient with systemic lupus erythematosus and progressive multifocal leukoencephalopathy. Neurol Sci 2001;22(1):17–20.
[159] Osorio S, De La Camara R, Golbano N, et al. Progressive multifocal leukoencephalopathy after stem cell transplantation, unsuccessfully treated with cidofovir. Bone Marrow Transplant 2002;30(12):963–6.

[160] Herrlinger U, Schwarzler F, Beck R, et al. Progressive multifocal leukoencephalopathy: cidofovir therapy in three patients with underlying hematological disease. J Neurol 2003; 250(5):612–4.

[161] Terrier B, Martinez V, Seilhean D, et al. Progressive multifocal leukoencephalopathy mimicking cerebral vasculitis in systemic granulomatosis. J Infect 2007;54(3):e133–5.

[162] Kerr DA, Chang CF, Gordon J, et al. Inhibition of human neurotropic virus (JCV) DNA replication in glial cells by camptothecin. Virology 1993;196(2):612–8.

[163] Vollmer-Haase J, Young P, Ringelstein EB. Efficacy of camptothecin in progressive multifocal leukoencephalopathy. Lancet 1997;349:1366.

[164] Yiannoutsos CT, Major EO, Curfman B, et al. Relation of JC virus DNA in the cerebrospinal fluid to survival in acquired immunodeficiency syndrome patients with biopsy-proven progressive multifocal leukoencephalopathy. Ann Neurol 1999;45(6):816–21.

[165] Taoufik Y, Gasnault J, Karaterki A, et al. Prognostic value of JC Virus load in cerebrospinal fluid of patients with progressive multifocal leukoencephalopathy. J Infect Dis 1998; 178(6):1816–20.

[166] Garcia De Viedma D, Diaz Infantes M, Miralles P, et al. JC virus load in progressive multifocal leukoencephalopathy: analysis of the correlation between the viral burden in cerebrospinal fluid, patient survival, and the volume of neurological lesions. Clin Infect Dis 2002;34(12):1568–75.

[167] Safdar A, Rubocki RJ, Horvath JA, et al. Fatal immune restoration disease in human immunodeficiency virus type 1-infected patients with progressive multifocal leukoencephalopathy: impact of antiretroviral therapy-associated immune reconstitution. Clin Infect Dis 2002;35(10):1250–7.

[168] Martinez JV, Mazziotti JV, Efron ED, et al. Immune reconstitution inflammatory syndrome associated with PML in AIDS: a treatable disorder. Neurology 2006;67(9):1692–4.

[169] Miralles P, Berenguer J, Lacruz C, et al. Inflammatory reactions in progressive multifocal leukoencephalopathy after highly active antiretroviral therapy. AIDS 2001;15(14):1900–2.

[170] Berger JR, Gallo B, Concha M. Progressive miultifocal leukoencephalopathy. In: Berger JR, Levy RM, editors. AIDS and the nervous system. Philadelphia: Lippincott-Raven; 1997. p. 569–94.

[171] Fong IW, Britton CB, Luinstra KE, et al. Diagnostic value of detecting JC virus DNA in cerebrospinal fluid of patients with progressive multifocal leukoencephalopathy. J Clin Microbiol 1995;33(2):484–6.

[172] Matthews B. Symptoms and signs of multiple sclerosis. In: Compston A, Ebers G, Lassmann H, editors. McAlpine's multiple sclerosis. London: Chruchill Livingstone; 1998. p. 145–90.

ELSEVIER
SAUNDERS

Neurol Clin 26 (2008) 855–864

NEUROLOGIC
CLINICS

Emerging Issues in Neurovirology: New Viruses, Diagnostic Tools, and Therapeutics

Richard T. Johnson, MD[a], Christopher Power, MD[b,*]

[a]*Department of Neurology, The Johns Hopkins Hospital, Pathology 627, Baltimore, MD 21287, USA*

[b]*Departments of Medicine (Neurology), Medical Microbiology and Immunology, and Psychiatry, University of Alberta, 6-11 Heritage Medical Research Centre, Edmonton, Alberta, T6G 2S2, Canada*

> "The middle of the 20th century can be viewed as the end of one of the most important social revolutions—the eradication of infectious diseases as a significant element in social life".
>
> —Sir Macfarlane Burnet, 1962

In the mid-twentieth century, the discovery of antibiotics, the control of syphilis, the development of poliovirus and measles vaccines, and the anticipation of world eradication of smallpox led many to predict an end of infectious diseases as a serious public health problem. Burnet's false prophesy was not alone; Robert Petersdorf [1] and others wrote obituaries for the specialty of infectious diseases and left the field.

Unforeseen was the emergence of new players, that is, new infections to replace and even dwarf the old. In the 1960s and 1970s, this emergence was foreshadowed by the appearance of new recombinant (duck-human) influenza viruses, legionnaires' disease, toxic shock syndrome, Lyme disease, and the neurovirulent La Crosse strains of California encephalitis virus. The crushing realization that emerging infections could pose an enormous global threat came in the 1980s with the emergence of HIV infections and AIDS [2].

Over the past 2 decades, significant outbreaks of disease due to "new" agents have occurred almost yearly (Box 1). Some have been due to the evolution of more virulent agents (eg, enterovirus 71, chikunqunya virus, and drug-resistant microbes), some to geographic relocation of agents (eg, Dengue type 3 in Sri Lanka and West Nile virus in North America), and some to

* Corresponding author.
E-mail address: chris.power@ualberta.ca (C. Power).

doi:10.1016/j.ncl.2008.04.003 *neurologic.theclinics.com*

Box 1. Emerging infections 1981–2003[a]

1981, **HIV** (worldwide)
1982, **Lyme borreliosis** identified (northeastern United States)
1984, Cryptosporidiosis (Texas)
1985, **Bovine spongiform encephalopathy** (United Kingdom)
1987, Multidrug-resistant tuberculosis (prisons in United States and Russia)
1989, Dengue 3, subtype III, virus (Sri Lanka)
1991, **Guanarito virus** (Venezuela)
1992, *Vibrio cholerae* 0139 (India)
1993, *Escherichia coli* 0157:H7 (western United States); Hantavirus pulmonary syndrome (southwestern United States)
1994, **Variant Creutzfeldt-Jacob disease** (United Kingdom); Hendra virus (Australia)
1995, Largest Ebola outbreak (Congo)
1996, **Reversion poliovirus vaccine virus** (Dominican Republic)
1997, Influenza virus (Hong Kong); vancomycin-resistant *Staphylococcus* (United States)
1998, **Enterovirus 71 fatal rhomboencephalitis** (Asia); **Nipah virus encephalitis** (Malaysia and Singapore)
1999, **West Nile virus encephalitis** (New York)
2000, Rift Valley fever (Saudi Arabia)
2002, SARS (Asia and Canada)
2003, Monkeypox (midwestern United States)
2006, **St. Louis encephalitis** (Argentina)

[a] Agents with a major impact on the central nervous system are in bold type.

contact with animals and crossing of species barriers (eg, bovine spongiform encephalopathy, variant Creutzfeldt-Jakob disease, Nipah virus, and the severe acute respiratory syndrome [SARS] virus). The majority of these agents have been viruses.

Added to this threat over the same time has been the growing concern over the weaponization and intentional release of infectious agents, culminating in the anthrax attacks in 2001. Millions have died and will die from the spontaneously emerging infections. In the three US terrorist acts, that is, the contamination of salad bars by a religious group in Oregon in 1984, the malicious contamination of donuts and muffins probably by a disgruntled worker in a Texas laboratory in 1996, and the anthrax mail contamination by an unknown assailant in 2001, only the latter resulted in deaths. The fear of far greater devastation from bioterrorism is real and likely to happen. The spontaneous emergence of new lethal agents is a certainty.

Origins of emerging infections

Several dramatic examples of new outbreaks of viral encephalitis have appeared in the past decade. The three examples summarized herein represent new clinical syndromes due to viruses appearing after evolutionary changes, geographic relocation, and zoonotic infection transmitted to humans.

Enterovirus 71

This enteric picornavirus, closely related to Coxsackie A 16, was originally recovered in California in 1969 and was associated with hand-foot-mouth syndrome, herpangina, occasional cases of viral meningitis, and a single case of fatal encephalitis [3]. Surprisingly, the virus appeared in Bulgaria and Hungary between 1975 and 1978 associated with a poliomyelitis-like syndrome, and isolates from these outbreaks caused flaccid paralysis after inoculation into monkeys [4]. The virus had evolved to simulate poliovirus disease. After another interval, the virus appeared in Southeast Asia in 1997 as the cause of fatal rhomboencephalitis with cardiopulmonary collapse in children under 5 years of age [5,6]. This brainstem localization of clinical findings, imaging studies, and neuropathologic findings indicate a further evolution of neurovirulence to cause a novel and highly lethal illness.

West Nile virus

West Nile virus was originally recovered from the blood of a febrile woman in Uganda in 1937 [7]. Subsequently, the virus was associated with a dengue-like illness in African children and later with occasional cases of aseptic meningitis in the Middle East. Ironically, the first documented cases of encephalitis due to West Nile virus were in New York City in 1954 after 95 patients with advanced cancer were inoculated with the virus on the premise that the "harmless" virus might selectively destroy tumor cells. Nine patients developed encephalitis, and virus was recovered from the cerebrospinal fluid of three patients [8]. Naturally occurring cases of encephalitis began to appear in the Mediterranean basin in the 1990s. A radical change in the epidemiologic pattern was then seen, with large epidemics of encephalitis with high mortality rates. The first epidemics occurred in Bucharest in 1996 and then in Volgograd in 1999. Clearly, a more virulent strain of the virus had evolved in this region [9].

In the summer of 1999, a cluster of patients with encephalitis and severe weakness were reported in the New York borough of Queens. They ranged in age from 58 to 87 years, had no common exposure, reported no family illnesses, and spent their evening hours out of doors where investigators found many mosquito larvae in stagnant water. The disease was correctly assumed to be arthropod borne but tentatively diagnosed as St. Louis encephalitis. A subsequent virus isolate from a bird dying with encephalitis in the neighboring Bronx was identified as West Nile virus, and reassessment of the human encephalitis cases showed them to be caused by West Nile virus [10]. Sequence analysis of the New York virus showed striking similarity to an isolate from a goose in Israel in the previous year [11].

Similar to recent European outbreaks, encephalitis was a common complication, and in 10% of patients there was flaccid paralysis indicating involvement of anterior horn cells of the spinal cord [12]. The virus may have been transported by a mosquito aboard an international flight or by a smuggled bird or viremic animal. Whatever the courier, the virus found a compatible habitat, successfully over wintered, and spread across the United States and Canada over the next 3 years [13]. The disease took on major public health importance in 2002 when it spread coast to coast, causing over 3000 human illnesses and 153 deaths. In 2003, 9862 cases of disease were recorded in humans; 2860 patients had meningitis or encephalitis, with over 200 deaths.

In 1999, many had predicted that the virus would not survive the winter or that future summers would present small geographically defined outbreaks similar to those caused by other arthropod-borne viruses in North America. The annual outbreaks and the spread from coast to coast, across Canada, and into Latin America have occurred, but cases also continue to occur annually in Eastern states. The wide range of transmission-competent avian and mosquito species portend a serious future health threat from this neurovirulent strain of West Nile virus now firmly established in its new habitat [14].

Nipah virus

In September of 1998, cases of encephalitis were reported among pig farmers and their families in Malaysia. Initially the disease was thought to be Japanese encephalitis, an endemic arthropod-borne disease in the region; however, the new disease affected primarily adults, occurred in household clusters, was associated with contacts with pigs that were ill, and affected some persons who had been immunized against Japanese encephalitis virus. Each of these factors suggested an alternative cause. In March of 1999, a paramyxovirus named Nipah virus was recovered and related to the disease [15]. An outbreak in a single abattoir in Singapore gave insight into the epidemiology of the infection, since only workers transporting or slaughtering pigs became infected. No family member or medical caregiver had evidence of infection, indicating a lack of human-to-human spread [16]. The slaughter of tens of thousands of pigs in Malaysia brought an end to the outbreak, but only after over 100 patients had died.

The clinical disease was unique, with multifocal neurologic signs such as cerebellar ataxia, brainstem signs, and segmental myoclonus. Imaging studies showed multiple small lesions scattered through the white matter. Pathologic studies showed widespread vasculitis with giant cell formation and occlusion of small vessels as well as neuronal infection [17]. Follow-up studies have shown that over 7% of survivors relapse [18].

The virus is related to the Hendra virus, a paramyxovirus associated with encephalitis in horses and a few handlers in Australia. The natural hosts of Hendra virus are fruit bats. Nipah virus has now been recovered from the saliva and urine of bats, including bats that hung in fruit trees overhanging the pig yards in Malaysia [19].

No further major epidemic has been reported, but scattered small outbreaks of Nipah encephalitis have been reported in Bengal and Bangladesh. These cases have not been related to exposures to sick pigs; therefore, some other modes of transmission from fruit bats to humans occur.

Viral infections of the nervous system on the horizon

Hepatitis C virus

The flavivirus hepatitis C virus (HCV) represents one of the leading chronic viral epidemics globally, with over 130 million infected individuals [20]. In many areas of the world, HCV infection parallels HIV-1 infection, with injection drug use or blood transfusion being the chief modes of transmission [21]. Like HIV-1, HCV is defined by multiple subtypes and immense molecular heterogeneity within subtypes with varying disease severity, largely evident as inflammatory liver disease, which progresses to a cirrhotic state. HCV-infected persons also report neurologic symptomatology, including neurocognitive symptoms (fatigue, mental slowing, poor concentration, forgetfulness, apathy) in the absence of liver disease and exhibit neurologic signs including low-grade encephalopathy (impaired attention and learning, psychomotor slowing) with an overall prevalence of 13% [22] and peripheral neuropathy (10%) [23] with or without cryoglobulinemia. Indeed, polymyositis, demyelinating polyneuropathies, and acute demyelinating disseminated encephalomyelitis have also been associated with HCV infection.

The underlying pathogenesis of HCV-related disease appears to be largely cellular immune-mediated liver damage in response to chronically infected hepatocytes. Multiple studies indicate that leukocytes are chronically infected, together with a persistent viremia. More recently, several studies suggest that HCV-encoded genome (both positive and negative strand RNA) and proteins are detectable in the cerebrospinal fluid and brain, often in the setting of HIV-1 coinfection. Both monocytoid (macrophages and microglia) and astrocytes can express selective viral proteins. The extent to which accompanying cellular (and humoral) immunity contributes to neural tissue injury as a bystander cytotoxic effect, or whether virus-encoded proteins are secreted and subsequently injure proximate cells (as proposed for HIV-1) remains uncertain to date. Nonetheless, quantitative neuropsychologic studies and neuroimaging including MRI and positron emission tomography (PET) disclose abnormalities in HCV-infected patients, which are compounded by HIV-1 coinfection. Although pegylated interferon-alpha is an effective therapy for some HCV subtypes, its impact on HCV-related neurologic disease remains uncertain. As this epidemic grows, the increased risk of neurologic disease requires vigilance and intervention as indicated.

Factors promoting emergent agents

The emergence of increasing numbers of new agents appears to be accelerating. This increase may be, in part, an artifact of better surveillance and reporting, but contemporary society has made changes that encourage the evolution of organisms (Box 2).

Widespread use and misuse of antibiotics and antiviral drugs has caused natural selection of drug-resistant agents. Increased preschooling and nursery care magnifies the spread of enteroviruses. Clearing forests for agriculture has led to encounters with new zoonotic agents. The building of dams has created ecological changes that alter the habitats of infectious agents. Contemporary animal husbandry with massive feedlots facilitates

Box 2. Societal changes that enhance the evolution and spread of neurotropic agents

- Providing an adequate pool of susceptibles
  - Increasing global population
  - Increasing human contacts (travel)
- Altering forms of human or animal contact
- Societal mores
  - Increased sexual contact
  - Day care with early exposure
  - Altering environment for urban expansion and recreation
  - Agricultural clearing or irrigation
  - Global movement of animals and animal products
- Medical practices
  - Blood transfusions
  - Immunosuppressive therapy
  - Organ transplants (infected donor)
  - Antimicrobial drugs (encourage resistance)

Data from Johnson RT. The Soriano Award Lecture. Emerging infections of the nervous system. J Neurol Sci 1994;124:3–14.

the spread of agents. Even medical advances in the administration of blood products and transplantation have led to the emergence of infections.

The two most important factors are the rapidly growing human population and its movement both in speed and volume [24,25]. In 1850, the world population was under 1 billion, and around-the-world travel under sail required over a year for the very few who ventured out. Now the world population has burgeoned to over 6 billion. The world can be circumnavigated by air in 24 hours. Furthermore, each year, over 500 million persons cross international borders on commercial airplane flights, 70 million work in other countries, and 50 million are refugees or internationally displaced persons [26]. The population can maintain agents that spread human-to-human, and the speed of travel allows any known infectious agent worldwide to be in a community within one incubation period.

The evolution of new agents, the relocation of viruses, and the crossing of species barriers are not new. The measles virus has only one host, humans, but to maintain the virus an interactive population of about 200,000 is required; otherwise, the virus disappears as it does on sparsely populated islands. The virus could not have persisted in early Neolithic or nomadic cultures. Indeed, the typical exanthem of measles was not described by Greek and Roman physicians but was first recorded by Rhazes of Baghdad. He dated its arrival in the Arab world to the sixth century. Contemporary sequence analysis indicates that the measles virus was probably derived from rhinderpest virus of African or Asian cattle, suggesting an early zoonotic spread across species barriers. The appearance of measles in France and Northern Europe has been dated to the eighth century after the Moorish invasion. Apparently, it was confined below the Pyrenees for centuries until an army crossed the mountains, maintaining the virus by soldier-to-soldier spread. Today, a similar virus could cross this geographic barrier in hours.

New diagnostic tools

With the advent of new high-throughput technologies including polymerase chain reaction (PCR) arrays, virus-specific cDNA and DNA microarrays, and the capacity to perform large-scale DNA sequencing [27], diagnostic tools for viral infections are rapidly advancing in complexity and in the spectrum of agents that can be detected or newly discovered. Indeed, multiple molecular tools have been used over the past decade to characterize West Nile virus, the SARS coronavirus, and Kaposi's sarcoma–associated herpesvirus (HHV-8) [28]. The recent development of this combined diagnostic approach was particularly rewarding because it led to the discovery of a new human arena virus causing fever and encephalitis. Both PCR and oligonucleotide microarray studies were unremarkable, but subsequent high-throughput DNA pyrosequencing of patient-derived cDNA, together with bioinformatics analyses of protein sequences to exclude human sequences, resulted in the identification of a new neurovirulent arena virus. This metagenomic pyrosequencing approach, albeit, cumbersome at a bioinformatics level, is unbiased in terms of identifying novel sequences and will be invaluable for the discovery and detection of new infectious agents of the nervous system [29].

In the past 5 years, new protein technologies, including matrix-assisted laser desorption ionization (MALDI) and stable isotope labeling with amino acids in cell culture (SILAC), have also advanced, permitting the detection of new proteins in body fluids including cerebrospinal fluid and the sequencing of individual proteins [30]. These technologies have immediate value as biomarker detection tools, facilitating the diagnosis of syndromic entities and the evaluation of response to specific treatments. Eventually, using nanotechnologies including microfluidic approaches coupled with conventional methods such as antibody detection or PCR, large-scale viral detection will be feasible in the field far from high-technology laboratories, permitting expedited diagnosis and intervention.

Responses to emergent neurologic infections

Little can be done to slow the emergence of new microbial threats. The challenge is to anticipate them, detect them early, and respond in effective ways that will contain them (Box 3). First, surveillance must be global, and international cooperation is crucial. It is easier to contain Ebola virus with improved infection control in African hospitals than to counter it at international airports. The SARS virus epidemic might have been restricted by quarantines more promptly had information and efforts been shared internationally. New methods of rapid identification and characterization of agents are being developed, but the capacity to perform these studies should be put in place worldwide with major strengthening of public health infrastructure. We cannot prevent the emergence of new microbes and diseases, but we can improve our knowledge of the ecology and molecular biology of these agents. We also can anticipate that every change we make in the environment may impact on the evolution, new encounters, or habitats of these agents.

Box 3. Control of emerging infections

Global surveillance and cooperation
Rapid identification methods
Characterization to determine origin
Development and deployment of vaccines and drugs
Strengthen the public health infrastructure worldwide

New therapeutic approaches are rapidly evolving for chronic and acute viral infections. Aside from new antiviral drugs that are being developed for a range of viruses such as tenofovir and others, strategies involving the use of therapeutic vaccines, the delivery of monoclonal antibodies, or the use of therapeutically primed dendritic cells represent highly novel approaches. The use of these agents has not yet reached neurovirology; however, other technologies, including the use of nanotechnology-derived liposomes for drug delivery to the brain, are now reality. Perhaps the greatest promise lies in the use of small molecules to prevent bystander-mediate injury of the brain. For example, the recent use of minocycline, beta-lactam antibiotics, and free radical scavengers as neuroprotective therapies raises exciting possibilities, albeit, with some caution given the variable outcomes from different studies.

In the current era of escalating globalization with rapid transport, changing climate, and an ever growing human population with associated changes in lifestyle, poverty, and war, the emergence of new neurologic infections is highly plausible. Understanding their origins using epidemiologic and molecular tools will contribute to improved control of agent spread throughout vulnerable populations. Although few interventions are effective in acute epidemics, the prompt identification of new infectious agents and the roll-out of vaccines together with new antiviral and neuroprotective drugs are promising for the management of future epidemics.

References

[1] Petersdorf R. The doctors' dilemma. N Engl J Med 1978;299:628–34.
[2] Johnson RT. Viral infections of the nervous system. 2nd edition. Philadelphia: Lippincott-Raven Publishers; 1998.
[3] Schmidt NJ, Lennette EH, Ho HH. An apparently new enterovirus isolated from patients with disease of the central nervous system. J Infect Dis 1974;129:304–9.
[4] Chumakov M, Voroshilova M, Shindarov L, et al. Enterovirus 71 isolated from cases of epidemic poliomyelitis-like disease in Bulgaria. Arch Virol 1979;60:329–40.
[5] Chen KT, Chang HL, Wang ST, et al. Epidemiologic features of hand-foot-mouth disease and herpangina caused by enterovirus 71 in Taiwan, 1998–2005. Pediatrics 2007;120: e244–52.
[6] Huang CC, Liu CC, Chang YC, et al. Neurologic complications in children with enterovirus 71 infection. N Engl J Med 1999;341:936–42.
[7] Smithburn K, Hughes T, Burke A, et al. A neurotropic virus isolated from the blood of a native of Uganda. Am J Trop Med Hyg 1940;20:471–92.
[8] Southam C, Moore A. Induced virus infections in man by Egypt isolates of West Nile virus. Am J Trop Med Hyg 1954;3:19–50.
[9] Johnson RT. West Nile virus in the US and abroad. Curr Clin Top Infect Dis 2002;22:52–60.
[10] Nash D, Mostashari F, Fine A, et al. The outbreak of West Nile virus infection in the New York City area in 1999. N Engl J Med 2001;344:1807–14.
[11] Lanciotti RS, Roehrig JT, Deubel V, et al. Origin of the West Nile virus responsible for an outbreak of encephalitis in the northeastern United States. Science 1999;286:2333–7.
[12] Johnson RT, Cornblath DR. Poliomyelitis and flaviviruses. Ann Neurol 2003;53:691–2.
[13] Johnson R, Irani D. West Nile virus encephalitis in the United States. Curr Neurol Neurosci Rep 2002;2:496–500.

[14] Davis LE, DeBiasi R, Goade DE, et al. West Nile virus neuroinvasive disease. Ann Neurol 2006;60:286–300.

[15] Chua KB, Goh KJ, Wong KT, et al. Fatal encephalitis due to Nipah virus among pig farmers in Malaysia. Lancet 1999;354:1257–9.

[16] Lee KE, Umapathi T, Tan CB, et al. The neurological manifestations of Nipah virus encephalitis, a novel paramyxovirus. Ann Neurol 1999;46:428–32.

[17] Goh KJ, Tan CT, Chew NK, et al. Clinical features of Nipah virus encephalitis among pig farmers in Malaysia. N Engl J Med 2000;342:1229–35.

[18] Tan CT, Goh KJ, Wong KT, et al. Relapsed and late-onset Nipah encephalitis. Ann Neurol 2002;51:703–8.

[19] Chua KB, Bellini WJ, Rota PA, et al. Nipah virus: a recently emergent deadly paramyxovirus. Science 2000;288:1432–5.

[20] Alter MJ. Epidemiology of hepatitis C virus infection. World J Gastroenterol 2007;13: 2436–41.

[21] Koziel MJ, Peters MG. Viral hepatitis in HIV infection. N Engl J Med 2007;356:1445–54.

[22] McAndrews MP, Farcnik K, Carlen P, et al. Prevalence and significance of neurocognitive dysfunction in hepatitis C in the absence of correlated risk factors. Hepatology 2005;41: 801–8.

[23] Santoro L, Manganelli F, Briani C, et al. Prevalence and characteristics of peripheral neuropathy in hepatitis C virus population. J Neurol Neurosurg Psychiatry 2006;77:626–9.

[24] Johnson RT. Emerging viral infections of the nervous system. J Neurovirol 2003;9:140–7.

[25] Johnson RT. The Soriano Award Lecture. Emerging infections of the nervous system. J Neurol Sci 1994;124:3–14.

[26] Wilson M. Ecological disturbances and emerging infections: travel, dams, shipment of goods, and vectors. In: Power C, Johnson RT, editors. Emerging Neurological Infections. Boca Raton: Taylor & Francis Group; 2005. p. 35–58.

[27] Holt RA, Warren R, Flibotte S, et al. Rebuilding microbial genomes. Bioessays 2007;29: 580–90.

[28] Lipkin WI. Pathogen discovery. PLoS Pathog 2008;4:e1000002.

[29] Delwart EL. Viral metagenomics. Rev Med Virol 2007;17:115–31.

[30] Tan SL, Ganji G, Paeper B, et al. Systems biology and the host response to viral infection. Nat Biotechnol 2007;25:1383–9.

ELSEVIER
SAUNDERS

Neurol Clin 26 (2008) 865–869

NEUROLOGIC
CLINICS

Index

Note: Page numbers of article titles are in **boldface** type.

A

B

C

doi:10.1016/S0733-8619(08)00082-0 *neurologic.theclinics.com*

I

L

M

S

T

V

W

Z